International Symposium on (e, 2e), Double Photoionization and Related Topics and 15th International Symposium on Polarization and Correlation in Electronic and Atomic Collisions 2009

Satellite Meetings of the XXVI ICPEAC 2009
Journal of Physics: Conference Series Volume 212

Lexington, Kentucky, USA
30 July – 1 August 2009

ISBN: 978-1-61738-265-9

Printed from e-media with permission by:

Curran Associates, Inc.
57 Morehouse Lane
Red Hook, NY 12571

Some format issues inherent in the e-media version may also appear in this print version.

Printed by Curran Associates, Inc. (2010)

For permission requests, please contact the Institute of Physics
at the address below.

Institute of Physics
Dirac House, Temple Back
Bristol BS1 6BE UK

Phone: 44 1 17 929 7481
Fax: 44 1 17 920 0979

techtracking@iop.org

Additional copies of this publication are available from:

Curran Associates, Inc.
57 Morehouse Lane
Red Hook, NY 12571 USA
Phone: 845-758-0400
Fax: 845-758-2634
Email: curran@proceedings.com
Web: www.proceedings.com

TABLE OF CONTENTS

PREFACE

From 30 July to 1 August 2009, over a hundred scientists from 18 countries attended the International Symposium on (e,2e), Double Photoionization and Related Topics and the 15th International Symposium on Polarization and Correlation in Electronic and Atomic Collisions which were held at the W T Young Library of the University of Kentucky, USA. Both conferences were satellite meetings of the XXVI International Conference on Photonic, Electronic and Atomic Collisions (ICPEAC) held in Kalamazoo, Michigan, USA, 21–28 July 2009.

These symposia covered a broad range of experimental and theoretical topics involving excitation, ionization (single and multiple), and molecular fragmentation, of a wide range of targets by photons and charged particles (polarized and unpolarized). Atomic targets ranged from hydrogen to the heavy elements and ions, while molecular targets ranged from H_2 to large molecules of biological interest. On the experimental front, cold target recoil ion momentum spectroscopy (COLTRIMS), also known as the Reaction Microscope because of the complete information it gives about a wide variety of reactions, is becoming commonplace and has greatly expanded the ability of researchers to perform previously inaccessible coincidence experiments. Meanwhile, more conventional spectrometers are also advancing and have been used for increasingly sophisticated and exacting measurements. On the theoretical front great progress has been made in the description of target states, and in the scattering calculations used to describe both simple and complex reactions. The international nature of collaborations between theorists and experimentalists is exemplified by, for example, the paper by Ren *et al* which has a total of 13 authors of whom the experimental group of six is from Heidelberg, Germany, one theoretical group is from Australia, with the remainder of the theoreticians coming from several different institutions in the United States. A total of 52 invited talks and 44 submitted posters covered recent advances in these topics. These proceedings present papers on 35 of the invited talks.

The Local Organizers gratefully acknowledge the financial support of the University of Kentucky College of Arts and Sciences, and the University of Kentucky Department of Physics and Astronomy. We also thank Carol Cotrill, Eva Ellis, Diane Yates, Sarah Crowe, and John Nichols, of the Department of Physics and Astronomy, University of Kentucky for their invaluable assistance in the smooth running of the conferences; Oleksandr Korneta for taking the group photograph; and Emily Martin for helping accompanying persons.

Nicholas L S Martin
University of Kentucky
Bruno A deHarak
Illinois Wesleyan University

International Scientific Organizing Committee

Co-Chairs

Don Madison (USA)

Klaus Bartschat (USA)

Members

Lorenzo Avaldi (Italy)
Jamal Berakdar (Germany)
Michael Brunger (Australia)
Greg Childers (USA)
JingKang Deng (China)
Alexander Dorn (Germany)
Jim Feagin (USA)
Nikolay Kabachnik (Russia)
Anatoli Kheifets (Australia)
George King (UK)
Tom Kirchner (Germany)
Azzedine Lahmam-Bennani (France)
Julian Lower (Australia)
William McCurdy (USA)
Andrew Murray (UK)
Bernard Piraux (Belgium)
Tim Reddish (Canada)
Roberto Rivarola (Argentina)
Michael Schulz (USA)
Anthony Starace (USA)
Giovanni Stefani (Italy)
Masahiko Takahashi (Japan)

Nils Andersen (Denmark)
Uwe Becker (Germany)
Igor Bray (Australia)
Nikolay Cherepkov (Russia)
Albert Crowe (UK)
Danielle Dowek (France)
Oscar Fojon (Argentina)
Tim Gay (USA)
Alexei Grum-Grzhimailo (Russia)
Friedrich Hanne (Germany)
Alan Huetz (France)
Morty Khakoo (USA)
Birgit Lohmann (Australia)
Bill McConkey (Canada)
Rajesh Srivastava (India)
Al Stauffer (Canada)
Jim Williams (Australia)
Akira Yagishita (Japan)
Peter Zetner (Canada)
Joachim Ullrich (Germany)
Erich Weigold (Australia)

PEER REVIEW STATEMENT

All papers published in this volume of *Journal of Physics: Conference Series* have been peer reviewed through processes administered by the proceedings Editors. Reviews were conducted by expert referees to the professional and scientific standards expected of a proceedings journal published by IOP Publishing.

International Symposium on (e,2e) and the 15th ISPCEAC IOP Publishing
Journal of Physics: Conference Series **212** (2010) 012001 doi:10.1088/1742-6596/212/1/012001

Multiple ionization of atoms with xuv attosecond pulses: two-photon double ionization of helium with 50 eV photons

H Bachau[1], E Foumouo[2], Ph Antoine[2], B Piraux[2], O Chuluunbaatar[3], Y. Popov[4] and R Shakeshaft[5]

(1) Centre Lasers Intenses et Applications, Université Bordeaux I-CNRS-CEA, 33405 Talence Cedex, France

(2) Laboratoire de Physique Atomique, Moléculaire et Optique, unité PAMO, Université Catholique de Louvain, 2 chemin du cyclotron, B-1348 Louvain-la-Neuve, Belgium

(3) Joint Institute for Nuclear Research, Dubna, Moscow region, 141980, Russia

(4) Nuclear Physics Institute, Moscow State University, Moscow, 119991, Russian Federation

(5) Physics Department, University of Southern California, Los Angeles, CA 90089-0484, USA

E-mail: bachau@celia.u-bordeaux1.fr

Abstract. We consider two-photon double ionization of helium by two xuv photons in the region around the sequential ionization threshold. We show that, on the attosecond timescale, the mechanism for double ionization is dominated by the absorption of one photon by each electron in the fundamental state $He(1s^2)$. We examine the dynamics of two-photon double ionization of helium for an averaged photon energy ω =50 eV, with a pulse duration of two optical cycles. The double ionization rate, energy and angular distributions are calculated by solving the time-dependent Schrödinger equation. Results are discussed on the basis of a model.

1. Introduction

The study of correlated multi-electron dynamics in atoms and molecules is a necessary step towards a better understanding of multi-electron ejection in complex systems interacting with intense lasers. In the xuv domain two-electron ejection *via* one-photon absorption has received considerable attention from both theoretical and experimental sides. This process is now well understood. In the non-linear interaction regime where more than one photon is absorbed, multi-electron ejection has been studied in the infrared frequency domain with intense fields. In this case, experimental evidence as well as theoretical analysis have favored a physical mechanism based on a rescattering picture for non-sequential double ionization [1]. Double ionization of helium by several XUV photons has a much lower probability to occur, and therefore requires xuv sources that are more intense than the conventional ones. The only available sources are high order harmonic generation (HOHG) and free electron lasers (FEL). Two-photon double ionization (TPDI) of helium has been measured using both types of sources. However, the level of uncertainty in the data is very high [2] and therefore precludes a reliable comparison with the large number of existing theoretical results. On the theoretical side, the TPDI has indeed received considerable attention (see [3] for a list of references, up to 2008), but there are large discrepancies between the results for the total and differential TPDI cross sections obtained with different models in the direct regime ($\omega < 2$ a.u.) [3]. Therefore TPDI is still clearly an open problem which may require considerable effort from both theorists and experimentalists to solve.

Over the last years we have thoroughly studied TPDI in both direct ($\omega < 2$ a.u.) and sequential ($\omega > 2$ a.u.) regimes. In particular, we investigated the dynamics of TPDI in case of sub-fs pulses by studying both the electron angular and energy distributions [4]. We have shown that the electrons are predominantly emitted back-to-back along the polarization axis in direct TPDI, revealing the important role of dynamical screening [3]. For sub-fs pulses, sequential TPDI tends to a transient regime and back-to-back emission becomes the dominant process [5], as in the direct regime. Here we investigate TPDI of helium with 50 eV photons in the limit of attosecond (as) pulses. With the help of a simple model we show that TDPI is dominated by the absorption of one photon by each electron of the initial state. Atomic units are used throughout this paper, unless stated otherwise.

2. Theoretical approach

2.1. Time-dependent perturbative and non-perturbative treatments

Our theoretical approach is based on the solution of the time-dependent Schrödinger equation (TDSE)

$$i\frac{\partial}{\partial t}\Psi(\vec{r}_1,\vec{r}_2,t) = \left[-\frac{1}{2}\nabla^2_{r_1} - \frac{1}{2}\nabla^2_{r_2} - \frac{2}{r_1} - \frac{2}{r_2} + \frac{1}{r_{12}} + D_G(t)\right]\Psi(\vec{r}_1,\vec{r}_2,t), \qquad (1)$$

where r_1 and r_2 are the radial coordinates of both electrons and $r_{12} = |\vec{r}_1 - \vec{r}_2|$ is the inter-electronic distance. $D_G(t)$ describes the dipole interaction of the system with the oscillating field either in the length gauge ($G \equiv L$) or in the velocity gauge ($G \equiv V$):

$$D_L(t) = \vec{E}(t)\cdot(\vec{r}_1 + \vec{r}_2), \qquad (2)$$
$$D_V(t) = -i\vec{A}(t)\cdot(\vec{\nabla}_1 + \vec{\nabla}_2). \qquad (3)$$

$\vec{A}(t)$ denotes the vector potential and $\vec{E}(t) = \hat{z}E_0 f(t)\sin(\omega t + \phi)$ the electric field which oscillates at the frequency ω with a phase ϕ and which is assumed linearly polarized along the z-axis. Here $f(t)$ is the pulse envelope given by:

$$f(t) = \begin{cases} \cos^2(\pi t/T);\ |t| \le \dfrac{T}{2}, \\ 0;\ |t| > \dfrac{T}{2}. \end{cases} \qquad (4)$$

In this expression, T represents the total duration of the pulse. We have:

$$\vec{A}(T/2) = -\int_{-T/2}^{T/2} dt\vec{E}(t) = 0, \qquad (5)$$

for any pulse duration and phase. This means that even for few-cycle pulses there is no static field component, and problems related to gauge invariance [6] do not arise.

Our method to solve the TDSE has been described in detail in [7, 8]. Briefly, we first expand the full wave packet of the system $\Psi(\vec{r}_1,\vec{r}_2,t)$ in terms of its field-free eigenstates. The wave functions associated with these eigenstates are calculated within a spectral method that consists of diagonalizing the atomic Hamiltonian in a basis of products of one-electron square integrable functions of the radial coordinates r_1 and r_2 and bipolar harmonics of the electron angular coordinates. The square integrable functions are either Coulomb Sturmian functions or B-splines.

After propagating the initial wave packet until such time that the interaction with the pulse has died out, we are left with the task of calculating the ionization probabilities, a task which poses a challenge in all theoretical approaches. We calculate the double ionization probability using two different approaches. The first one is a simple projection of the wave packet onto a product of Coulomb functions, at $t \geq T/2$. The second one, developed in the context of Sturmian functions, uses the Jacobi-matrix method [10] to generate in the Coulomb Sturmian basis, a multichannel scattering wave function that describes accurately the single continuum of He while incorporating the correct asymptotic conditions. Projecting the final wave packet $\Psi(\vec{r}_1, \vec{r}_2, t)$ at the end of the pulse on this function provides a tool to disentangle the single- and double-ionization components [8].

In order to probe how electron correlations in the ground state of He affect the double-escape process, we have developed a simple model in which electron correlations enter only in the ground state of He. This model is based on lowest order time-dependent perturbation theory. This is justified since in the high frequency and low intensity regimes considered here the ponderomotive shift of the electrons is negligible compared to the photon energy. For two-photon double ionization, the probability amplitude reads:

$$U^{(2)} = -\sum_{\alpha} \langle \Psi_f | z_1 + z_2 | \Psi_\alpha \rangle \langle \Psi_\alpha | z_1 + z_2 | \Psi_i \rangle K(E_0, \omega, \phi, E_i, E_\alpha, E_f), \tag{6}$$

where the function K is given by:

$$\begin{aligned} K(E_0, \omega, \phi, E_i, E_\alpha, E_f) &= \int_{-T/2}^{T/2} d\tau_1 E_0 f(\tau_1) \sin(\omega\tau_1 + \phi) e^{i\omega_{f\alpha}\tau_1} \\ & \int_{-T/2}^{\tau_1} d\tau_2 E_0 f(\tau_2) \sin(\omega\tau_2 + \phi) e^{i\omega_{\alpha i}\tau_2}. \end{aligned} \tag{7}$$

$\omega_{f\alpha} = E_f - E_\alpha$ and $\omega_{\alpha i} = E_\alpha - E_i$ with E_i, E_α and E_f the initial, the intermediate, and the final state energy respectively. E_0 is the maximum field amplitude. The wave function of the initial state Ψ_i is written as follows:

$$\Psi_i(\vec{r}_1, \vec{r}_2) = \sum_{l,\nu,n} \phi^{l,l}_{\nu,n} \mathcal{A} F^{l,l}_{\nu,n}(r_1, r_2) \Lambda^{0,0}_{l,l}(\Omega_1, \Omega_2), \tag{8}$$

where the coefficients $\phi^{l,l}_{\nu,n}$ are calculated by diagonalizing the atomic Hamiltonian of He. $\mathcal{A}$ is the antisymmetrization operator and $\Lambda^{0,0}_{l,l}(\Omega_1, \Omega_2)$ is a bipolar harmonic in the angular coordinates of both electrons. The radial function $F^{l,l}_{\nu,n}(r_1, r_2)$ is written as a product of two hydrogenic bound states of principal quantum number ν and n respectively:

$$F^{l,l}_{\nu,n}(r_1, r_2) = \psi_{\nu,l}(r_1) \psi_{n,l}(r_2). \tag{9}$$

Let us stress that the above expression (8) for Ψ_i is not correct because this expansion does not contain continuum states. However, it contains some radial and angular correlation. In this calculation of $\Psi_i(\vec{r}_1, \vec{r}_2)$, we have included 4 pairs (ℓ, ℓ) of the electron angular momenta ((0,0), (1,1), (2,2) and (3,3)) and taken both ν and n varying from $1+\ell$ to $4+\ell$. The resulting estimate of the ground state energy is -2.84 a.u. The intermediate states Ψ_α and the final state Ψ_f are defined as an antisymmetrized product of bound or/and Coulomb states. We have tested this model by comparing the results with those obtained by the approach described at the beginning of this section. The agreement is good both qualitatively and quantitatively [5]. In addition, this model also gives reasonable results in the direct regime.

Before discussing the TDSE results we briefly present several models which include electron correlations in various degrees of approximation. Their pertinence is discussed with respect to the pulse duration; they illustrate the strong dependence of the TPDI dynamics on T.

2.2. The zero-order approximation and the independent electron model

We recall first a previous study where the role of electron correlation was treated in lowest order perturbation theory [9]. In this approach, the He ground state is described by the $1s^2$ configuration, i.e., by a product of two hydrogenic orbitals $1s$. The other states are the single continuum state He(1skp) and double continuum He(kpk'p), also represented by hydrogenic orbitals (all orbitals are calculated with $Z = 2$). In this approximation the energy $E^0_{1s^2}$ of the ground state would then be given by $2E^0_{1s} + < 1s^2|1/r_{12}|1s^2 >$ where E^0_{1s} is the hydrogenic energy of $1s$ and $< 1s^2|1/r_{12}|1s^2 >$ the electron interaction energy. For long pulse durations, it is possible to derive an analytical expression for the photoelectron energy spectrum, it reads

$$|U_{k,k'}(T \to \infty)|^2 \propto \frac{|\langle kp|z_1|1s\rangle\langle k'p|z_2|1s\rangle|^2}{[(E_k + E_{k'} - E^0_{1s^2} - 2\omega)^2 + \frac{1}{4}\gamma^2_{He}][(E_k - E^0_{1s} - \omega)^2 + \frac{1}{4}\gamma^2_{He^+}]}, \qquad (10)$$

where γ_{He} and γ_{He^+} are the photoionization widths of He and He$^+$, respectively. $\langle kp|\mu|1s\rangle$ represents the dipole coupling between the $1s$ orbital and the kp continuum. E_k and $E_{k'}$ are the energies of the ejected electrons. The electron spectrum shows two peaks separated by the energy $E_I =< 1s^2|1/r_{12}|1s^2 >$. E_I is the electron interaction energy exchanged during the ionization process. Here it is worth noticing that the effective widths of the peaks is the result not of a single Lorentzian but of a convolution.

If now we fully neglect the electron interactions, i.e. if we adopt the independent electron model, it is easy to show that the expression (6) splits into the product of two terms representing the absorption of one photon from each $1s$ orbital. $U^{(2)}$ is written

$$U^{(2)} = -\sqrt{2}\langle kp|z_1|1s\rangle K_M(E_{1s}, E_k)\langle k'p|z_2|1s\rangle K_M(E_{1s}, E_{k'}), \qquad (11)$$

where

$$K_M(E_i, E_f) = \int_{-T/2}^{T/2} d\tau E_0 f(\tau) sin(\omega\tau + \phi) e^{i(E_f - E_i)\tau}. \qquad (12)$$

It is worth noticing that the final state has a unique angular configuration of the electron pair — it is (1,1). The corresponding angular distribution peaks when both electrons move in the same direction with the same energy, a feature which can occur in the absence of electron correlation [3].

2.3. A simple model to investigate TPDI with ultrashort pulses

We return now to equation (6). Our aim is to develop a simple model to describe TPDI in the limit of ultrashort pulses. The initial state is now represented by a simple product of 1s orbitals with a screening charge of 27/16. For ultrashort pulses (e.g., two optical cycles with a photon energy of 50 eV), the function $K(E_0, \omega, \phi, E_i, E_\alpha, E_f)$ shows a slow dependence on the intermediate energy E_α. Therefore, we replace the value E_α by the resonant energy $E_r = E_i + \omega$. It is straightforward to show that

$$U^{(2)} = -\langle\Psi_f|(z_1 + z_2)^2|\Psi_i\rangle K(E_0, \omega, \phi, E_i, E_r, E_f). \qquad (13)$$

Here it is interesting to note that the matrix element $\langle\Psi_f|(z_1 + z_2)^2|\Psi_i\rangle$ exhibits two types of couplings. The first one, $z_1 z_2$, is associated with the absorption of one photon by each electron in Ψ_i, as illustrated in equations (10) and (11). The second term, z_i^2 (with $i = 1, 2$), represents the absorption of two photons by one electron, the second electron being ionized through electron interactions. This latter term does not contribute when the electron interactions are fully neglected. These two terms have a crucial impact on angular distributions in the final state; this is discussed in the following section.

3. Results and discussion

We consider TPDI with 50 eV photons and a pulse duration of two optical cycles (2c), the intensity is 10^{13} W/cm^2. Note that one optical cycle (1c) has a duration of 82 as. We have checked that three photon absorption is negligible at the intensity considered.

First we have compared our model (see section 2.3) with TDSE calculations; the agreement is excellent. We have checked that the coupling $2z_1z_2$ dominates over z_1^2 and z_2^2 (although the latter are non-negligible).

L	(l_1, l_2)	$0c$	$2c$	$4c$
2	(1,1)	$0.18\ 10^{-8}$	$0.17\ 10^{-8}$	$0.16\ 10^{-8}$
0	(1,1)	$0.82\ 10^{-9}$	$0.60\ 10^{-9}$	$0.48\ 10^{-9}$
2	(0,2)	$0.71\ 10^{-10}$	$0.24\ 10^{-9}$	$0.34\ 10^{-9}$
0	(0,0)	$0.49\ 10^{-10}$	$0.17\ 10^{-9}$	$0.24\ 10^{-9}$

Table 1. Dominant angular pair contributions in double continuum produced by two-photon absorption at $t = T/2$ (i.e., at the end of the pulse), $t = T/2 + 2c$ and $t = T/2 + 4c$. The laser parameters are the same than in figure 1.

After solving the TDSE, angular distributions have been extracted by projecting the total wavefunction $\Psi(\vec{r}_1, \vec{r}_2, t)$ (see equation 1) onto an antisymmetrized product of Coulomb functions (see equations (12) and (13) in [5]). Using the latter functions, calculated with $Z = 2$, we neglect the electron interactions in double continuum but these interactions are included in TDSE and free propagation. Table 1 shows the dominant angular pair contributions calculated at different times after the end of the pulse. The table clearly shows that the pair $(1, 1)$ dominates at $t = T/2$ (i.e., at the end of the pulse). We see that the contributions of other angular pairs, in particular (0,2), increases at $t = T/2 + 2c$ and $t = T/2 + 4c$. Here it is worth mentioning that when a three particle system disintegrates the term 'angular momentum distribution' becomes meaningless. Indeed, it is in principle necessary to take into account more and more angular momenta as the particles separate further, the only exception being the collinear configuration [11].

Figure 1 shows the angular distribution in the case were both electrons share a similar energy (0.4 a.u.). The first electron is emitted along the z axis with an angle of $\theta = 0$. As in the precedent case the information is extracted from $\Psi(\vec{r}_1, \vec{r}_2, t)$ at various times t. At $t = T/2$ the figure shows that the second electron is preferentially emitted in the opposite direction ($\theta = \pi$) but there is a non-negligible probability that both electrons are emitted in the same direction. At the end of the short pulse both electrons are close to the nucleus and strongly interact. After four cycles (i.e., at $t = T/2 + 4c$) the latter contribution is negligible, the distribution clearly shows that the electron are emitted in opposite directions. The results agree with the analysis of table 1, in fact, the presence of the single angular pair (1,1) in the double continuum would lead to a symmetric distribution [3].

In conclusion, for a pulse with attosecond duration and photon energy 50 eV the double

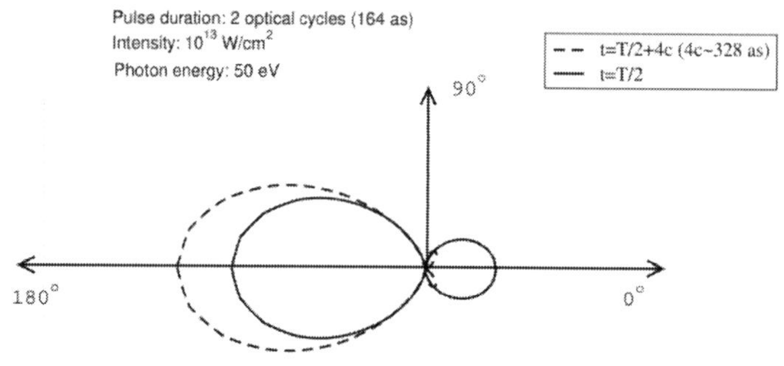

Figure 1. TPDI angular distribution, in polar coordinates. Both electrons have an energy of 0.4 a.u., one of them being emitted along the z axis ($\theta = 0$). The calculations are performed at the time $T/2$ (i.e., at the end of the pulse, full line) and at $T/2 + 4c$ (i.e., four cycles after the end of the pulse, dashed line). The laser parameters are indicated in the figure.

ionization process is dominated by the absorption of one photon by each electron. This leads, at the end of the pulse, to a wave packet with strongly interacting electrons where the double continuum is dominated by an angular pair $(1, 1)$. Subsequently, the strong electron interaction in the double ionization channel generates other angular pairs, resulting physically in electron emission in opposite directions. Therefore, in the limit of ultrashort pulse durations, it is possible to disentangle the effect of the laser coupling and electron correlations. In the attosecond regime the angular and energy distributions show unexpected features; they will be analyzed in details in a forthcoming publication.

Acknowledgments
Work partially supported by European COST Action CM0702. Y. P. thanks the Université Bordeaux I for hospitality and financial support.

References
[1] Corkum P B 1993 *Phys. Rev. Lett.* **71** 1994
[2] Antoine Ph, Foumouo E, Piraux B, Shimizu T, Hazegawa H, Nabekawa Y and Midorikawa K 2008 *Phys. Rev.* A **78** 023415
[3] Foumouo E, Antoine Ph, Piraux B, Malegat L, Bachau H and Shakeshaft R 2008 *J. Phys. B* **41** 051001
[4] Foumouo E, Laulan S, Piraux B and Bachau H 2006 *J. Phys. B* **39** S427
[5] Foumouo E, Antoine Ph, Bachau H and Piraux B 2008 *New J. Phys.* **10** 025017
[6] Madsen L B 2002 *Phys. Rev.* A **65** 053417
[7] Laulan S and Bachau H 2003 *Phys. Rev.* A **68** 013409
[8] Foumouo E, Lagmago Kamta G, Edah G and Piraux B 2006 *Phys. Rev.* A **74** 063409
[9] Bachau H and Lambropoulos P 1991 *Phys. Rev.* A **44** R9
[10] Heller E J and Yamani H A 1974 *Phys. Rev.* A **9** 1201
[11] Gailitis M 1990 *J. Phys. B* **23** 85

Journal of Physics: Conference Series **212** (2010) 012002 doi:10.1088/1742-6596/212/1/012002

Inner shell excitation, ionization and fragmentation of pyrimidine

P. Bolognesi[1], P. O'Keeffe[1], V. Feyer[2], O. Plekan[2,8], K. Prince[2], M. Coreno[1,3], G. Mattioli[4], A. Amore Bonapasta[4], W. Zhang[5,6,7], V. Carravetta[5], Y. Ovcharenko[1,8] and L. Avaldi[1,3]

1) CNR-IMIP, Area della Ricerca di Roma 1, Monterotondo Scalo, Italy

2) Sincrotrone Trieste, Area Science Park, Basovizza (Trieste), Italy

3) CNR-INFM-TASC, Gas Phase Beamline@Elettra, Area Science Park, Trieste, Italy

4) CNR-ISM, Area della Ricerca di Roma 1, Monterotondo Scalo, Italy

5) CNR-IPCF, via Moruzzi 1, Pisa, Italy

6) Hefei National Laboratory for Physical Science at the Microscale, University of Science and Technology of China, Hefei, China

7) Department of Theoretical Chemistry, School of Biotechnology, Royal Institute of Technology, Stockholm, Sweden

8) Permanent address: Institute of Electron Physics, Uzhgorod, Ukraine

E-mail: paola.bolognesi@imip.cnr.it

Abstract. The inner shell excitation and ionisation of pyrimidine have been studied at the carbon K edge by near-edge X ray absorption fine structure (NEXAFS) and X ray photoelectron (XPS) spectroscopies. The theoretical predictions of density functional theory (DFT) provide a satisfactory assignment of the complex spectra of this polyatomic molecule. The fragmentation following the $C(1s^{-1})\pi^*$ excitation has been investigated by resonant Auger electron-ion coincidence spectroscopy, which allows a site and state selective study.

1. Introduction

Pyrimidines are an important class of organic molecules because the pyrimidine ring forms the base structure of three nucleic acids (uracil, cytosine and thymine) and their halogenated substituted bases have found applications as radiosensitizers in radiotherapy [1]. A photoabsorption process in a biological system may cause repairable and/or unreparaible changes/damages in a biological cell. The macroscopic damage can be tracked down to a microscopic scale, where the initial processes on the elementary constituents are the same as those studied in molecular physics and photochemistry. Therefore it is expected that gas phase experiments of model molecules can provide insight in the physical and chemical properties of biological molecules. This has lead to an extensive investigation of the structure and dynamics of biomolecules and their solvated complexes [2- 5].

A recent theoretical study of the electronic structure and chemical properties of DNA/RNA bases [6] has shown that pyrimidine bases inherit certain properties from their parent pyrimidine, while the purine bases do not. Moreover the observation of high relative biological effectiveness of soft X ray in cells [7] has raised the question whether this is due to a specific physical/chemical effect or to the preferential localization of the ionizations in the DNA bases. In the former case this implies a fragmentation process induced by the radiation, while the latter implies that the bond breaks are due to the electron swarms produced following the absorption of the soft X-ray. To answer these questions one has to measure the electron spectra of the inner shell ionized and excited molecules as well as their site selected fragmentation patterns. For this purpose we have undertaken a detailed investigation of the inner shell photoabsorption, photoionization and photofragmentation of pyrimidine and some of its halogenated substitutes. Here some results of the XPS, NEXAFS and resonant Auger measurements of pyrimidine together with the analysis of the fragmentation following inner shell excitation are reported.

2. Experimental

All the measurements have been performed at the Gas Phase Photoemission beamline of the Elettra synchrotron radiation source, Trieste [8]. The 100% linearly polarised radiation from the undulator (period 12.5 cm, length 4.5 m) is deflected to the variable-angle-spherical grating monochromator, which covers the energy region 13-1000 eV with five interchangeable gratings.

The XPS photoemission and resonant Auger spectra were measured using a commercial 6-channel, 150 mm hemispherical electron energy VG analyzer. The axis of the electron analyzer is set at the magic angle in the parallel plane, i.e. in the plane defined by the electric vector of the light and the photon propagation direction, at 54.7° with respect to the electric vector of the light. In this geometry the measurements are insensitive to the photoelectron asymmetry β parameter. The C (1s) spectrum of pyrimidine (Figure 1) was measured at about 100 eV above ionization threshold, using an energy resolution of the analyzer of about 100 meV [9]. This collection energy in the photoelectron spectrum guarantees that PCI effects [10] can be neglected in the data analysis. The energy spectra were calibrated using a mixture of the molecule under study and CO_2, whose XPS peak (C(1s), v=0) is at 297.69 eV [11]. The C NEXAFS spectrum of pyrimidine (Figure 2.a) was measured using a channeltron multiplier close to the interaction region to detect the ions produced by the decay of the excited molecular states [12]. The ion detector has been placed at about the 'magic angle' with respect the photon beam polarization axis, so that the measured cross section is not affected by asymmetry parameters. The NEXAFS spectrum has been obtained scanning the photon energy over the range 283-300 eV. The spectrum was then normalized to the variation of the photon beam intensity due to time and monochromator transmission efficiency. The energy resolution was approximately 75 meV. For the measurements of the site and state selected fragmentation the same hemispherical analyser, set at 90° with respect to the electric vector, and a time-of-flight ion analyzer mounted opposite to it have been used. The characteristic of this set-up as well as the procedure adopted in the measurements have been described recently [13] and will not be repeated here. An important peculiarity of the resonant Auger-ion coincidence spectroscopy is the ability of the technique to select both the initial state and site (by setting the photon energy on a specific NEXAFS peak) and the final ionic state (by selecting the kinetic energy of a particular resonant Auger electron).

Pyrimidine is in the form of liquid at room temperature, with enough vapour pressure to produce the gas density needed for gas phase experiments. The sample was purchased from Sigma-Aldrich, with purity always higher than 99 % and used without further purification. The compound was kept in a glass vial outside the vacuum chamber and introduced in the interaction region via a leak valve.

3. Results

The C(1s) XPS spectrum of pyrimidine is shown in figure 1. The spectrum is composed of three main peaks, corresponding to the ionization of the three non equivalent carbon atomic sites in the molecule, the C2, C4/C6 and C5. Each one of these XPS peaks is most likely the convolution of an undefined number of overlapping and unresolved vibrational progressions. The resulting asymmetric lineshape of each peak has been approximated in the fitting procedure by an asymmetric Gaussian function with the position of the centroid, the area and the low/high energy side widths used as free parameters. This procedure provides a good description of the spectra, with only small discrepancies on the tails of the peaks. The centroid of each peak has been assumed as the binding energy of the state and is reported in Table 1 together with the estimated areas from the fitting. In the same table the results of a DFT calculation [9] and previous calculations of F. Wang *et al.*[6] are reported. The different binding energies of the non equivalent C atoms are mainly due to the presence of the two electron-attractive N atoms, that have the overall effect of shifting the binding energies towards higher values with respect to the benzene case [14]. Indeed the C2 carbon atom located between the two N atoms is affected by the larger binding energy shift, while the C5 being the farthest one has the smallest shift. These observations can be rationalized in terms of the inductive and resonant models of the charge screening in aromatic molecules.

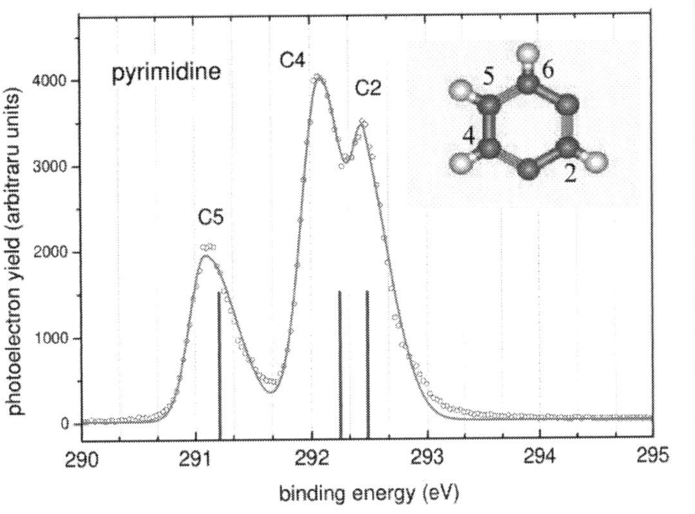

Figure 1. C (1s) XPS spectrum of pyrimidine: experimental results (open dots) fitted with asymmetric Gaussian lineshape (full line) and compared to the DFT theoretical calculations (vertical bars).

Table 1. C(1s) experimental binding energies and relative intensities of pyrimidine. The theoretical values of [9,6] have been arbitrarily aligned to the C2 site of pyrimidine for comparison with the experimental results. The uncertainty in the theoretical calculation is estimated to be around 0.1 eV.

C(1s) state	Binding energy (eV) Experiment	Relative Intensity experiment	Theory (eV) Ref [9]	Ref [6]
CO_2(C1s, v=0)	297.69 ± 0.02	--		
	291.09 ± 0.12	0.52 ± 0.04	291.20 (C5)	290.98 (C5)
pyrimidine	292.08 ± 0.02	1.00 ± 0.05	292.24 (C4)	292.11 (C4)
	292.48 ± 0.03	0.53 ± 0.07	292.48 (C2)*	292.48 (C2)*

* zero of the theoretical binding energy scale

In the inductive model an electron attractive substituent removes charge density from the neighboring sites, while the weaker resonant effect involves the π conjugated bonds along the ring. It is interesting to note for example that an analysis based on the difference electron density maps, obtained by subtracting the valence electronic density of a "standard" calculation from the valence electronic density of the corresponding "core hole" calculation, shows that a contribution to the screening of the C2 hole comes from the farthest C5 atom [9]. This contribution is explained by a resonant effect involving the C5 atom in *para* position with respect to the C2 hole.

The portion of the C(1s) NEXAFS spectrum of pyrimidine between 284-288 eV is shown in Figure 2.a. The spectrum is dominated by the strong π^* resonance representing the transition of the 1s electron to the lowest unoccupied, antibonding, molecular orbital. The peaks labeled 1′, 1″ and 1‴ have been assigned by a DFT calculation [12] to the transition from the three inequivalent carbon sites, C5,C4/C6 and C2 respectively. At higher photon energy the NEXAFS spectrum shows peaks due to a mixed contributions of σ^* and π^* orbitals corresponding to excitation of Rydberg or mixed valence-Rydberg orbitals [12]. The present experimental and theoretical spectra are in good agreement with previous results [15], even though the higher resolution of the present measurement allows the observation of some extra features on the π^* that can be assigned to vibrational progressions, as already observed in benzene [16]. The decay of the inner shell excited pyrimidine via resonant Auger electron emission has been studied at the energies which correspond to the excitation to the π^* orbital of the three non equivalent sites. The results are collected in figures 2.b,c and d. The final states in the resonant Auger decay are singly charged ion states, thus the resonant Auger spectra have been compared with the PES spectrum (figure 2.e) of pyrimidine recently reported and assigned [17]. In figures 2.b-d the three resonant Auger spectra display different intensity among themselves and with respect to the PES spectrum.

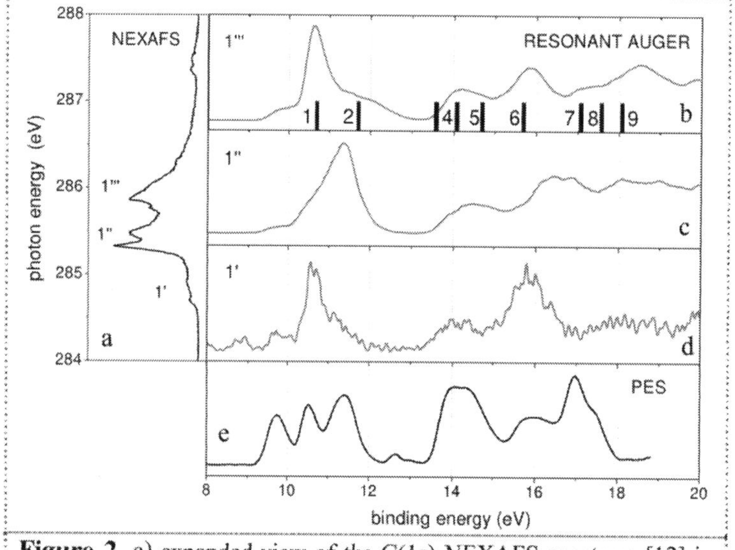

Figure 2. a) expanded view of the C(1s) NEXAFS spectrum [12] in the region of the π^* transition; transitions labeled 1′, 1″ and 1‴ have been assigned to the excitation of the C5, C4/C6 and C2 carbon sites into the π^* orbital; b-d) the resonant Auger electron spectra measured at photon energies corresponding to the excitation of the 1′, 1″ and 1‴ NEXAFS features are compared to the photoelectron spectrum measured at 21.22 eV photon energy [17].

The intensity of a resonant Auger transition depends on the overlap of the wave functions of the intermediate and final states, i.e. the $C(1s^{-1})\pi^*$ and the ionized valence orbitals. Interestingly, the resonant Auger decay seems to favor the population of final ionic states with similar localization of the charge distribution as in the intermediate state. For example, according to the Mulliken population analysis [17], the HOMO state of pyrimidine receives a major contribution from the lone pair orbitals of the nitrogen atoms. Very little superposition with the $C(1s^{-1})\pi^*$, core hole state is expected, and indeed observed (see figures 2.b-d at about 9.7 eV binding energy). On the contrary, the HOMO-1 orbital, with a stronger contribution from the C5 and C2 atoms, shows a preferential population from the resonant Auger decay of the 1′ and 1‴ NEXAFS states, localized on the C5 and C2 site, respectively, according to our calculations [12].

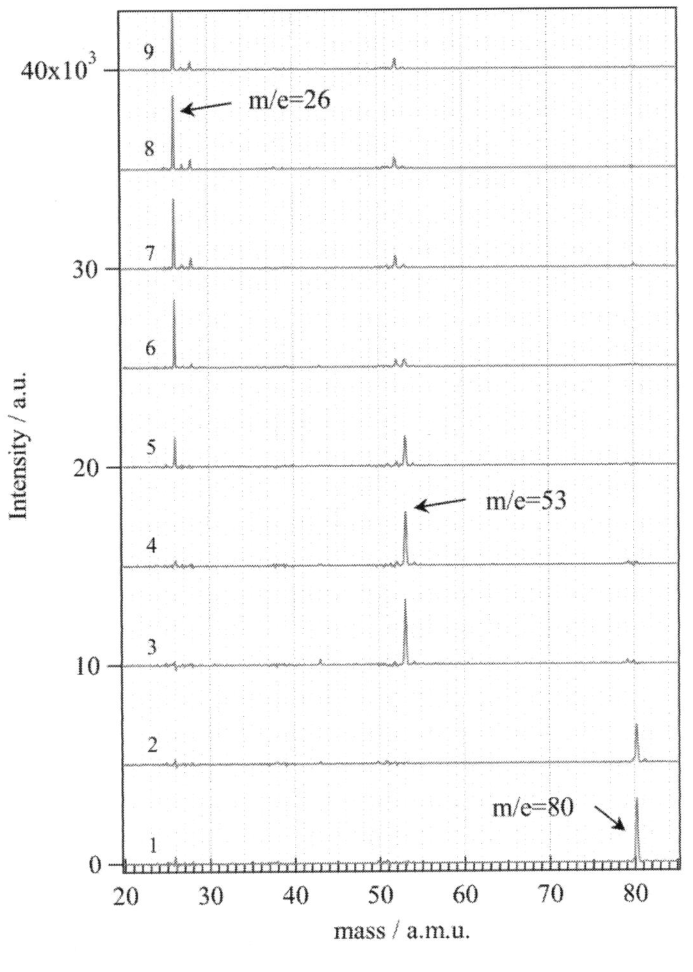

Figure 3. The mass spectra of pyrimidine measured in coincidence with resonant Auger electrons at selected kinetic energies, labeled 1 to 9 in figure 2.b.

At a few selected energies of the resonant Auger electron (labelled 1-9 in figure 2.b), which correspond to several different final states of the singly charged ion, the fragmentation patterns of pyrimidine have been measured. In figure 3 the fragmentation patterns measured at the 1‴ position of the NEXAFS spectrum is shown. Three groups of charged fragments of decreasing masses appear in the spectra of figure 3 as the kinetic energy of the resonant Auger electrons is decreased, i.e ion states of increasing binding energy are selected. The parent molecule corresponds to the peak at m/e=80, while the fragments due to the loss of one or two HCN groups appear at m/e=53 and 26, respectively. Within the statistics and sensitivity of the present experiments, no fragmentation of the molecule is observed for binding energies <11.5eV, while at binding energies >12 eV the full fragmentation occurs. This is consistent with the picture proposed by Schwell et al. [18], who set at a binding energy >11 eV the appearance potential for the loss of a neutral C_2H_2 fragment, which represents the lowest and weak fragmentation channel. In the same work, those authors set at 12.27 and 14.2 eV the thresholds for the appearance of the $C_3H_3N^+$ and $HCCH^+$ fragments, corresponding to the peaks at m/e=53 and 26, respectively. The observations reported in figure 3, can be easily understood in terms of the increasing residual energy available for the fragmentation of the singly charged molecule as ion states of increasing binding energy are selected. No substantial differences are observed in the fragmentation patterns by moving the photon energy to the 1″ and 1′ NEXAFS states, i.e. making a selective excitation of different molecular sites. This allows us to conclude that the fragmentation of pyrimidine following inner shell excitation depends only on the final state of the singly charged ion reached in the relaxation process.

4. Conclusions

The electron spectra produced by C(1s) ionization and excitation with synchrotron radiation have been studied. The spectra have been assigned by comparison with DFT calculations [9]. The measured values of the chemical shifts of the three inequivalent C atoms in the XPS has been explained in terms of both inductive and resonant effects. The resonant Auger spectrum shows a selective population of the final states of the singly charged ion and the site and state selected fragmentation patterns appear to depend only on the final state of the singly charged ion. This set of data represents the benchmark for an extensive study with the same techniques of the halogenated pyrimidines, which aims to provide information useful to disentangle the mechanisms of the radiation damage of soft X-ray in cells.

Acknowledgments

The authors thank the Elettra support program for Italian users.

[1] Dewey W C and Humprey R M 1965 *Radiat. Res.* **26** 538
[2]Wang F 2005 *J. Mol. Struct.* **728** 31
[3]Fujii K *et al* 2004 *J. Phys. Chem. B* **108** 8031
[4] Sanche L 2002 *Mass Spectrom. Rev.* **21** 349
[5] Plekan O *et al.* 2008 *Chem. Phys.* **347** 360
[6] Wang F *et al.* 2008 *J. Synchr. Rad.* **15** 624
[7] Fayard B *et al.* 2002 *Radiat. Res.* **157** 538
[8] Prince K C *et al.* 1998 *J. Synchotron Rad.* **5** 565
[9] Bolognesi P *et al* 2009 *J. Phys. Chem.* submitted
[10] van der Straten P *et al*1988 *Z. Phys.* **D8** 35
[11] Hatamoto T et al. 2007 *J. Electron Spectrosc. Relat. Phenom.* **155** 54
[12] Bolognesi P *et al* 2009 *J. Chem. Phys.* to be submitted
[13] Plekan O *et al.* 2008 *Phys. Scr.* **78** 058105
[14] Jolly W L *et al* 1984 *Atom. Data and Nucl. Tables* **31** 433
[15] Vall-llosera G *et al* 2008 *J. Chem. Phys.* **128** 044316
[16] Puttner R *et al* 2004 *Chem. Phys. Lett.* **393** 361
[17] O'Keeffe P *et al* 2009 *Mol. Phys*, accepted for publication
[18] Schwell M *et al* 2008 *Chem. Phys.* **353** 145

Three-dimensional cross sections for electron impact ionization of atoms and molecules

X. Ren[1]*, A. Senftleben[1], T. Pflüger[1], M. Holzwarth[1], A. Dorn[1]*, K. Bartschat[2], I. Bray[3], D.V. Fursa[3], J. Colgan[4], M.S. Pindzola[5], O. Al-Hagan[6], D.H. Madison[6] and J. Ullrich[1]

[1] Max-Planck-Institute for Nuclear Physics, Saupfercheckweg 1, 69117 Heidelberg, Germany

[2] Department of Physics and Astronomy, Drake University, Des Moines, Iowa 50311, USA

[3] ARC Centre for Antimatter-Matter Studies, Curtin University, GPO Box U1987 Perth, Western Australia, Australia

[4] Theoretical Division, Los Alamos National Laboratory, Los Alamos, New Mexico 87545, USA

[5] Department of Physics, Auburn University, Auburn, Alabama 36849, USA

[6] Physics Department, Missouri University of Science and Technology, Rolla, Missouri 65409, USA

E-mail: Xueguang.Ren@mpi-hd.mpg.de, Alexander.Dorn@mpi-hd.mpg.de

Abstract. Using a multi-particle momentum spectrometer (reaction microscope), three-dimensional and fully differential cross sections (FDCS) for electron impact ionization are obtained, providing benchmark data for comprehensive tests of theoretical calculations. Since all final-state particles, including the scattered projectile were detected, a good momentum transfer resolution was obtained also for heavy targets like Ar. Results for ionization of the Ar 3*p*- and He 1*s*-orbitals by 200 eV electron impact are presented. The cross section patterns display rich structure, which is partially reproduced by theory, although differences persist out-of-the scattering plane. Kinematically complete experiments for electron impact ionization of simple diatomic molecules have attracted increased attention concerning molecular structure effects. (e, 2e) on He and H_2 is studied at equivalent collision kinematics to explore the differences of atomic and molecular ionization. Here FDCS were obtained covering the whole solid angle.

1. Introduction

Electron impact single ionization, the so-called (e, 2e) reaction, on atoms or molecules represents one of the fundamental dynamical few-body problems in quantum mechanics, and plays a key role in many naturally occurring and technical plasma phenomena. Since the first pioneering work on single ionization of He [1] a large amount of cross sections has been measured over a wide range of kinematics and geometries, on the other hand most of the dynamic studies were performed in the scattering plane, where the ejected, the incoming and the scattered projectile were in a common plane. The agreement between state-of-the-art theories and experimental data has been significantly improved especially for simple targets such as atomic hydrogen and He [2–4]. For heavier or more complex many-electron atoms, the agreement is not as good particularly at low and intermediate collision energies [5–8].

Recently three-dimensional images of the complete electron emission pattern have been obtained for the single ionization of He by the impact of ions [9] and electrons [10] due to the development of many-particle imaging techniques, the so called "reaction microscope". An unexpected structure in the fully differential cross sections (FDCS) out of the scattering plane was observed for the single ionization of He [9, 10], and has so far not been fully understood, (for more details see [11, 12]). Full three-dimensional (3D) studies of ionization from the heavier targets, such as the case of Ar, would be interesting because of the rich structures that they can display. Furthermore, it would be important to explore the experimental images of ***p***-orbital ionization (e.g. Ar 3***p***) into not only the scattering plane but also the full solid angle. This work is to reveal the features of single ionization on He 1***s*** and Ar 3***p*** orbitals at 200 eV impact energy by using the three-dimensional images. In addition, experiments for electron impact ionization of simple diatomic molecules have attracted increased attention recently, since strong molecular structure effects were observed for electron emission in the plane perpendicular to the incoming beam direction [13, 14]. In order to study these effects more comprehensively we have performed comparative (e, 2e) experiments on the two-electron targets He and H_2 at low impact energies (16 eV excess energy) covering the full solid angle of electron emission.

2. Experimental Setup

The experiments were performed by using a highly efficient reaction microscope which was especially designed for electron impact experiments. The experimental setup of the reaction microscope and experimental procedure have been described in detail elsewhere [15, 16] (Figure 1).

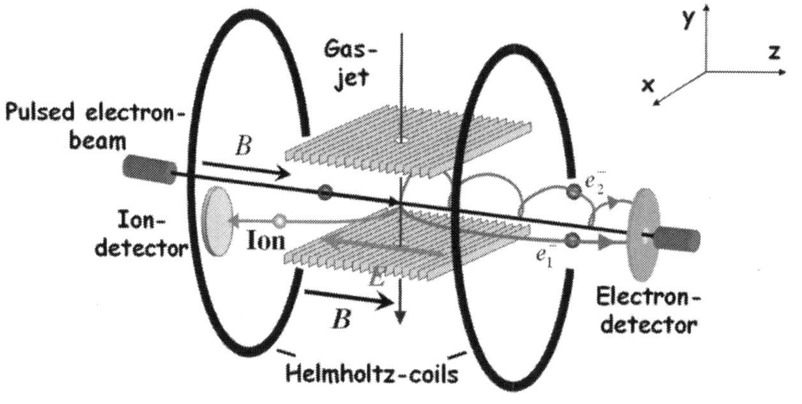

Figure 1. Schematic view of the experimental setup.

Briefly, a well focused (1 mm), pulsed electron beam (pulse length:1–2 ns, repetition rate: 180 kHz), produced by a standard thermo cathode gun, crosses and ionizes on atomic or molecular gas target (1mm diameter, 10^{12} atoms/cm^3), which is produced in a two-stage supersonic gas jet. All three final-state particles, i.e., the scattered projectile electron, the emitted target electron, and the recoiling ion, were extracted into the opposite directions, respectively, along the projectile beam axis (the longitudinal direction defines the z-axis) by a homogeneous electric field over 11 cm and projected onto two-dimensional position-sensitive multi-hit detectors. A uniform longitudinal magnetic field was applied to confine the transverse motion of the electrons, such that all electrons with energies below 25 eV were collected with the full solid angle. While the ion-detector is equipped with a wedge and strip anode, the electron detector employs a delay-line anode for position read-out with increased multihit capability. From the hitting positions and the time-of-flight the vector momenta of the particles can be reconstructed. It is noted that, since the projectile beam axis is adjusted exactly parallel to the electric and magnetic extraction fields, there is a central bore in the forward electron detector to avoid the complete saturation of the electron detector by the unscattered projectiles ($\sim 10^8$ s^{-1}) and allow for the passage of the non-deflected electrons (see Figure 1). In this way a large part of the full solid angle is covered, 100% for the detection of recoil-ion and 80% for electrons below 25 eV. For the present measurements, the scattered projectile electrons were detected together with ejected electrons, as well as the residual ion, such that the triple coincidence of all final-state continuum particles delivered superior background suppression. Furthermore, the experimental electron momentum resolution is independent of the recoil ion momentum resolution (and therefore independent of the target temperature). That makes it feasible to measure heavier targets with the reaction microscope. The electron momentum resolution depends on how well the time and position of the ionizing collision can be determined from the pulse width and the focus diameter of the projectile beam in the target. Note that for our present (e, 2e) measurements on the light targets He and H_2 at low impact energies (16 eV excess energy). One of the outgoing electrons and the recoil ion are detected in coincidence, and the momentum of the second outgoing electron is recalculated by using the momentum conservation.

3. Results and discussion

3.1. ionization of He and Ar at intermediate energy

Figure 2 exhibits the experimental and theoretical three-dimensional electron emission patterns for the (e, 2e) cross section of He by 200 eV electron impact at a scattering angle of $\theta_1 = 10°$. The energy of the emitted electron is fixed at $E_2 = 15$ eV.

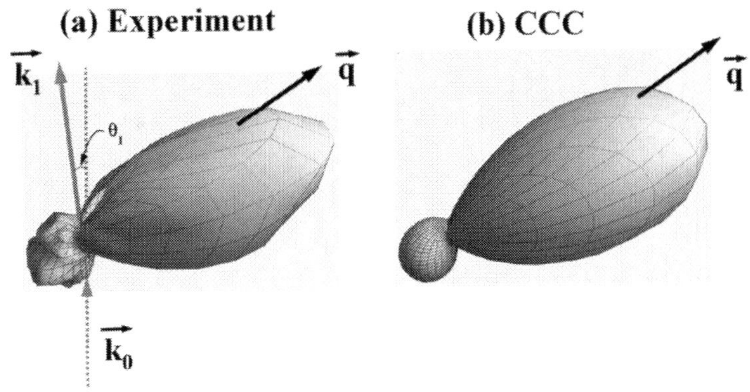

Figure 2. 3D cross sections of the ejected electron (15 eV) for single ionization of He by 200 eV electrons. (a) experiment, (b) CCC calculation

In such a 3D plot, the FDCS for a particular direction is given as the distance from the origin of the plot (also the collision point) to the point on the surface which is intersected by the ionized electron's trajectory. The cross section patterns are governed by the well-known double-lobe structure: the binary lobe in the direction of momentum transfer $\boldsymbol{q}$ corresponding to electrons emitted in a single binary collision with the projectile and the recoil lobe in the direction of $-\boldsymbol{q}$, where the target electron initially going along the $\boldsymbol{q}$ direction backscatters in the ionic potential. Due to the long-range nature of the Coulomb-force, the emitted electron is repelled by the outgoing projectile. For this reason, the binary and recoil lobes do not point exactly along the directions of $\boldsymbol{q}$ and $-\boldsymbol{q}$, respectively. Instead they are tilted away from the outgoing projectile direction. The importance of such post-collision interaction (PCI) effect increases with decreasing relative velocity of the two outgoing electrons.

For the present asymmetric kinematics, the angular electron emission pattern has been well studied only in coplanar experiments. The three-dimensional representation in figure 2 allows for a more complete investigation. Also shown in Figure 2 is the corresponding cross sections predicted by the convergent close coupling (CCC) method. In a global view, good agreement between experimental data and CCC calculation is visible. The magnitudes and the angular position of the binary and recoil lobes are nicely reproduced. More detailed comparisons show discrepancies in the region bridging the angular range between the binary and recoil lobes, which is the plane perpendicular to the scattering plane including the incoming projectile. In this perpendicular plane higher intensity is observed in experiment than the calculation. Such phenomenon was also clearly visible in the (e, 2e) data on He at 1keV impact energy [17].

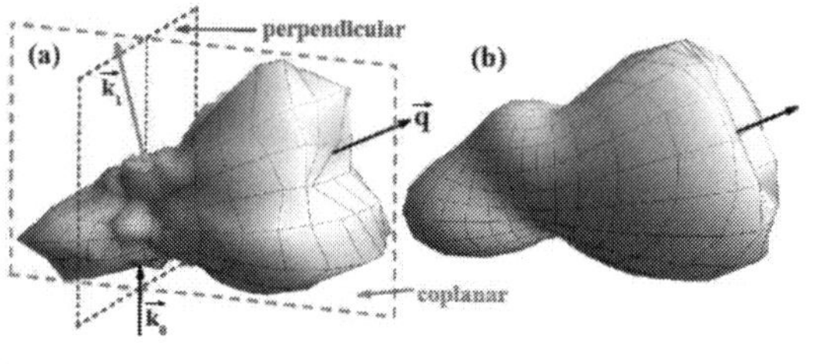

Figure 3. 3D images of the ionized electron (10 eV) emission pattern for single ionization of Ar (3***p***) by 200 eV electrons. (a) experiment, (b) DW2–RM calculation.

For the heavier target of Ar, the (e 2e) differential cross section behaviour is more complex than He, because of its many-electron property. As an example, Figure 3 presents the 3D FDCS for Ar at a projectile scattering angle of θ_1 = 10° in comparison with predictions from a hybrid second-order DWBA + R-matrix model [5]. The energy of the emitted electron is fixed at E_2 = 10 eV. From the global view of the 3D plots, we see the qualitative agreements between the experiment and the theory. The patterns are again governed by the binary lobe and the recoil lobe. In both experimental and theoretical plots the binary lobe shows a minimum in the momentum transfer direction. This minimum is due to the momentum profile of the Ar 3***p***-orbital. Which is in contrast to the He **1*s***-orbial momentum profile showing a maximum. The experimental data also highlight some discrepancies compared with the calculations, especially in the region, in between the binary and the recoil lobes. The experiment shows a double-peak structure similar to the binary peak feature in the scattering plane, instead the calculation only shows one peak. This might be a signature of the importance of projectile-nucleus scattering interaction.

3.2. ionization of He and H_2 at low energy

(e, 2e) on simple diatomic molecules has attracted increased attention recently since strong molecular structure effects were observed for electron emission in the plane perpendicular to the incoming beam direction [13, 14]. In order to study these effects more comprehensively we have performed comparative (e, 2e) experiment on the two-electron targets He and H_2 covering the full solid angle of electron emission at 16 eV excess energy.

In Figure 4, we compare our experimental results on He with previous experiments and CCC calculations, where one slow electron (E_1 = 4eV) is fixed to θ_1 = 80° in Figure 4(a) and θ_1 = 147° in Figure 4(b) and the cross section is plotted as a function of the second electron emission angle θ_2 in the scattering plane. The agreement is perfect between the two experiments and the CCC calculations. One of the advantages of our new experiments is that the full angular range is reached not only for the angular distributions in the scattering plane but also for the full 4π solid angle.

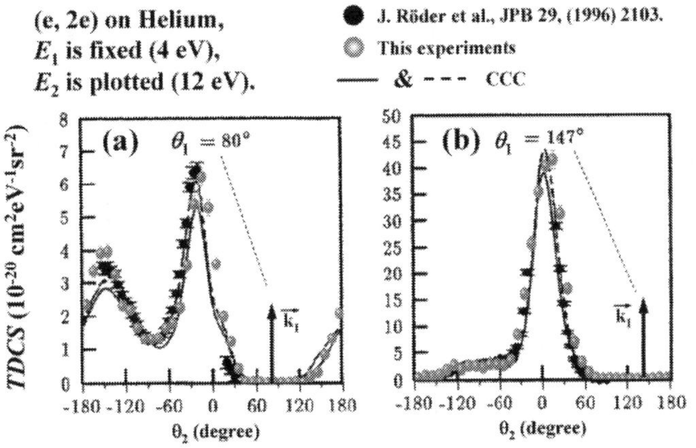

Figure 4. (e, 2e) on He at 16 eV excess energy, one of the outgoing electrons ($\boldsymbol{E}_1$ = 4eV) is fixed to $\boldsymbol{\theta}_1$ = 80° for (a) and $\boldsymbol{\theta}_1$ = 147° for (b), and the second outgoing electron angle ($\boldsymbol{E}_2$ = 12eV) is plotted in the scattering plane. Also included are the previous experimental and theoretical results in [18].

The full 3D image of the second electron emissions is shown in Figure 5. The (e, 2e) cross sections for H_2 and He are plotted as three-dimensional polar plots representing the emission pattern for the fast electron (**$\boldsymbol{E}_2$ = 12 eV**). The slow electron (**$\boldsymbol{E}_1$ = 4 eV**) is fixed to $\boldsymbol{\theta}_1$ = 70°.

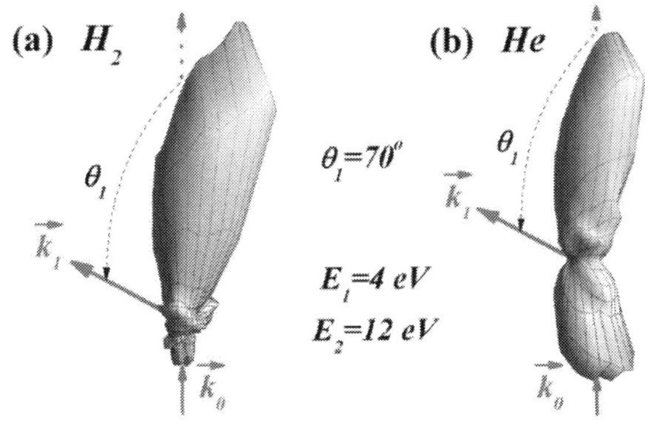

Figure 5. (e, 2e) cross sections as a 3D polar plot for the electron emission (12 eV) from H_2 (a) E_0=31.5 eV and He (b) E_0=40.5 eV.

It can be seen from the 3D images that the cross sections are governed by the well-known double lobe structure: the binary lobe and the recoil lobe. Distinct differences with a much stronger recoil intensity (bottom lobe) for He can be seen which can be attributed to the stronger nuclear potential in comparison with H_2 where the two positive charges are 1.4 a.u. apart from each other. The 3D images also show richer structure in the cross section of H_2 target than He in the range perpendicular to the incident beam direction [13, 14]. More detailed comparisons will be discussed in [19].

4. Summary

Kinematically complete experiments of (e, 2e) on He and Ar at intermediate impact energy of 200 eV and single ionization of He and H_2 by lower energy electrons (16 eV excess energy) have been performed. Full 3D images of the electron emission are obtained. For the intermediate projectile energy and heavier atomic collisions, strong out-of-plane intensity is observed. The discrepancies between experiments and theories are possibly due to approximations in the description of the projectile-target and projectile-ejected electron interactions. For low energy collisions, distinct differences between atomic and molecular ionization are visible, where much stronger recoil intensity for He is observed in comparison with H_2. Furthermore H_2 shows richer structures outside the scattering plane than He.

5. References

[1] Ehrhardt H, et al., 1969 *Phys. Rev. Lett.* **22** 89
[2] Resigno T N, et al., 1999 *Science* **286** 2474
[3] Bray I 2002 *Phys. Rev. Lett.* **89** 273201
[4] Colgan J, et al., 2006 *Phys. Rev. A* **73** 042710
[5] Bartschat K, et al., 2005 *Phys. Rev. A* **72** 022728
[6] Prideaux A, et al., 2005 *Phys. Rev. A* **72** 032702
[7] Naja A, et al., 2008 *J. Phys. B* **41** 085205
[8] Stevenson M A, et al., 2008 *Phys. Rev. A* **77** 032708
[9] Schulz M, et al., 2003 *Nature (London)* **422** 48
[10] Dürr M, et al., 2006 *Phys. Rev. Lett.* **96** 243202
[11] Schulz M and Madison D H 2006 *Int. J. Mod. Phys.* **21** 3649
[12] Schulz M, et al., 2007 *Phys. Rev. A* **76** 032712
[13] Al-Hagan O, et al., 2009 *Nature Physics* **5** 59
[14] Colgan J, et al., 2008 *Phys. Rev. Lett.* **101** 233201
[15] Ullrich J, et al., 2003 *Rep. Prog. Phys.* **66** 1463
[16] Dürr M, et al., 2006 *J. Phys. B* **39** 4097
[17] Dürr M, et al., 2008 *Phys. Rev. A* **77** 032717
[18] Röder J, et al., 1996 *J. Phys. B* **29** 2103
[19] Ren X, et al., 2009 in preparation.

Journal of Physics: Conference Series 212 (2010) 012004 doi:10.1088/1742-6596/212/1/012004

Recent theoretical progress in treating electron impact ionization of molecules

[1]Ola Al-Hagan, Chuangang Ning[2], Kate Nixon[3], Andrew Murray[3], Christopher Colyer[4], Mark Stevenson[4], Birgit Lohmann[4] and Don Madison[1]

[1]Department of Physics, Missouri University of Science & Technology, Rolla, MO 65409, USA
[2]Department of Physics and Key Laboratory of Atomic and Molecular NanoSciences of MOE, Tsinghua University, Beijing 100084, People's Republic of China
[3]School of Physics & Astronomy, University of Manchester, Manchester M13 9PL, United Kingdom
[4]ARC Centre for Antimatter-Matter Studies, University of Adelaide, Adelaide, SA 5005, Australia

E-mail:oan6f@mst.edu

Abstract. (e,2e) ionization differential cross sections are presented for several molecules. We will compare experimental results with theoretical calculations using the molecular three body distorted wave (M3DW) approximations for H_2 , N_2 , H_2O and Formic Acid (FA) using better wave-function for the molecules than we had in previous works. Generally, good agreement is found between the M3DW approach and experiments.

1. Introduction
There has been impressive progress in the area of theoretical treatments of charge particle collisions with atoms and molecules in the last decade. There have been many (e, 2e) studies for ionization of atoms and this area is now fairly mature. There have been some experimental and theoretical studies performed for the (e, 2e) processes with molecular targets but most of these studies have been performed either for high incident energies or for small molecules [1-4]. Most recently, low to intermediate incident energies have been reported for relatively simple molecular systems [5-6]. For these cases the dynamics of the ionization collisions become important and therefore more sophisticated models are needed to get good agreement with the experimental data.

In this paper, we will use the molecular three-body distorted wave (M3DW) approximation method coupled with the orientation averaged molecular orbital (OAMO) approximation. We apply this treatment to calculate the triple differential cross section (TDCS) for a variety of electron angles and energies for H_2, N_2, H_2O and HCOOH (Formic Acid - FA) using better wave-function for the molecules than we had in previous works.

2. Theory
The molecular 3-body distorted wave (M3DW) approximation has been presented by our group in previous publications [7-9] so only a brief outline of the theory will be presented. The triple differential cross section (TDCS) for the M3DW is giving by:

$$\frac{d^5\sigma}{d\Omega_a\Omega_b dE_b} = \frac{1}{(2\pi)^5}\frac{k_a k_b}{k_i}|T|^2 \tag{1}$$

Where $\vec{k}_i$,$\vec{k}_a$, and $\vec{k}_b$ are the wave vectors for the initial, scattered and ejected electrons. The amplitude is given by:

$$T = \left\langle \chi_a^-(\vec{k}_a,\mathbf{r}_1)\chi_b^-(\vec{k}_b,\mathbf{r}_2)C_{scat-eject}(r_{12})\,|\,V-U_i\,|\,\phi_j^{OA}(\mathbf{r}_2)\chi_i^+(\vec{k}_i,\mathbf{r}_1)\right\rangle \tag{2}$$

Where r_1 and r_2 are the coordinates of the incident and the bound electrons, χ_i, χ_a, and χ_b are the distorted waves for the incident, scattered, and ejected electrons respectively, and $\phi_j^{OA}(r_2)$ is the initial bound-state wave-function which is approximate as the orientation averaged molecular wave-function for the molecular orbital of interest. The molecular wave function was calculated by Ning using density functional theory (DFT) along with the standard hybrid B3LYP [10] functional by means of the ADF 2007 (Amsterdam Density Functional) program [11] with the TZ2P (triple-zeta with two polarization functions) Slater type basis sets. The present molecular wave-functions are better than the ones we used in previous works. The factor $C_{scat-eject}(r_{12})$ is the Coulomb-distortion factor between the two final state electrons, V is the initial state interaction potential between the incident electron and the neutral molecule, and U_i is a spherically symmetric distorting potential which used to calculate the initial-state distorted wave for the incident electron $\chi_i^+(\vec{k}_i,\mathbf{r}_1)$.

The molecular distorted waves are calculated using a spherically averaged distorting potential as described in previous works [7-9]. The Schrödinger equation for the incoming electron wave-function is given by:

$$(T + U_i - \frac{k_i^2}{2})\chi_i^+(\vec{k}_i, r) = 0 \tag{3}$$

where T is the kinetic energy operator. The initial state distorting potential contains three components $U_i = U_s + U_E + U_{CP}$, where U_s is the initial state spherically symmetric static potential which is obtained from the molecular charge density averaged over all angular orientations, U_E is the exchange potential of Furness-McCarthy [12] which approximates the effect of the continuum electron exchanging with the passive bound electrons in the molecule, and U_{CP} is the correlation-polarization potential of Perdew and Zunger [13,14].

The final state for the system is approximated as a product of distorted waves for the two continuum electrons times the average Coulomb-distortion factor. The final state distorted waves are calculated as the initial state except that the final state spherically symmetric static distorting potential for the molecular ion which is used for U_s.

3. Results and discussion

3.1. Molecular hydrogen(H_2)

Our recent study using the M3DW method yielded good agreement with the experimental measurements for triply differential cross sections (TDCS) for ionization of both H_2 and He by electron impact in a plane perpendicular to the incident beam direction with symmetric final state energies [15]. Figures 1 and 2 contain a comparison between our calculations and some recent experimental data [16] for ionization of H_2 taken by Andrew Murray and Christian Kaiser at Manchester University. The Manchester apparatus is designed such that the angle between the incident beam direction and the detection plane (defined as ψ) can be varied. The scattering plane corresponds to $\psi=0^0$ and the perpendicular plane corresponds to $\psi=90^0$. The TDCS results in figures 1 and 2 are plotted as a function of the half-angle between the two final state electrons in the detection plane (i.e. 2ξ is the angle between the electrons in the detection plane). For low incident electron energies, we have found that using the full Coulomb-distortion factor $C(r_{12})$ in M3DW calculations overestimates the effect of the final state electron-electron repulsion, normally called the post collision interaction (PCI), while the Ward-Macek average C-factor $C(r_{12}^{ave})$ [17] yields better agreement with experimental data so we have used the Ward-Macek approximation.

Although the experimental data are not absolute, only one normalization factor is needed for the different ψ angles and we have chosen to normalize experiment to theory for $\psi=90^0$. Two different M3DW calculations are presented – one including the correlation polarization potential and one excluding it. As can be seen from figures 1 and 2, there is good agreement between the experiment and the theory for large values of ψ especially in the perpendicular plane when the correlation-polarization potential is included. At low ψ values, the agreement between the experiment and the theory is not as satisfactory. The largest experimental cross sections for both equal ($E_a=E_b=10$ eV) and unequal ($E_a=18$ eV, $E_b=2$ eV) energy sharing were not in the scattering plane but rather in a plane where $\psi=45^0$. The M3DW also predicts the largest cross sections for the 45^0 plane if correlation and polarization is included in the calculations.

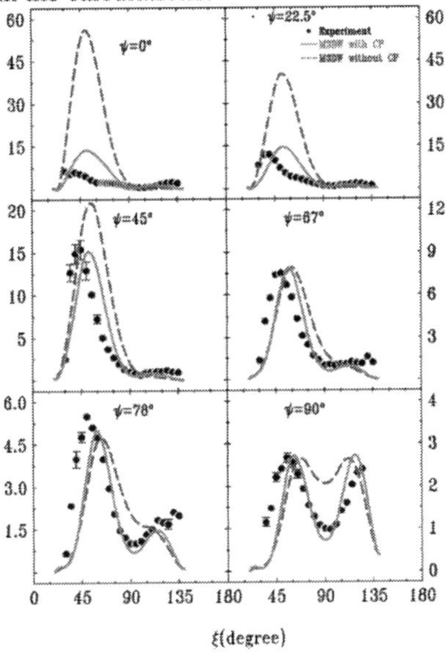

Figure 1. TDCS for electron impact ionization of H_2 for equal final state energies $E_a=E_b=10$ eV. See text for definition of angles. The measurements are compared with M3DW calculations obtained with and without the correlation-polarization potential.

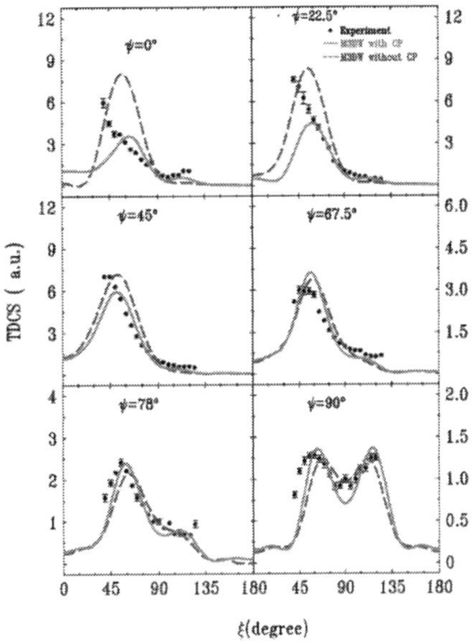

Figure 2. TDCS for the electron impact ionization of H_2 for unequal final state energies $E_a=18$ eV and $E_b=2$ eV. See text for definition of angles.

3.2. Molecular nitrogen (N_2)

N_2 measurements are of particular interest due to the possibility of observing the effects of 2-center Young's-type interference terms in the cross sections [18]. Gao et al. [19] predicted a very strong Young's type interference effect for ionization of the $3\sigma_g$ state of N_2 for small projectile scattering angles when the ejected electron comes out at 180^0 (i.e. the backward beam direction) but this prediction is yet to be verified experimentally. This prediction resulted from a M3DW calculation using a polarization potential containing arbitrary cut-off parameters and a fairly elementary molecular orbital. We repeated these calculations using the M3DW method with an improved correlation-polarization potential [13-14] and improved molecular orbitals. The M3DW with the improved polarization potential and original molecular orbital is shown as the blue dotted line in figure 3 and the agreement with experiment improved but there was a predicted peak near 100^0 which is not seen in the experimental results. Then we did another M3DW calculation and this time we used a better wave-function calculated by Ning. The M3DW with Ning's wave-function is shown also as the solid red line. As can be seen from figure 3, the new calculation is in even better agreement with experimental data and the theory still predicts a Young's type interference peak around 180^0. Since the agreement between theory and the experiment is fairly good, we are encouraged to think that the predicted 180^0 peak may be real. Until now, the existing experimental data is inconclusive concerning the existence of Young's interference effects for N_2.

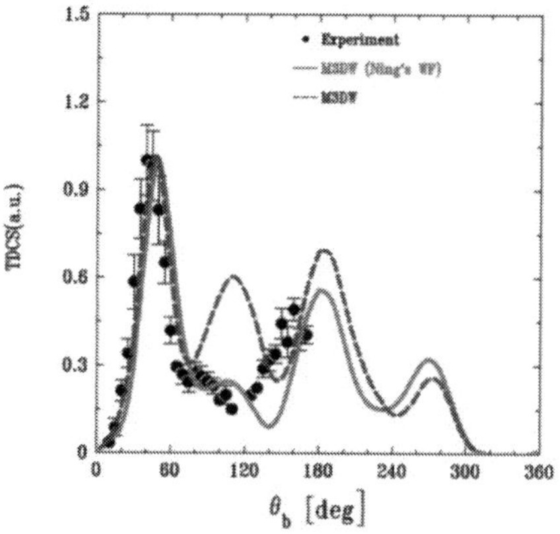

Figure 3. TDCS for the $3\sigma_g$ state of N_2 with E_0=75.6 eV, E_a=E_b=30 eV and θ_a=22^0. The experiment data are compared to two sets of M3DW. The dotted blue line is the M3DW using an old wave-function and the solid red line is the M3DW using an improved wave-function. The experimental data are those of Murray et al. [20].

3.3. Water (H_2O)

A couple years ago we compared the results of the M3DW method with experimental results for ionization of the $1b_1$ state of H_2O [21] and we found qualitative agreement with experiment but the results were somewhat disappointing. We now believe that the disappointing results stemmed from the OAMO being invalid for the $1b_1$ state. Kate Nixon and Andrew Murray have very recently measured triple differential cross sections for low incident energy electron-impact ionization of the $3a_1$ molecular state of H_2O and the OAMO approximation should be much better for this state. They used the same experimental apparatus as for H_2. Figure 4 shows the experimental and theoretical TDCS for H_2O in the symmetric coplanar geometry with excess energy of 10 eV and 20 eV. There is a relatively good agreement between the experimental data and the M3DW (including the correlation-polarization potentials) and the DWBA calculations which is the same calculation as the M3DW except the PCI term is not included in the calculations. The DWBA without PCI has unphysically large cross sections for 20 eV excess energy when the two electrons leave the collisions in the same direction and this is a common failure of the DWBA. The agreement between experiment and theory found here for the $3a_1$ state is better than we previously found for the $1b_1$ state indicating that the OAMO approximation is much better for this state.

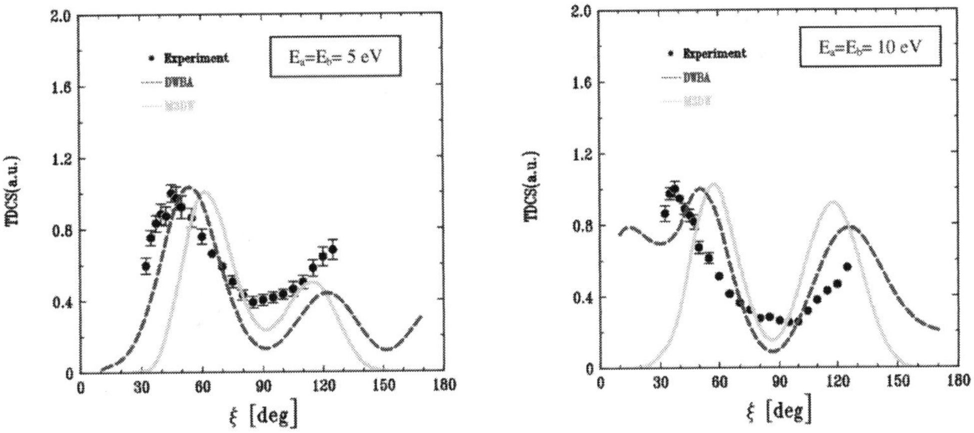

Figure 4. TDCS for electron impact ionization of H_2O in symmetric coplanar geometry as a function of ξ (2ξ is the angle between the two outgoing electrons). The cross sections are presented for excess energies of 10 eV and 20 eV.

3.4. Formic acid (HCOOH)

Birgit Lohmann's group at the ARC Center of Excellence for Antimatter-Matter studies at the University of Adelaide, Australia have recently measured (e,2e) ionization differential cross sections ionization of formic acid (HCOOH) for an incident electron energy of 100 eV and an ejected electron energy of 10 eV. This is a planar molecule with carbon near the center of mass which is of biological interest. The HOMO (highest occupied molecular orbital) is the 10a′ (ionization potential of 11.6 eV) and the next state is the 2a″ (ionization potential of 12.45 eV) and these two states cannot be resolved in the experiment so the experimental data represent a sum of the 10a′ and 2a″ states. Unfortunately the OAMO approximation is not valid for the 2a″ state so we can only calculate results for the 10a′ HOMO state.

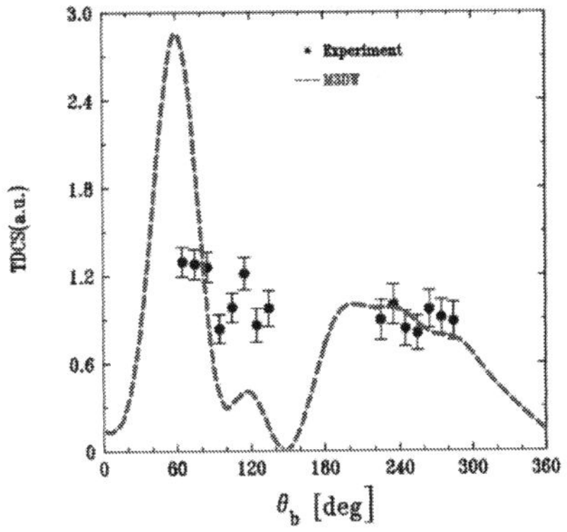

Figure 5. Triple differential cross section of ionization of Formic Acid with E_0=100 eV, E_b=10 eV and $\theta_a=10^0$ as a function of the ejected electron angle. The Experimental measurements represent a sum of the 10a′ and 2a″ states while the M3DW results are for the 10a′ state only.

Figure 5 shows a comparison between the experimental data and the M3DW results without the correlation-polarization potential. The experimental data have been normalized to theory in the recoil region. Although the M3DW agrees well with the shape of the recoil peak, the theory predicts a larger and more pronounced binary peak than found in the experimental data. Since the effect of the 2a″ is unknown, it would be highly desirable to have experimental results which resolved the 10a′ state to ascertain how well the M3DW works for a larger molecule such as this.

4. Conclusions

In this paper, we presented TDCS for electron impact ionization of different molecules and compared the experimental results with the M3DW. Overall the theory is in reasonably good agreement with the experiments. Including the correlation polarization potential in the M3DW improved the agreement with the experiment for H_2, N_2, and H_2O. Replacing our old wave-function with Ning's wave-function has also improved the agreement with experiment for N_2 (the H_2 results did not change). We looked at two larger molecules – water and formic acid. We found better agreement with experiment for the $3a_1$ state of H_2O than we had previously found for the $1b_1$ state. For formic acid, we found good agreement with the shape of the recoil peak but not the binary peak. However, the experimental data represented a sum of the 10a′ and 2a″ states while we were only able to calculate results for the 10a′ state so validity of the M3DW method using OAMO for large molecules has not been adequately tested.

Acknowledgment

This work supported by the National Science Foundation under Grant. No. PHY-0757749 and the Australian Research Council Centre of Excellence for Antimatter-Matter Studies and the American NSF Also, we acknowledge the British Council for funding Kate Nixon, the University of Manchester for funding Christopher Kaiser, and the Saudi Ministry of Higher Education's King Abdullah Bin Abdul-Aziz Scholarship for funding Ola Al-Hagan. The author Chuangang Ning would like to acknowledge the support of the National Natural Science Foundation of China under contract No. 10704046.

References

[1] McCarthy I E and Weigold E 1976 *Phys. Rep.* C **27** 275.
[2] McCarthy I E and Weigold E 1988 *Rep.Prgo. Phys.* **51** 299.
[3] Stia C R,Fojon O A, Weck P F, Hanssen J, joulakian B J, and Rivarola R D 2002 *Phys. Rev.* A **66** 052709.
[4] Gao J, Madison D H, and Peacher J L 2006 *J. Phys.* B **39** 1275.
[5] Milne-Brownlie D S,Cavanagh S J, Lohmann B, Champion C, Hervieux P A, and Hanssen J 2004 *Phys. Rev.* A **69** 032701.
[6] Murray A J, 2005 *J. Phys.* B **38** 1999.
[7] Gao J, Madison D H, Peacher J L 2005 *J. Chem. Phys.* **123** 204314.
[8] Gao J, Madison D H, Peacher J L 2005 *Phys. Rev.* A **72** 032721.
[9] Gao J, Madison D H, Peacher J L 2005 *J. Chem. Phys.* **123** 204302.
[10] C. Lee, W. Yang, and R. G. Parr 1988 Phys. Rev. B **37**, 785.
[11] C. F. Guerra *et al.* 1998 Theor. Chem. Acc. **99**, 391.
[12] Furness J B and McCarthy I E 1973 *J.Phys. B* **6** 2280.
[13] Perdew J P and Zunger A 1981 *Phys. Rev.* B **23** 5048.
[14] Padial N Tand Norcross DW 1984 *phys. Rev.* A **29** 1742.
[15] Al-Hagan O, Kaiser C, Madison D H, and Murray A J 2008 *Nat. Phys* **5** 59.
[16] Colgan J, Al-Hagan O, Madison D H ,Kaiser C, Murray A J and Pindzola M S 2009 *Phys. Rev.* A**79** 052704.
[17] Ward S J and Macek J H 1994 *Phys. Rev.* A **49** 1049.
[18] Stia C R, Fojon O A, Weck P F, Hanssen J and Rivarola R D 2003 *J. Phys.* B **36** L257.
[19] Gao J, Madison D H and Peacher J L 2005 Phys. Rev. A **72** 032721.
[20] Murray A J, Hussey M J, Gao J, and Madison D H 2006 *J. Phys* B **39** 3945.
[21] Kaiser C, Spieker D, Gao J, Hussey M, Murray A J, and Madison D H 2007 *J. Phys. B: At. Mol. Opt. Phys.* **40** 2563.

Fully Differential Molecular-Frame Electron Impact Ionization Measurements

S Bellm[1,3], J Lower[1], D Mueller[2] and E Weigold[1]

[1]Research School of Physics and Engineering, The Australian National University, Canberra, ACT, 0200, Australia

[2]Department of Physics, University of North Texas, Denton, Texas, 76203, USA

E-mail: susan.bellm@adelaide.edu.au, erich.weigold@anu.edu.au

Abstract. (e,2e) measurements in the molecular frame using a new spectrometer are described. This work requires a triple coincidence experiment in which (e,2e+ion) events are detected. Preliminary results are presented for a nitrogen target, which illustrate the performance of the apparatus. The nitrogen data clearly shows that the (e,2e) cross section depends sensitively on the orientation of the molecule relative to the incident direction.

1. Introduction

(e,2e) experiments are a powerful tool for providing information on the electronic structure of atoms and molecules and on the mechanisms underlying electron impact ionization. However, until recently the information that could be obtained from these kinds of experiments for molecular targets has been limited because measurements have been averaged over the molecular orientations. The orientation of a molecule can be inferred from the momenta of ion fragments produced following dissociative ionization. The method relies on fragmentation occurring within the axial recoil approximation, that is that the molecule fragments in a time interval considerably faster than its rotational period [1], i.e. that the energy of dissociation is much greater than the rotational energies.

While numerous molecular-frame experiments have been reported for photoionization [2], only two studies have been reported for electron-impact ionization [3, 4], and these have been severely limited by poor statistics. These experiments are more difficult for electron-impact induced ionization than for the analogous photoionization measurements because an extra charged particle must be detected in coincidence and its kinematics fully determined. The photoionization experiments frequently employ charged particle imaging techniques such as cold target recoil ion momentum spectroscopy COLTRIMS [5] and velocity map imaging (VMI) [6], which can provide detailed kinetic energy and angular distributions of fragment ions. Again these techniques are simpler to implement for photoionization measurements as the electric and magnetic fields required for imaging can be used without interfering with the ionizing projectile.

Takahashi and co-workers performed the first (e,2e) measurement for H_2 molecules fixed in space [3]. Our group also performed a preliminary study to assess the feasibility of obtaining orientation

[3] Present Address: ARC Centre for Antimatter-Matter Studies, The University of Adelaide, Adelaide, SA, 5005, Australia

resolved data [4]. While these measurements showed that obtaining orientation-resolved data was indeed possible, they also showed large improvements in count rate would be required to develop a dependable technique. We have recently modified our high efficiency electron spectrometer to considerably improve both electron and ion detection efficiencies. In our preliminary study only ions emitted within an approximately 30 degree emission-cone were detected on a position-sensitive detector located above the reaction plane, corresponding to measurement of about 10% of the total solid angle. We have since introduced a pulsed ion extraction scheme so that all ions emitted over the full 4π solid angle are detected and their momenta determined.

Theory predicts very strong orientation dependent effects should be observed. Calculations performed on both N_2 and H_2 targets show that the cross sections are expected to be quite different for different molecular orientations [7, 8, 9]. To date there is no experimental data to test these predictions. In order to test the theoretical models and to shed new light on the process of electron impact dissociative ionization, we have developed an apparatus and the associated techniques to enable fully differential (e,2e+ion) cross sections to be measured in the molecular frame.

2. Experiment

2.1. Overview

As the electron analyzers have been described in detail in a previous publication [10] only an overview will be given here which will focus on details relevant to the incorporation of the new ion spectrometer into this apparatus. The electron analyzers and ion spectrometer are shown schematically in figure 1.

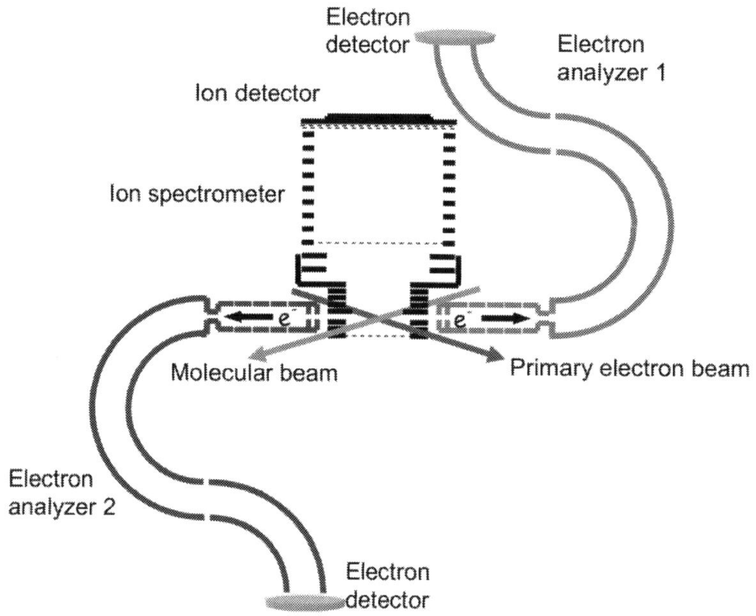

Figure 1. Schematic of the apparatus.

A primary beam of spin polarized electrons is produced from a strained GaAs photocathode by illuminating it with circularly polarized laser light. The light at 850 nm is produced from a laser diode and first passed through an acousto-optic modulator, which periodically switches the light on and off and accordingly turns the electron beam on and off. The electron beam is deflected though 90 degrees and transported at high energy through a differential pumping stage to the main collision chamber where it is decelerated to the collision energy of interest. Deceleration from the transport energy of around 700 eV occurs in two stages using a 7-element electrostatic lens system.

The electron beam then intersects a molecular target beam which is formed by the effusion of the target gas through a 0.5 mm internal diameter titanium capillary, mounted orthogonal to the electron beam. Both the electron beam and molecular target beam are oriented such that they are parallel to the face of the ion detector and lie within the scattering plane, which is defined by the momentum vectors of the incident and measured emitted electrons. Electrons emitted within the scattering plane are momentum and energy analysed in two separate toroidal sector electrostatic energy analysers. Each analyser consists of an entrance lens system, double toroidal sector electrodes and a time- and position-sensitive delay line detector [11].

When an electron pair is detected in coincidence, the extraction electrodes are switched on immediately and ions are extracted and imaged onto the ion detector. This is a third time- and position-sensitive detector, again consisting of an 80 mm active area microchannel plate pair and delay line anode [11]. It takes ~165 ns from the moment of ionization for the extraction field to be turned on. This delay is comprised of the time for an electron to travel through the spectrometer, the time to process the signal in the detection electronics and the time required to switch the extraction voltage from 0 to ~210 V. The momenta of the ions are reconstructed from their arrival times and positions on the detector by matching these to simulated arrival times and positions which are calculated using SIMION software [12].

2.2. Modifications to electron analysers

As a means of increasing the (e,2e) count rate for these molecular frame ionization experiments, the electron analysers have been reconfigured to allow for more asymmetric kinematics to be measured. For a molecular nitrogen target, Avaldi *et al* [13] demonstrated that a factor of 10 increase in data collection speed can be achieved if asymmetric rather than symmetric kinematics are employed. One analyser presently covers the range θ_1= -10° to -50° and the other θ_2 = 30° to 130°, where θ_1 and θ_2 are measured in the scattering plane with respect to the primary electron beam.

The electron analyzer entrance optics for both analyzers had to be modified to allow the ion extraction electrodes to be inserted around the interaction region. In addition, to help maximize the triple coincidence count rate, the electron analyzer lenses have been adapted to enable a greater electron transmission through the analyzers whilst still maintaining good energy resolution. With the modified lens system, electrons emitted with angles of up to +/-3° out of the scattering plane are detected.

2.3. Ion Spectrometer

The ion spectrometer is shown in detail in figure 2. The spectrometer consists of a series of titanium rings which are electrically isolated from each other. The first seven rings have an outer diameter of 78 mm and an inner diameter of 50 mm with the exception of the 4th ring which has a smaller inner diameter of 40 mm and is designed to block and therefore to reduce the number of background ions that reach the detector. The remaining rings have an outer diameter of 120 mm, the first few of these have an inner diameter of 80 mm and the next rings, which form a drift region, have a larger inner diameter of 100 mm.

The first ten rings are joined together with resistors which form a voltage divider chain and which are used to create a two-stage electric field. The fast pulsed extraction voltage of 212 V is applied to the first electrode. The gas capillary located midway between the 3rd an 4th rings has a separate pulsed voltage applied, which is set to match the potential at the interaction region when the ion extraction field is turned on. The high transmission mesh (~90% transmission) labeled M1 is held at ground potential. The rings between M2 and the ion detector are joined together and held at a constant potential of 81 V creating a drift region. Two more high transmission meshes are located immediately in front of the ion detector. These meshes are also both held at 81 V and are to shield the drift region from the high potential applied to the front of the ion detector. The first two rings at the end of the spectrometer are both filled with a high transmission copper mesh to allow the interaction region to be pumped more efficiently and the purpose of the double mesh is to prevent electric field penetration from outside the spectrometer so as to ensure a uniform extraction field at the interaction region.

A possible complication in performing these measurements can arise from the buildup of low-energy ions in the interaction region which will all be extracted when an (e,2e) event is identified and may result in false triple coincidence events being detected. To reduce this buildup we have introduced a cleaning and measuring cycle which is repeated every 10 microseconds. During the cleaning time a potential of 178 V is applied to the gas capillary to sweep all the positively charged ions from the interaction region and the electron beam is turned off to prevent the production of further ions. At the commencement of the measurement time the potential on the needle is reduced to 0 V and the electron beam is switched on for 2 microseconds. The electron beam current, which affects the number of ionization events, is set at a level such that the ratio of ions detected to (e,2e) events is independent of beam current and only true triple coincidence events are measured.

Also shown in figure 2 are N^+ ion trajectories modeled using the program SIMION. The N^+ ions, all with the same energy of 10 eV, are started from five different positions separated by 1 mm. Twelve ions are started at each position with different initial angles which are stepped in 30° intervals. In the simulation the ions are allowed to drift field free for 165 ns before potentials are applied to the extraction electrodes and the ions are imaged onto the detector.

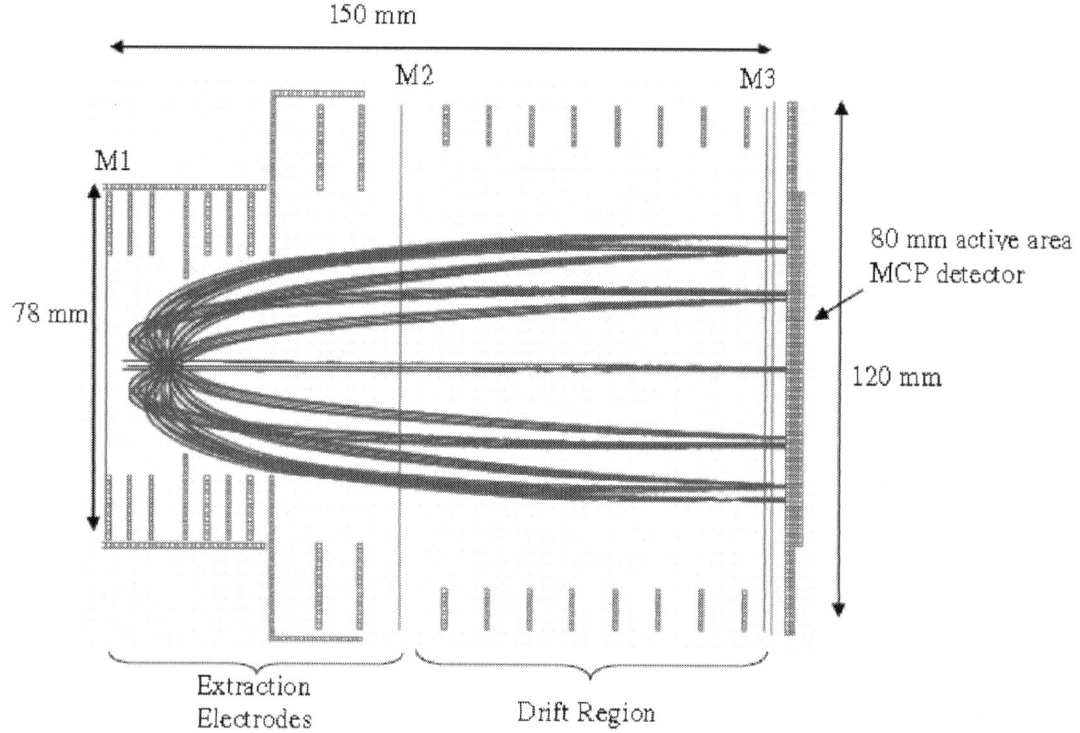

Figure 2. Simion simulation showing a cross section through the centre of the ion spectrometer. The ion trajectories are for 10 eV N^+ ions. See text for details.

Ions with different initial spatial positions but the same initial energy and angle of emission are focused together on the detector.

Experiments to test the spectrometer were performed with a nitrogen target at an incident electron energy of 178 eV. At this energy around 20% of ionization events lead to dissociation [14]. The scattered projectile (fast) and ejected target (slow) electrons were detected with mean energies of E_1=100 eV and E_2= 40 eV respectively and in the angular range θ_1 = -10° to -50° degrees and θ_2 = 30° to 70°. Fragment ions emitted with energies up to 10 eV and over 4π solid angle are detected.

3. Results

An N^+ ion time of flight spectrum for nitrogen measured using the apparatus is presented in figure 3. The velocity component of the ion's momentum perpendicular to the extraction direction (i.e. in the scattering plane) determines the position it arrives at on the two dimensional detector, zero energy ions arriving in the centre of the detector. The ion spectrometer has been designed so that information about the ions' velocity component parallel to the ion extraction direction is preserved, which enables the full 4π angular and momentum distribution of fragment ions to be determined. This is in contrast to the velocity map imaging technique [6], in which the mass resolution is maximized by time focusing so that ions with the same mass arrive in a narrow time interval and information about the momenta of the ions along the spectrometer axis is lost. There is a narrow peak occurring at 4.8 microseconds in the time-of-flight spectrum corresponding to N^+ fragment ions with essentially zero kinetic energy and also to very low energy N_2^{2+} ions. N_2^+ ions all arrive at a later time due to their higher mass to charge ratio and can therefore be easily eliminated.

Figure 4 shows the ion kinetic energy distribution of N^+ fragment ions reconstructed from their arrival times and positions on the detector. Only the momenta of ions with energies between 0.2 eV and 10 eV are reconstructed, as the very low energy ions may include contributions from N_2^{2+} ions.

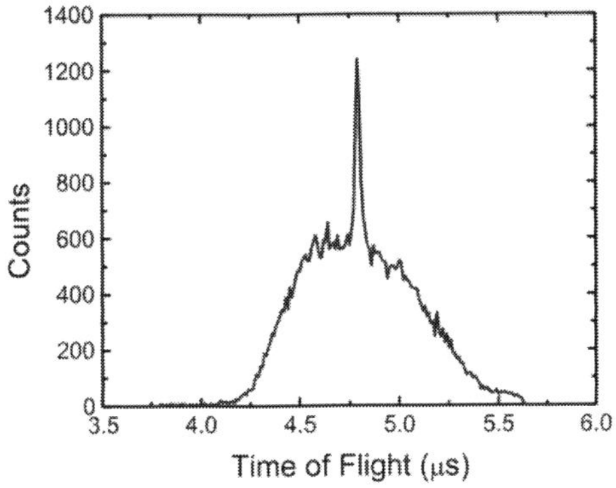

Figure 3. A typical ion time of flight spectrum from nitrogen. See text for details.

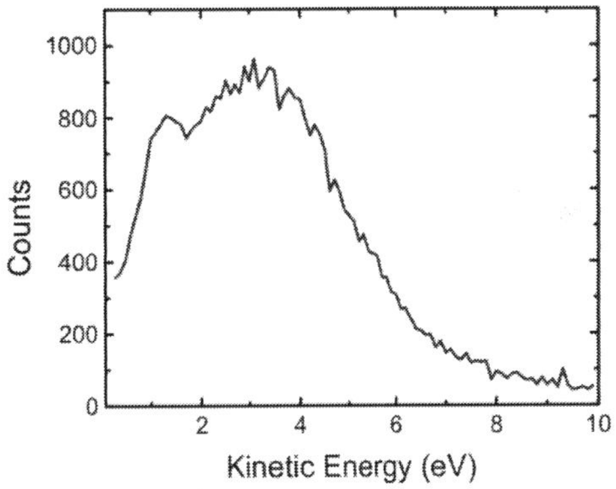

Figure 4. Ion kinetic energy distribution of N^+ ions emitted over all angles. See text for details.

Figure 5 shows the correlation between N^+ ion kinetic energy and electron binding energy. The intensity lies along lines with slope 0.5 corresponding to different dissociation channels where the N^+ and N fragments are formed in different internal energy states. Approximately 2200 triple coincidence events per day are obtained with a nitrogen target.

The orientation dependence of fragmentation of molecular nitrogen measured using the new spectrometer is shown in figure 6. Summed ejected (slow) electron counts measured between 30° and 70° are plotted as a function of the scattering angle of the fast electron for ions in the two bands in figure 5. Two orientations of the molecules perpendicular to the incident electron beam have been chosen by selecting triple coincidence events in which the ion fragments were emitted within 30°

emission cones oriented either perpendicular to the scattering plane or lying within the scattering plane.

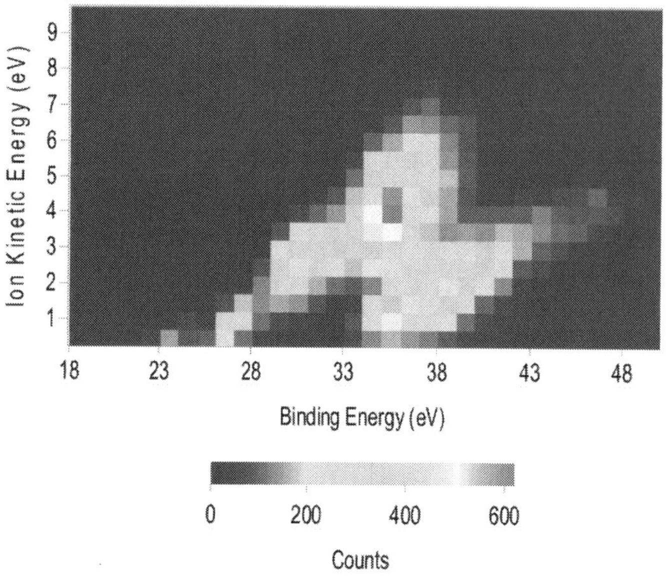

Figure 5. Ion kinetic energy as a function of electron binding energy.

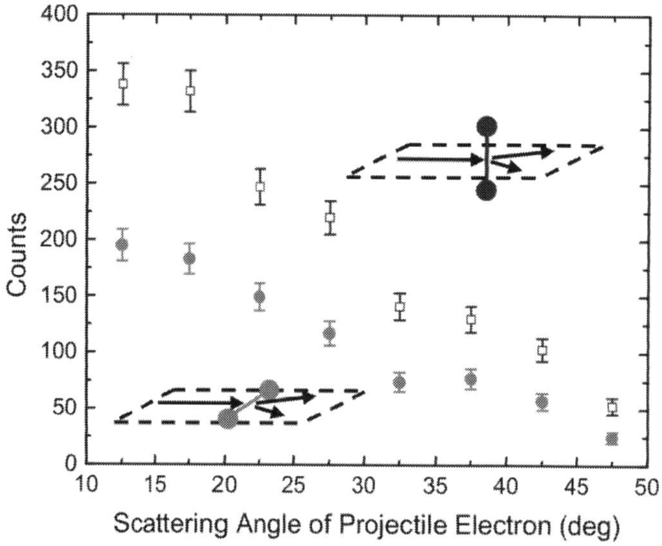

Figure 6. Orientation dependence of dissociative fragmentation for molecular nitrogen resulting from an (e, 2e) ionization event. Ejected electron counts are plotted as a function of the projectile scattering angle. The curves shown are for different orientations of the nitrogen molecules. Solid circles: molecules lying within the scattering plane, but perpendicular to the incident electron beam direction. Open squares: molecules oriented perpendicular to scattering plane.

When the molecule is oriented perpendicular to the scattering plane the cross section is significantly larger than when the molecule lies within the scattering plane. There are currently no calculations available for comparison with our measurements. The calculations of Gao *et al* [9] for nitrogen are for transitions to the ground state of N_2^+, which does not dissociate.

The present data has only undergone preliminary analysis. Further analysis of the data, including investigating the dissociative fragmentation of different excited states and additional molecular orientations, will appear in a future publication.

4. Conclusion

Here we have described progress in the development of a new (e,2e+ion) spectrometer for use in studies of the electron-impact-induced dissociative ionization of molecules. We have shown that the apparatus can be used to provide the angular dependencies of molecular fragmentation processes and

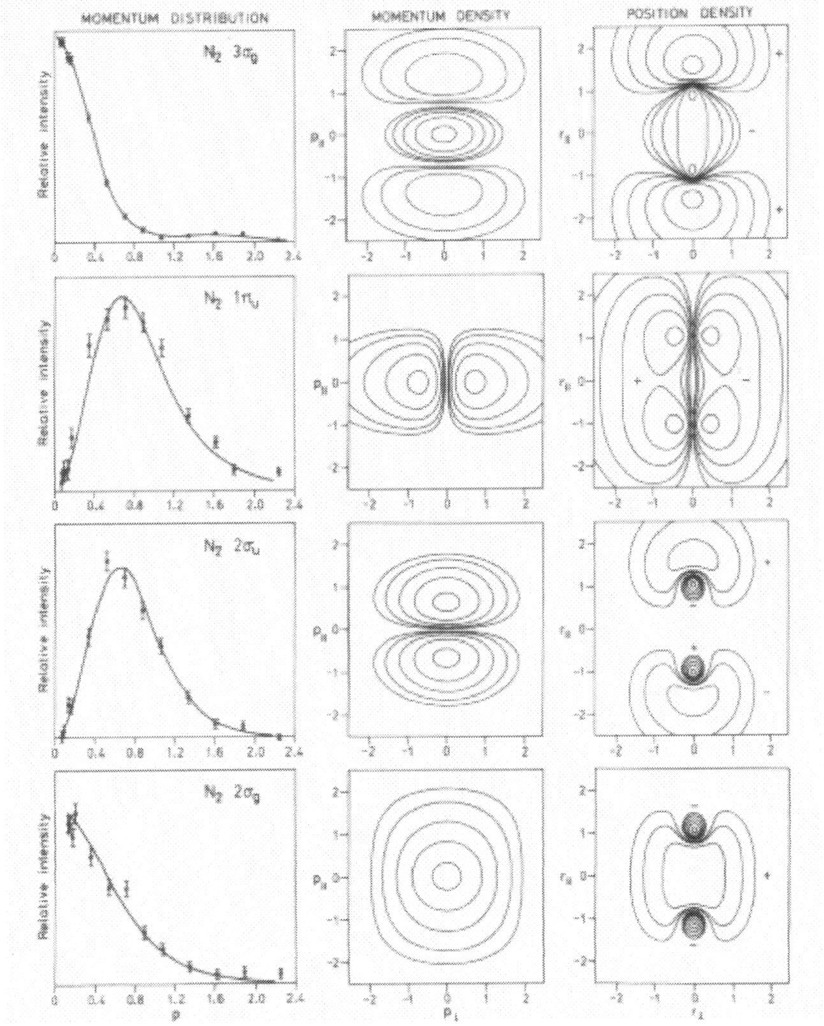

Figure 7. Experimental [16] and calculated (SCF) spherically averaged orbital momentum densities (left hand panels) and momentum and position densities for molecular nitrogen.

thus enables (e,2e) measurements to be performed within the molecular frame. The technique promises to provide a greater understanding of electron-impact-induced molecular fragmentation processes.

The work reported here also opens up the possibility of measuring the momentum distributions of molecular orbitals in the molecular frame using electron momentum spectroscopy (EMS) [15]. In the past EMS has been limited to measuring spherically averaged orbital momentum distributions. Figure 7 shows EMS data [16] for the valence orbitals of nitrogen compared with calculated spherically averaged orbital momentum densities as well as contour maps of corresponding position and momentum densities. This indicates the extra information that can be obtained for measurements in the molecular frame.

Acknowledgements
This work was supported by the Australian Research Council.

References

[1] Zare R N 1972 *Mol. Photochem* **4** 1
[2] Weber T *et al.* 2004 *Nature* **431** 437
[3] Takahashi M, Watanabe N, Khajuria Y, Udagawa Y and Eland J H D 2005 *Phys. Rev. Lett.* **94** 213202
Takahashi M, Watanabe N, Khajuria Y, Nakayama K, Udagawa Y and Eland J H D 2004 *J. Electron Spectrosc. Rel. Phenom.* **141** 83
Takahashi M and Udagawa Y 2007 Toward three-dimensional electron momentum distributions *Nanoscale Interactions and their Applications: Essays in Honour of Ian McCarthy*, ed F Wang and M J Brunger (Transworld Research Network India) Chapter 15 pp 157-168
[4] Lower J, Baxendell J and Bellm S 2007 (e,2e) Measurements in the Molecular Frame *Nanoscale Interactions and their Applications: Essays in Honour of Ian McCarthy*, ed F Wang and M J Brunger (Transworld Research Network India) Chapter 9 pp 95-106
[5] Dörner R *et al.* 1997 *Nucl. Instrum. Methods* B **124** 225
[6] Eppink A T J B and Parker D H 1997 *Rev. Sci. Inst.* **68** 3477
[7] Al-Hagen O, Kaiser C, Madison D and Murray A 2009 *Nature Physics* **5** 59
[8] Colgan J, Pindzola M, Robicheaux F, Kaiser C, Murray A J and Madison D H 2008 *Phys. Rev. Lett.* **101** 233201
[9] Gao J, Madison D H and Peacher J L 2005 *Phys. Rev. A* **72** 032721
[10] Lower J, Panajotovic R, Bellm S and Weigold E 2008 *Rev. Sci. Inst.* **78** 111301
[11] RoentDek GmbH, Kelkheim, Germany
[12] SIMION, ion source software, P.O. Box 2726 Idaho Falls, ID 83404, USA
[13] Avaldi L, Camilloni R, Fainelle E and Stefani G 1987 *J. Phys. B. At. Mol. Phys.* **20** 4163
[14] Tian C and Vidal C R 1998 *J. Phys. B* **31** 5369
[15] Weigold E and McCarthy I E 1999 *Electron Momentum Spectroscopy* (Kluwer Acacdemic/Plenum Publishers New York)
[16] Cook J P D *et al* 1990 *Chem. Phys.* **141** 211

Single-photon double K-shell ionization of low-Z atoms

J Hoszowska[1], A S Kheifets[2], J-Cl Dousse[1], I Bray[3], W Cao[1], K Fennane[1], Y Kayser[1], M Kavčič[4], J Szlachetko[1] and M Szlachetko[1]

[1]Department of Physics, University of Fribourg, CH-1700 Fribourg, Switzerland
[2]Research School of Physical Sciences, Australian National University, ACT 0200 Canberra, Australia
[3]ARC Centre for Matter-Antimatter Studies, Curtin University, WA 6845 Perth, Australia
[4]J. Stefan Institute, SI-1001, Ljubljana, Slovenia

E-mail: `joanna.hoszowska@unifr.ch`

Abstract. The photon energy dependence of the double K-shell ionization of light atoms is reported. Experimental double-to-single photoionization cross section ratios for Mg, Al, Si and Ca were obtained from measurements of high-resolution x-ray emission spectra. The double photoionization (DPI) cross-sections for K-shell hollow atom production are compared to convergent close-coupling calculations (CCC) for neutral atoms and He-like ions. The relative importance of the initial-state and final-state electron-electron interactions to the K-shell DPI in many-electron atoms and two-electron ions is addressed. Physical mechanisms and scaling laws of the K-shell double photoionization are examined. A semiempirical universal scaling of the DPI cross sections with the effective nuclear charge for neutral atoms $2 \leq Z \leq 47$ is established.

1. Introduction

The double photoionization (DPI) process has been a subject of intensive research both on the experimental and theoretical side and is still attracting considerable attention [1–11]. Because the photon interacts with only one of the bound electrons, two-electron ejection resulting from single-photon absorption is driven solely by many-electron interactions. Electron correlation effects lie in the heart of understanding of atomic structure. Yet, their quantitative description is far from complete or satisfactory. In the quest for an exact theoretical treatment of electron correlations in many-body systems the investigation of the photon energy dependence of single-photon double ionization of atoms is of fundamental importance.

The K-shell DPI of neutral elements produces K-shell hollow atoms i.e., atoms with an empty innermost shell and outer shells occupied. Thanks to the high brilliance and energy-tunability of x-ray synchtrotron radiation sources, experiments on the photon energy dependence of the DPI in neutral atoms became feasible. However, due to the low probability ($\sim 10^{-2}$-10^{-6}) for creating double K-shell vacancies, experimental investigation of the K-shell DPI in neutral atoms is quite challenging [10–16]. We report on the photon energy dependence of the double K-shell ionization of Mg, Al, Si and Ca [17, 18] for a wide range of photon energies from the DPI threshold up and beyond the maximum of the double-to-single photoionization ratios P_{KK}. So far, experimental results for P_{KK} far beyond the broad maximum are only available for He [19]. The experimental double-to-single photoionization cross-section ratios are compared to convergent close-coupling calculations (CCC) for neutral atoms [18] and their corresponding two-electron ion counterparts. The relative contribution of the initial-state correlations and final-state electron-electron interactions to the K-shell DPI is addressed. Universal scaling of the DPI cross sections and the double-to-single photoionization ratios with the effective nuclear charge for neutral atoms is examined.

2. Experimental method

Experiments were performed at two undulator beam lines ID21 and ID26 at the European Synchrotron Radiation Facility (ESRF), using monochromatic synchrotron radiation and the Fribourg von Hamos Bragg-type curved crystal x-ray spectrometer [20]. The experimental method was based on the measurements of the high-resolution hypersatellite x-ray spectra following the radiative decay of the K-shell double vacancy states. For the studied low-Z elements, the $K\alpha_2^h$ hypersatellite ($^1P_1 \rightarrow {}^1S_0$) was predominant, the spin-flip $K\alpha_1^h$ transition ($^3P_1 \rightarrow {}^1S_0$) being dipole forbidden in the LS coupling scheme. As for light elements the K hypersatellites are partly overlapping with the L-satellites of the $K\beta$ transitions, the use of high-resolution was mandatory. The double-to-single photoionization cross section ratios P_{KK} were derived from the relative intensities of the resolved hypersatellite $K\alpha^h$ ($1s^{-2} \rightarrow 1s^{-1}2p^{-1}$) to the diagram $K\alpha$ ($1s^{-1} \rightarrow 2p^{-1}$) x-ray transitions using the expression $P_{KK}=\frac{I_{K\alpha^h}}{I_{K\alpha}}\frac{\omega_K}{\omega_{KK}}$, where ω_K and ω_{KK} are the fluorescence yields for the single- and double- hole states [21], respectively. The DPI cross sections were determined employing the relation $\sigma^{2+} = P_{KK}\sigma^{+}$, where σ^{+} stands for the single K-shell photoionization cross section from [22]. The $K\alpha^h$ and $K\alpha$ x-ray transitions were measured at different incident beam energies from 2.746 to 8.0 keV for Mg, from 3.122 to 7.0 keV for Al, from 3.6 to 10.0 keV for Si and for Ca in the 8.5 to 28.0 keV photon energy range. The intensities were corrected for sample self-absorption and absorption of the incident x-rays, the photon flux, the spectrometer solid angle, as well as for the detector quantum efficiency. The energy resolution of the spectrometer was comparable to the natural line widths of the measured K x-ray transitions. The incident photon flux was $\sim$1-3$\times 10^{12}$ ph/s at both beam lines.

3. Theoretical model

The single and double photoionization cross sections were calculated using the convergent close-coupling (CCC) method. In the CCC formalism, the DPI is treated as a two step process. The first is the full absorption of the photon energy by one electron. The second is the interaction of this electron with the nucleus and the remaining electron which results in the promotion of the remaining electron into the continuum. To describe the first step, the 20-term Hylleraas ground state wave-function due to Hart and Herzberg [23] was employed for Mg^{10+}. For Al^{11+} and Si^{12+} more accurate results were obtained with a 9-term multi-configuration Hartree-Fock (HF) ground state calculated using the computer code by Dyall *et al.* [24]. The final states with two electrons in the continuum were described by a close-coupling expansion on the basis of channel functions. Calculations in the three gauges of the electromagnetic operator produced convergent results. More information on the photoionization CCC formalism can be found in Refs. [25, 26].

For neutral atomic targets Mg and Ca the CCC calculations of the K-shell DPI were based on the frozen-core model with two active ns^2 electrons which was used previously for the valence shell DPI [27]. Different ground-state wave functions of various degrees of correlation were used. Results of these new *ab initio* nonperturbative calculations for many-electron systems, including gauge convergence issues and difficulties, are reported by Kheifets *et al.* [18]. In the present paper, we compare the experimental data with the CCC calculations performed in the velocity gauge using the Hartree-Fock ground-state wave functions. These results were found to be closest to the experimental ones.

Journal of Physics: Conference Series **212** (2010) 012006 doi:10.1088/1742-6596/212/1/012006

4. Results and Discussion

4.1. Double-to-single photoionization cross-section ratios

Single-photon double ionization is discussed in terms of two dominant mechanisms, namely, *shakeoff* (SO) and *knockout* (KO) [2–4, 9, 11, 28, 29]. Shake-off corresponds to a sudden ejection of the photoelectron following the single photon absorption and a subsequent change of the atomic field which may lead to a removal of the remaining electron to the continuum. To date, the shakeoff process is well defined in the asymptotic high-energy limit, but at finite energies no unique definition exists [30]. Despite this fact, the different SO expressions lead to quite similar numerical results. In case of KO, the outgoing photoelectron knocks out the second 1s electron in an (e,2e)-like electron impact half-collision. Initial ground-state correlations are important in SO, while final-state electron interactions govern KO. Although the two mechanisms have different interaction times and very different photon energy dependences, the separation of KO and SO with photon energy and the quantification of inteferences is not straightforward.

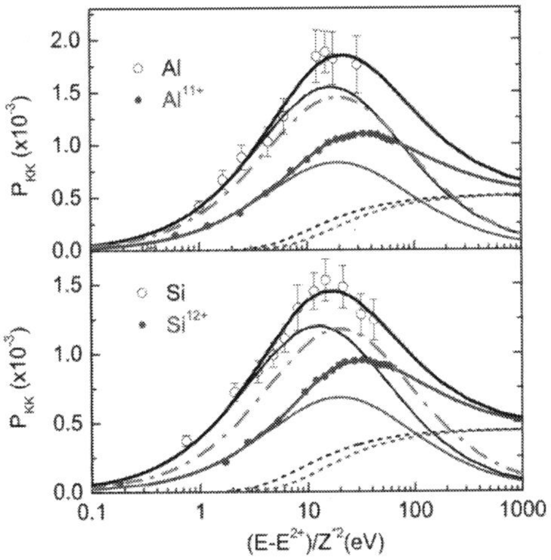

Figure 1. Double-to-single K-shell photoionization ratios for neutral atoms (black open circles) compared to the CCC calculations (velocity gauge) for the corresponding He-like ions (blue closed circles) as a function of the scaled excess energy. The best fits to our data (see text) are represented by solid thick lines, the KO by thin solid lines, and the SO by dashed lines. Black and blue lines correspond to the neutral atoms and ions, respectively. Red dot-dashed lines show the KO curve for the He-like ions scaled by Z^4/Z^{*4}.

To bring new insights as to the role played by outer shell electrons in the K-shell DPI, the phototoionization cross-sections for hollow atom production are compared to the CCC results of their corresponding two-electron ion counterparts. The experimental P_{KK}^{max} values for neutral atoms were found in good agreement with the $1/Z^{1.6}$-dependence suggested by Kanter *et al.* [10, 12] but systematically higher with respect to those of the He-like like ions. Because the influence of passive electrons on the ground-state correlations is small [31], one can assume that the photoabsorption asymptotic SO values for both neutral atoms and He-like ions are almost the same. Thus, the observed higher values of P_{KK}^{max} suggest that the electron scattering contribution to DPI is more important for neutral atoms compared to the He isoelectronic series. A clue to this finding resides in the proportionality of electron-impact ionization of a H-like ion to the KO term of the double photoionization of the corresponding He-like ion [3, 4, 28, 29]. Since along the hydrogen isoelectronic sequence the ionization cross sections σ_e scaled with the square of the ionization potentials I and plotted against the normalized electron energy E_e/I yield a universal curve [32], the KO contribution should scale with the square of the ionization potentials for a He-like ion and for the same neutral atom, respectively.

To corroborate the above statement, we have fitted the double-to-single K-shell ionization ratios for neutral and two-electron targets adopting an approach based on an incoherent summation of SO and KO terms. For $P_{KK}^{SO}(E)$ we have used the SO expression of Thomas [33] and for $P_{KK}^{KO}(E)$ the analytical form of the universal shape function for electron impact ionization of H-like ions of Aichele *et al.* given by Eq. (5) in Ref. [32]. The second ionization

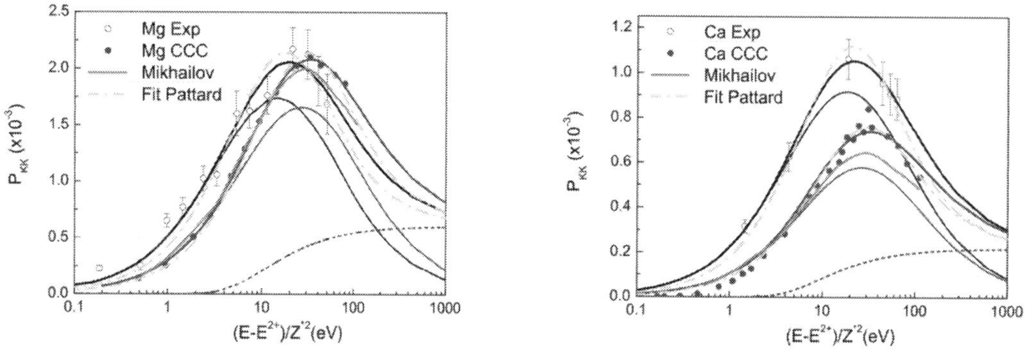

Figure 2. Double-to-single photoionization cross-section ratio of Mg and Ca versus the scaled excess energy. Experimental data are depicted with open black circles and the CCC calculations (HF, velocity gauge) with closed blue circles. The line styles of the best fits to our data are the same as in figure 1. The fit with the analytical shape function of Pattard (see text) is shown by green dot-dashed lines. The thick solid magenta line represents the scaled LOPT calculations of Mikhailov *et al.* [7].

potential E^+, calculated with the GRASP code [24], the asymptotic SO ratios from [34] and for He-like targets also the DPI threshold energies E^{2+} were kept fixed in the fitting procedure. The results of best fits to Al and Si data are depicted in figure 1. Indeed, the KO terms of He-like ions multiplied by the Z^4/Z^{*4} ratios are close in magnitude to the ones for neutral atoms. The effective nuclear charge Z^* was deduced using the hydrogenic formula $E^+=Z^{*2}$Ry, where Ry = 13.6 eV, and for the He-like targets $Z^*=Z$. These results suggest that KO dominates near threshold and for intermediate excitation energies.

The experimental and theoretical CCC double-to-single photoionization cross-secions of Mg and Ca versus the reduced excess energy are presented in figure 2. The two sets of data were also fitted with our empirical SO-KO model. It can be seen that for Mg the overall shape and magnitude of CCC calculations is in excellent agreement with the experimental P_{KK} values, while for Ca the calculations underestimate the experimental data. For both Mg and Ca, a relative shift of the CCC curves towards the $(E\text{-}E^{2+})/Z^{*2}$ value for two-electron targets can be observed. For comparison, the predictions within the lowest-order perturbation theory (LOPT) by Mikhailov *et al.* [7] are also depicted in figure 2. The plotted LOPT curves were obtained by scaling the suggested universal ratio of the double-to-single K-shell photoionization cross-sections for neutral atoms from Ref. [7]. Note, that Mikhailov *et al.* deduce the effective nuclear charge Z^* from the first ionization potential (K-shell binding energy) and not E^+. The LOPT curves are found to be quite close to the CCC calculations for Mg, and slightly lower to those for Ca.

It also instructive to fit the data with the universal shape function for single-photon multiple ionization proposed by Pattard [35]:

$$\frac{\sigma^{2+}}{\sigma^{+}} = R_{\infty}\frac{E^{\alpha}(\Delta E + E_0)^{7/2}}{(\Delta E + E_1)^{\alpha+7/2}}, \tag{1}$$

where ΔE is the excess energy, α stands for the Wannier threshold exponent (α varies smoothly from 1.056 (Z=2) to 1.0 in the limit $Z\rightarrow\infty$), R_∞ is the asymptotic double-to-single photoabsorption ratio, and E_0 and E_1 are fit parameters. This shape function is based on the proposed analytical parametrization for ionization by bare projectiles [36], and it satisfies the Wannier law near threshold and the Bethe-Born theory at high-energies. The fitted curves for

Mg and Ca are shown as dot-dashed lines in figure 2. In the fitting procedure R_∞ was fixed at values from Ref. [34]. While for He-like ions $E_0 \approx 2E_1$, the E_0 paramaters were found to be $\sim 2.5E_1$ for the experimental P_{KK} of Mg, Al and Si and $\sim 2.7E_1$ for Ca, respectively. The fits to CCC calculations yield $\sim 2.4E_1$ for both Mg and Ca. These values are in contrast with the ones reported by Huotari *et al.* [11] who find E_0 almost equal to E_1. Indeed, only when letting R_∞ free in the fit we find $E_0 \sim E_1$, but then R_∞ takes the value of $\sim P_{KK}^{max}$. On inspection of figure 2 it can be seen that the fit with the universal shape function by Pattard works quite well for both the measured data and the CCC calculations, and is comparable to the curves from the SO-KO model fit.

4.2. Scaling laws

It is worthwhile to ask whether the scaling properties of DPI cross sections and the double-to-single photoionization cross-section ratios for two-electron systems suggested by Kornberg and Miraglia [37], hold for neutral atoms. To this end, first power-law fits to the maximum values of σ^{2+} as a function of Z^* were performed. As expected, for He-like ions a $Z^{*-4.08(3)}$ fall-off was found, however, for neutral atoms a $Z^{*-3.68(11)}$ dependence was determined (see figure 3b). Indeed, the scaled double photoionization cross sections for Mg, Al, Si and Ca in reduced coordinates $\sigma^{2+}Z^{*3.68}$ against $(E\text{-}E^{2+})/Z^{*2}$ collapse onto a single curve (see figure 3a). Further, the $\sigma^{2+}Z^{*3.68}$ data for neutral atoms in the range $2 \leq Z \leq 47$, within the experimental uncertainties, exhibit a universal scaling behavior and coincide with the $\sigma^{2+}Z^{*4}$ for the He isoelectronic series. A hint to the difference for the scaling exponents of σ^{2+} is already given in

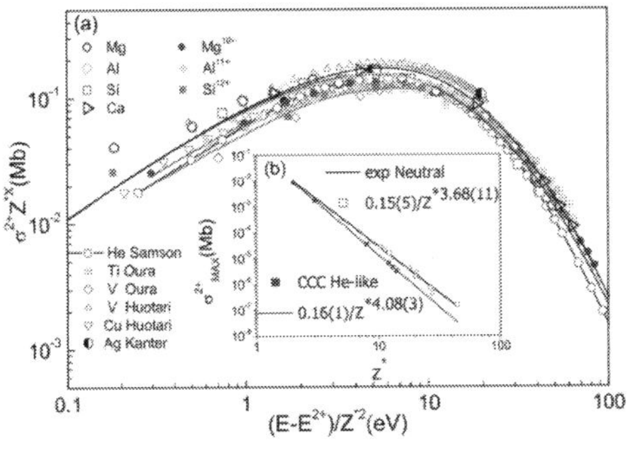

Figure 3. (a) Scaled experimental DPI cross sections for Mg, Al, Si and Ca compared to the scaled CCC calculations for He-like ions and experimental data for higher Z elements as a function of the scaled excess energy. The curves were deduced from the fitted values of P_{KK}. (b) Shown are the power-law fits to the maximum values of σ^{2+} as a function of Z^*.

the weaker Z dependence of P_{KK}^{max} for neutral atoms of $Z^{-1.61(5)}$ [10, 12]. In this perspective, one should not expect the double-to-single photoionization cross-section ratios for neutral atoms to obey the scaling relation with Z^{*2} which is valid for the He-isoelectronic sequence [3]. Indeed, figure 4 illustrates that the scaled double-to-single photoionization cross-section ratios for neutral elements do not fall on one curve.

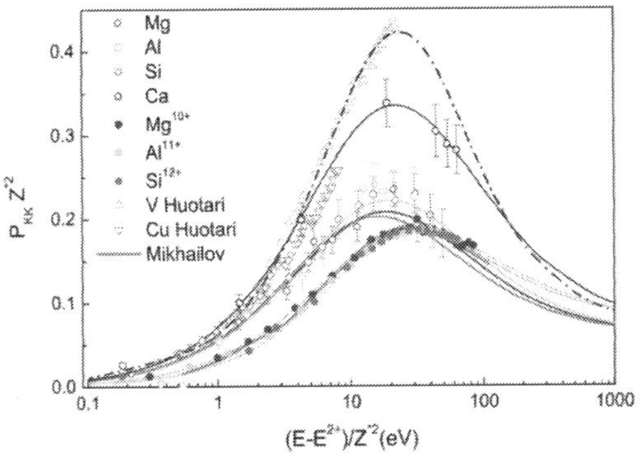

Figure 4. Scaled double-to-single photoionization cross-section ratios versus the scaled excess energy. The curves represent the fit with the SO-KO model. Solid lines correspond to our experimental results and dashed lines to CCC calculations for He-like ions, respectively. Experimental data for V and Cu are from Huotari *et al.* [11], where the dot-dashed line shows the fit with Pattard's shape function. The thick solid magenta line represents the calculations of Mikhailov *et al.* [7].

5. Concluding Remarks

The comparison of our results with other experimental data and CCC calculations suggest that the *post*-photoabsorption electron-electron correlation is different for neutral atoms and He-like ions. An empirical model to separate the knock-out (KO) and shake-off (SO) mechanisms of the K-shell DPI process is proposed. It was demonstrated that the K-shell double-to-single photoionization cross-section ratios of neutral atoms do not obey the Z^{*2} scaling. A semiempirical universal scaling of the double photoionization cross sections with the effective nuclear charge for neutral atoms in the range $2 \leq Z \leq 47$ was established.

Acknowledgments

The support of the Swiss National Science Foundation and the ESRF is acknowledged. Resources of the National Computational Infrastructure (NCI) Facility were used.

References

[1] Hino K, Hino K I, Ishihara T, Shimizu F, Toshima N and McGuire J H 1993 *Phys. Rev. A* **48** 1271
[2] Schneider T, Chocian P L and Rost J M 2002 *Phys. Rev. Lett.* **89** 073002
[3] Schneider T and Rost J M 2003 *Phys. Rev. A* **67** 062704
[4] Kheifets A S 2001 *J. Phys. B: At. Mol. Opt. Phys.* **34** L247
[5] Briggs J S and Schmidt V 2000 *J. Phys. B: At. Mol. Opt. Phys.* **33** R1
[6] Avaldi L and Heutz A 2005 *J. Phys. B: At. Mol. Opt. Phys.* **38** S861
[7] Mikhailov A I, Mikhailov I A, Moskalev A N, Nefiodov A V, Plunien G and Soff G 2004 *Phys. Rev. A* **69** 032703
[8] Colgan J, Pindzola M S and Robicheaux F 2001 *J. Phys. B: At. Mol. Opt. Phys.* **34** L457
[9] Knapp A *et al.* 2002 *Phys. Rev. Lett.* **89** 033004
[10] Kanter E P, Ahmad I, Dunford R W, Gemmell D S, Krässig B, Southworth S H and Young L 2006 *Phys. Rev. A* **73** 022708
[11] Huotari S, Hämäläinen K, Diamant R, Sharon R, Kao C C and Deutsch M 2008 *Phys. Rev. Lett.* **101** 043001
[12] Kanter E P, Dunford R W, Krässig B and Southworth S H 1999 *Phys. Rev. Lett.* **83** 508
[13] Southworth S H, Kanter E P, Krässig B, Young L, Armen G B, Levin J C, Ederer D L and Chen M H 2003 *Phys. Rev. A* **67** 062712
[14] Oura M *et al.* 2002 *J. Phys. B: At. Mol. Opt. Phys.* **35** 3847
[15] Diamant R, Huotari S, Hämäläinen K, Sharon R, Kao C C and Deutsch M 2009 *Phys. Rev. A.* **79** 062511
[16] Diamant R, Huotari S, Hämäläinen K, Sharon R, Kao C C, Honkimäki V, Buslaps T and Deutsch M 2009 *Phys. Rev. A.* **79** 062512
[17] Hoszowska J *et al.* 2009 *Phys. Rev. Lett.* **102** 073006
[18] Kheifets A S, Bray I and Hoszowska J 2009 *Phys. Rev. A* **79** 042504
[19] Samson J A R, Stolte W C, He Z X, Cutler J N, Lu Y and Bartlett R J 1998 *Phys. Rev. A* **57** 1906
[20] Hoszowska J, Dousse J Cl, Kern J and Rhême Ch 1996 *Nucl. Instr. Meth. Phys. Res. A* **376** 129
[21] Chen M H 1991 *Phys. Rev. A* **44** 239
[22] http:/physics.nist.gov/xcom
[23] Hart J F and Herzberg G 1957 *Phys. Rev.* **106** 79
[24] Dyall K G, Grant I P, Johnson C T, Parpia F A and Plummer E P 1989 *Comp. Phys. Comm.* **55** 425
[25] Kheifets A S and Bray I 1998 *Phys. Rev. A* **57** 2590
[26] Kheifets A S and Bray I 1998 *Phys. Rev. A* **58** 4501
[27] Kheifets A S and Bray I 2007 *Phys. Rev. A* **75** 042703
[28] Samson J A R 1990 *Phys. Rev. Lett.* **65** 2861
[29] Pattard T and Burgdörfer J 2001 *Phys. Rev. A* **64** 042720
[30] Pattard T, Schneider T and Rost J M 2003 *J. Phys. B: At. Mol. Opt. Phys.* **36** L189
[31] Mitnik D M and Miraglia J 2005 *J. Phys. B: At. Mol. Opt. Phys.* **38** 3325
[32] Aichele K *et al.* 1998 *J. Phys. B: At. Mol. Opt. Phys.* **31** 2369
[33] Thomas T D 1984 *Phys. Rev. Lett.* **52** 417
[34] Forrey R C, Sadeghpour H R, Baker J D, III J D M and Dalgarno A 1995 *Phys. Rev. A* **51** 2112
[35] Pattard T 2002 *J. Phys. B: At. Mol. Opt. Phys.* **35** L207
[36] Rost J M and Pattard T 1997 *Phys. Rev. A* **55** R5
[37] Kornberg M A and Miraglia J E 1994 *Phys. Rev. A* **49** 5120

Journal of Physics: Conference Series **212** (2010) 012007 doi:10.1088/1742-6596/212/1/012007

Localization of inner shell photoelectron emission and interatomic Coulombic decay in neon dimers

K. Kreidi[1,2], T. Jahnke[1], Th. Weber[3], T. Havermeier[1], R. E. Grisenti[1,4], X. Liu[5], Y. Morisita[6], S. Schössler[1], L. Ph. H. Schmidt[1], M. Schöffler[1], M. Odenweller[1], N. Neumann[1], L. Foucar[1], J. Titze[1], B. Ulrich[1], F. Sturm[1], C. Stuck[1], R. Wallauer[1], S. Voss[1], I. Lauter[1], H. K. Kim[1], M. Rudloff[1], H. Fukuzawa[5], G. Prümper[5], N. Saito[6], K. Ueda[5], A. Czasch[1], O. Jagutzki[1], H. Schmidt-Böcking[1], S. K. Semenov[7], N. A. Cherepkov[7] and R. Dörner[1]

[1] Institut für Kernphysik, J. W. Goethe Universität, Max-von-Laue-Str. 1, 60438 Frankfurt, Germany
[2] DESY, Notkestrasse 85, 22607 Hamburg, Germany
[3] Lawrence Berkeley National Laboratory, Berkeley CA 94720, USA
[4] Gesellschaft für Schwerionenforschung, Planckstr. 1, 64291 Darmstadt, Germany
[5] Institute of Multidisciplinary Research for Advanced Materials, Tohoku University, Sendai 980-8577, Japan
[6] National Metrology Institute of Japan, AIST, Tsukuba 305-8568, Japan
[7] State University of Aerospace Instrumentation, 190000 St. Petersburg, Russia

E-mail: kreidi@atom.uni-frankfurt.de

Abstract. By using the COLd Target Recoil Ion Momentum Spectroscopy (COLTRIMS) we have investigated the 1s photoionization of neon dimers. Interatomic Coulombic Decay (ICD) takes place in the asymmetric charge breakup Ne2+/Ne1+. This breakup is used to determine whether the inner shell vacancies, resulting out of the photoionization of the dimer, and the valence shell vacancies, resulting out of the ICD, are localized at one of the atoms or delocalized over the two equivalent sites of the neon dimer.

The question if vacancies in a homonuclear diatomic molecule are localized at one of the two indistinguishable atoms or delocalized over both sites has been discussed since more than 30 years now (e.g. [1, 2, 3, 4, 5, 6, 7, 8, 9, 10, 11]). Here we want to answer this question for the van-der-Waals bound Ne_2. Whereas the geometry of Ne_2 suggests the localized picture especially for the inner shell orbitals, the symmetry of the system indicates to describe the orbitals in Ne_2 by delocalized molecular wave functions of well-defined *gerade* (g) or *ungerade* (u) symmetry. This delocalized picture is also used in todays state-of-the-art theory concerning the photoionization and the decay of Ne_2 [12]. There have been different works using different observables to address this question. In the first theoretical claim Bagus and Schaefer have used the photoionization energy [2]. Their Hartree-Fock calculation showed a lower energy for the symmetry broken - i.e. localized - basis and reached a better agreement with the experiment. It was later argued that this was a peculiarity of the Hartree-Fock calculation itself [3]. In contrast the possible measurement of the energy splitting of $1\sigma_g$ und $1\sigma_u$ in N_2 [4, 5] and the agreement of this splitting with todays theories [5] is taken as a clear evidence for the delocalization of the inner shell vacancies. In the present work we used a quantum mechanically meaningful and an experimentally accessible way to probe the localization of the vacancies in the neon dimer. The question is addressed here by investigating the angular distribution of the emitted electrons in the molecular frame [13, 14, 15]. That means for a fixed orientation of the molecular axis we determine the angular distribution of the inner shell photoelectrons and of the valence shell electrons which are emitted via Interatomic Coulombic Decay (ICD) [16] following Auger decay [17]. If the electron is emitted from a delocalized orbital the angular distribution will be symmetric with respect to the *right* and the *left* atomic site of the dimer. If in contrast the electron is emitted from a localized source, i.e. from only one atom of the dimer, the angular distribution can show a clear *right/left* asymmetry due to the scattering of the electron wave in an asymmetric molecular potential. However for a homonuclear molecule like Ne_2 it is only possible to observe such an asymmetry in the angular distribution if both sites of the dimer can be distinguished experimentally. In this work this is done by the different charges of both neon atoms in the asymmetric charge breakup Ne^{2+}/Ne^{1+}. The first experiment for the breakup N^{2+}/N^{1+} found no asymmetry in the angular distribution [18]. We want to investigate this question here on Ne_2 because of fact the that the binding energy in this van-der-Waals bound system is only 3 meV and with this the internuclear distance of 5.86 a.u. between both neon atoms is about 10 times bigger than the radius of the outer shell of one neon atom. In addition the energy splitting between the 1s ionized g and u states is negligibly small compared to the natural line width. All these arguments suggest to think of two neon atoms with independent shells so that the localized basis should be the right one to describe the neon dimer. However the opposite is suggested by the symmetry of the system. For Ne_2 as well as for N_2 this symmetry indicates to describe the system by delocalized molecular wave functions. The relaxation mechanism which results in the asymmetric charge breakup Ne^{2+}/Ne^{1+} is named ICD following Auger decay. The whole relaxation pathway is shown in figure 1. The first step is the 1s photoionization which is followed by an Auger decay into the dicationic one-site state $Ne^{2+}(2s^{-1}2p^{-1})/Ne$. After this ICD takes place ending up in the tricationic two-site state $Ne^{2+}(2p^{-2})/Ne^{1+}(2p^{-1})$. In the final step the neon dimer fragments in a Coulomb-explosion. This localized sketch in figure 1 is only a demonstrative way to explain the relaxation in Ne_2. All actual calculations, e.g [17], use the molecular wavefunctions of well-defined g/u symmetry to describe the ICD in Ne_2 in good agreement with experimental results [12, 19].

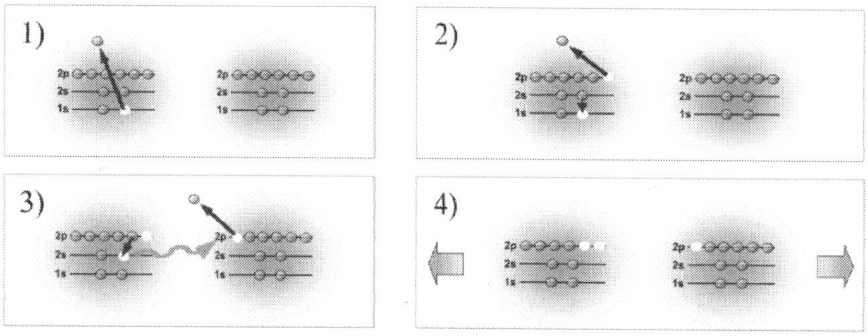

Figure 1. $1s$ photoionization (1) leads to an Auger decay resulting in a dicationic one-site state of Ne_2 (2). The Auger decay is followed by interatomic Coulombic decay (3). Ending up in a state where both atoms are charged positively, the dimer fragments back-to-back in a Coulomb-explosion (4).

The experiment was performed at BL UE56/1 SGM at BESSY in Berlin using the COLTRIMS technique [20]. As a source for Ne_2 a supersonic jet that was precooled to 180 K was employed. It was crossed with the photon beam. Products from the photoreaction were guided by an electric field of 20 V/cm and a magnetic field of 6 Gauss towards two channel plate detectors with delayline readout [21]. Electrons up to 12 eV and ions up to 10 eV were detectable with a solid angle of 4π and in coincidence .

Due to the fact that at the end of the reaction shown in figure 1 we can distinguish both neon atoms by their charge, we can determine the source of the emitted photo- and ICD-electrons. With this we are able to observe a possible asymmetry in the angular distribution. In figure 2 the angular distribution of the 1s photoelectrons in the molecular frame is shown for different orientations of the molecular axis. One can clearly see a $right/left$ asymmetry which switches from figure 2a) to 2b) with the orientation of the axis. In the small panels at the lower right of figure 2 a Frozen Core Hartree-Fock calculation is shown for a fully localized 1s vacancy in the neon dimer. The localization of the vacancy after photoionization is introduced by a linear combination of the symmetry adapted $1\sigma_g$ and $1\sigma_u$ wavefunctions: $|r\rangle = (|1\sigma_g\rangle + |1\sigma_u\rangle)/\sqrt{2}$ and $|l\rangle = (|1\sigma_g\rangle - |1\sigma_u\rangle)/\sqrt{2}$. One can find a good qualitative agreement between the experimental and the theoretical results whereas the theory overestimates the asymmetry in the angular distribution.

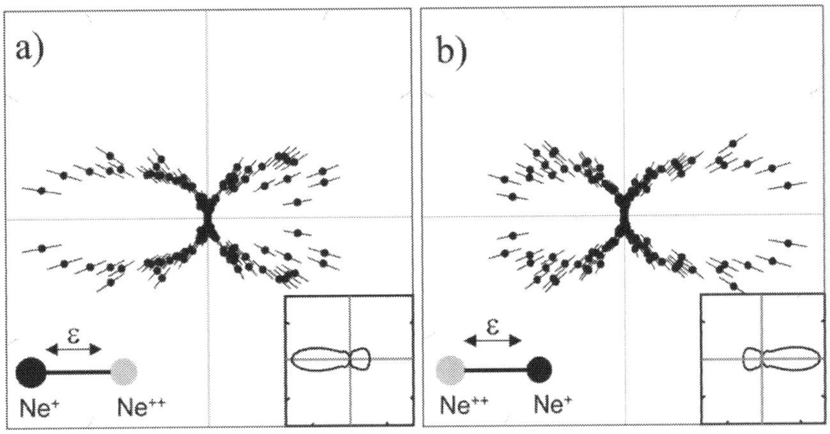

Figure 2. Angular distribution of the 10 eV 1s photoelectron in dependence of the orientation of the dimer axis and the direction of the polarization vector ε (horizontal). In (a) and (b) the dimer is aligned parallely to the polarization vector within an angle of $\pm$ 18°. The asymmetry, being a result of core hole localization, is clearly visible in the experimental data (circles). Solid line: frozen core Hartree-Fock calculation assuming a localized emission.

To investigate the localization of the vacancies in the valence shell of the neon dimer it is necessary to take a closer look on the ICD process in which the ICD-electrons, representing the electrons from the valence shell, are emitted. In figure 3 the two different contributions of the ICD are shown. A) represents the 'direct' contribution of the electron-electron Coulomb matrix element. Here ICD takes place via the transfer of a virtual photon. B) shows the 'exchange' contribution of the matrix element. Here ICD happens via an electron transfer. In this case the spatial overlap of the involved wavefunctions is the crucial contribution to the decay probability. Therefore pathway B) is suppressed at large internuclear distances where pathway A) is still open [22]. Both pathways can be distinguished by the different sum energy of the ICD electron and both ions [22]. The important difference between A) and B) for investigating the angular

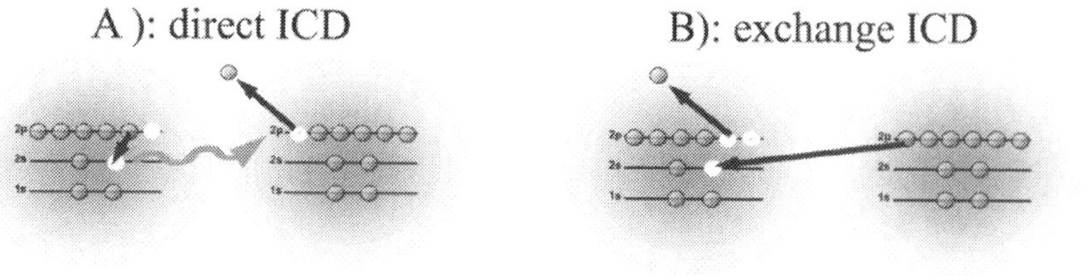

Figure 3. Both contributions of the interatomic Coulombic decay. In A) the ion's excitation energy is emitted via a virtual photon transfer and in B) via an electron transfer. From A) to B) the source of the emitted ICD electron changes from the singly ionized to the doubly ionized neon atom.

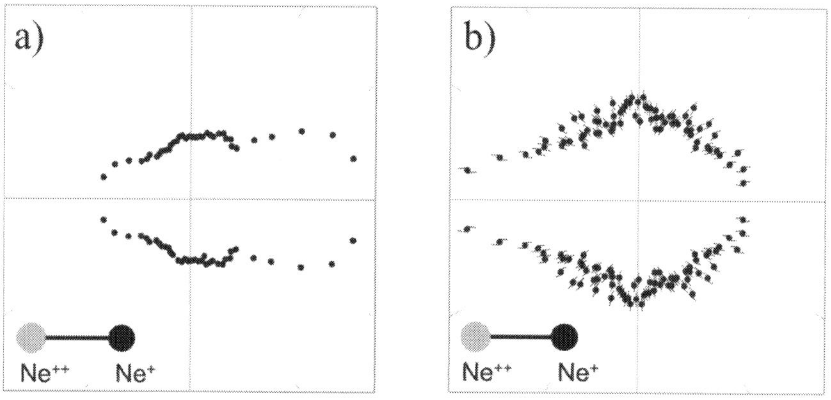

Figure 4. Angular distribution of the ICD electron in the dimer frame. The dimer is aligned horizontal and it is integrated over the orientation of the polarization vector. (a) ICD electrons which are created by virtual photon exchange (channel A in figure 3), (b) angular distribution of the ICD electrons emitted after transfer of an electron (channel B in figure 3). As suggested by the sketch in figure 3 the direction of the asymmetry switches from (a) to (b) depending on the decay path.

distribution of the ICD electrons in the molecular frame is the change of their source from the singly to the doubly ionized neon atom.

The angular distributions of the ICD electrons emitted in pathway A) and in pathway B) are shown in figure 4a) and b) respectively. A clear *right/left* asymmetry is visible in both cases and the preferred direction of electron emission switches depending on the source (singly or doubly ionized neon atom) of the ICD electron.

In conclusion we have shown a clear asymmetry in the molecular frame angular distribution of the inner shell photoelectrons and the ICD electrons being emitted from the valence shell of the neon dimer [23]. That means, in conflict with the use of delocalized wavefunctions in todays theory to describe homonuclear diatomic molecules, the inner shell vacancies as well as the vacancies in the valence shell of the neon dimer show strong features of localization.

[1] L. C. Snyder 1971 *J. Chem. Phys.* **55** 95
[2] P. S. Bagus 1972 *J. Chem. Phys.* **56** 224
[3] J. A. Kintop, W. V. M. Machado, and L. G. Ferreira 1991 *Phys. Rev. A* **43** 3348
[4] A. Thiel, J. Schirmer, and H. Köppel 2003 *J. Chem. Phys.* **119** 2088
[5] M. Ehara *et al* 2006 *J. Chem. Phys.* **124** 124311
[6] D. Dill *et al* 1978 *Phys. Rev. Lett.* **41** 1230
[7] P. Glans *et al* 1996 *Phys. Rev. Lett.* **76** 2448
[8] M. Matsumoto *et al* 2006 *J. Phys. B* **39** 375
[9] O. Björneholm *et al* 2000 *Phys. Rev. Lett.* **84** 2826
[10] C. T. Chen, Y. Ma, and F. Sette 1989 *Phys. Rev. A* **40** 6737
[11] D. Rolles *et al* 2005 *Nature* **437** 711
[12] S. Scheit *et al* 2004 *J. Chem. Phys.* **121** 8393
[13] A.V. Golovin *et al* 1997 *Phys. Rev. Lett.* **79** 4554
[14] N. Saito *et al* 2005 *J. Phys. B* **38** L277
[15] J. Adachi *et al* 2007 *J. Phys. B* **40** F285
[16] L. S. Cederbaum, J. Zobeley, and F. Tarantelli 1997 *Phys. Rev. Lett.* **79** 4778
[17] R. Santra, and L. S. Cederbaum 2003 *Phys. Rev. Lett.* **90** 153401 R. Santra, and L. S. Cederbaum 2005 *Phys. Rev. Lett.* **94** 199901(E)
[18] T. Weber *et al* 2001 *J. Phys. B* **34** 3669
[19] T. Jahnke *et al* 2004 *Phys. Rev. Lett.* **93** 083002
[20] J. Ullrich *et al* 2003 *Rep. Prog. Phys.* **66** 1463
[21] O. Jagutzki *et al* 2002 *Nucl. Instr. Meth. A* **477** 244
[22] T. Jahnke *et al* 2007 *Phys. Rev. Lett.* **99** 153401
[23] K. Kreidi *et al* 2008 *J. Phys. B* **41** 101002.

Journal of Physics: Conference Series **212** (2010) 012008 doi:10.1088/1742-6596/212/1/012008

Vortices in atomic wave functions

J. H. Macek

Department of Physics and Astronomy, University of Tennessee, Knoxville, Tennessee 37496 and Physics Division, Oak Ridge National Laboratory, Oak Ridge, Tennessee 37831

E-mail: jmacek@utk.edu

Abstract. Vortices in atomic wave functions are shown to have observable consequences. It is shown that time-dependent electron wave functions in coordinate space go over to electron emission amplitudes in the limit that time becomes infinite. This relation between wave functions and emission amplitudes is called the imaging theorem. According to this theorem vortices in wave functions at small times when particles interact in a complex way appear as vortices in electron momentum distributions where they produce observable features. Conversely, some observable structures may be interpreted in terms of vortices no matter how they are seen or computed. We verify that previously unexplained features in (e,2e) triply differential cross sections can be interpreted in terms of vortices.

1. Introduction

Structures in triply differential cross sections (TDCS) are often markers for a particular type of motion represented in the quantum theory by time-dependent Schrödinger wave functions [1]. Cusps and some zeros in TDCS have been interpreted in terms of particle-particle interactions and the interference of Schrödinger waves that underlie the momentum distributions of observed electrons. For example, double slit type interferences are predicted and observed for electrons ejected from diatomic molecules. In these instances the interference produces a distribution that vanishes along a nodal surface [1].

Almost all structures can be interpreted in the terminology given above, however, isolated zeros in TDCS have been observed [2] which do not fit into standard dynamical pictures of atomic processes [1]. It has been known since the early days of quantum mechanics that isolated zeros appear in complex single-particle wave functions, and the relation of these zeros to vortices has been explored by earlier workers. The imaging theorem discussed below suggests that isolated zeros in TDCS may be related to vortex structures in atomic wave functions.

2. The imaging theorem

One usually associates electron momentum distributions $P(\boldsymbol{k})d^3k = |A(\boldsymbol{k})|^2d^3k$ with Fourier transforms $\tilde{\psi}(\boldsymbol{k})$ of free-particle coordinate-space wave functions $\psi(\boldsymbol{k})$ at large times where the bound state components are explicitly subtracted off. As usually happens the free-particle waves have evolved from a bound or otherwise tightly confined wave packet at earlier times. In this case, the wave function incorporates an "explosion factor" equal to $t^{-3/2}\exp(imr^2/2\hbar t)$ so that

$$\psi(\boldsymbol{r},t) = \left(\frac{m}{\hbar t}\right)^{-3/2} \exp\left(i\frac{mr^2}{2\hbar t}\right) f(\boldsymbol{r},t) \tag{1}$$

where f is a "reduced" wave function. Upon taking the Fourier transform of $\psi(\boldsymbol{r},t)$ using the method of stationary phases for large times one sees that

$$\tilde{\psi}(\boldsymbol{k},t) = f\left(\frac{\hbar \boldsymbol{k} t}{m}, t\right) \exp\left(-i\frac{k^2 t}{2m} + 3i\frac{\pi}{4}\right) = \left(\frac{\hbar t}{m}\right)^{3/2} \psi\left(\frac{\hbar \boldsymbol{k} t}{m}, t\right) \exp\left(-i\frac{k^2 t}{m} + 3i\frac{\pi}{4}\right) \tag{2}$$

in the limit of large times. This identity shows that

$$\lim_{t\to\infty} (m/\hbar t)^3 |\psi(\boldsymbol{k}t,t)|^2 d^3r = |A(\boldsymbol{k})|^2 d^3k, \tag{3}$$

which is one version of the imaging theorem. In the remainder of this report atomic units with $m = \hbar = e = 1$, where m is the electron mass, will be used.

The derivation of the imaging theorem applies to a single electrond in external time-dependent fields which effectively shut off after some finite time $t_{\rm off}$. It is readily extended to processes where more than one electron is ejected in the final state. The theorem also applies to (e,2e) processes where it is understood that the incident electron a is confined to a wave packet of microscopic dimensions. Only after colliding with an atom and ejecting a second electron b do we apply the identity of Eq. (3) to both electrons. In this case $\boldsymbol{k}_a, \boldsymbol{k}_b$ in the amplitudes $A(\boldsymbol{k}_a, \boldsymbol{k}_b)$ are understood to satisfy energy conservation, so that the $\boldsymbol{k}_a, \boldsymbol{k}_b$ variable space is actually 5 dimensional.

3. Vortices in TDCS for proton impact

Computations of electron momentum distributions using a modified Lattice-Time-Dependent Schrödinger Equation method (LTDSE) for proton impact on atomic hydrogen found unusual structures in the electron momentum distributions also called the Triply Differential Cross Sections (TDCS) [3,4]. The results for the amplitude $|A(k_y = 0, k_x, k_z)|$ are shown in in Fig. (1) where isolated holes are present near the continuum capture and direct ionization cusps. The

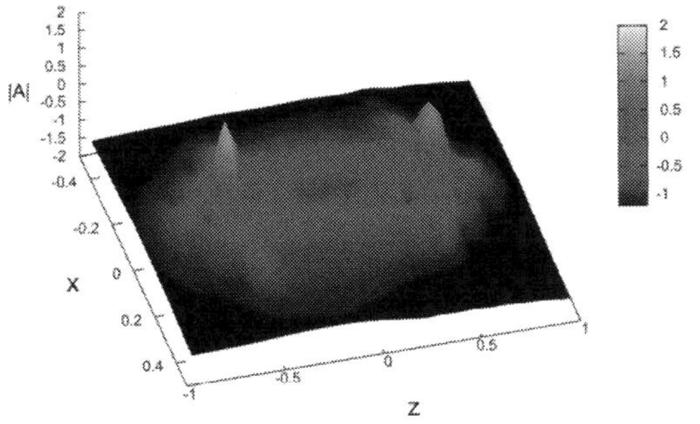

Figure 1. Contour plot of the absolute value of the ionization amplitude $|A(\boldsymbol{k})|$ for 5 keV proton impact on atomic hydrogen at unit impact parameter. The plot shows pronounced minima near the direct ionization and continuum capture cusps.

computational method showed that these holes emerged from similar holes in the time-dependent coordinate space wave function. The local probability current $\mathrm{Im}\frac{\boldsymbol{\nabla}\psi}{\psi}$ circulates around these holes and integration of the current around the holes gives a value of 2π, indicating that they correspond to isolated first order zeros in $A(\boldsymbol{k})$, *i. e.* to vortices [5]. Because the vortices produce holes in the TDCS, they are observable features. The first indication, to our knowledge, that vortices in atomic wave functions eventually appear as vortices in the TDCS were the "holes" in momentum distributions reported in Ref. [3].

The theory of Ref. [4] also relates these vortices to angular momentum transfer from the relative motion of target and projectile to angular momentum of continuum electrons in the final state. By angular momentum transfer we imply that the vector $\boldsymbol{L}$ becomes non-zero. Bound

Journal of Physics: Conference Series **212** (2010) 012008 doi:10.1088/1742-6596/212/1/012008

states with non-zero $\boldsymbol{L}$ are said to be oriented [6]. Vortices in continuum functions indicates that continuum states may also be oriented, thus the imaging theorem shows that exact zeros in momentum distributions correspond to oriented electron. Because exact zeros may be seen experimentally, the observation of vortices is also an observation of continuum state orientation. Unusual minima may be present in some observations of TDCS in ion-atom collisions, however, such minima have never been identified with vortices in time-dependent wave functions, thus observations of such features is necessary in order to confirm that vortices are present in electron momentum distributions.

4. Vortices in electron (e,2e) impact

The imaging theorem applies to ionization by electron impact. In the case of ionization by fast electron impact, one can consider that the fast electron a ejects electron b with a velocity that is comparable to the mean velocity in the initial state, which is much less than the velocity of the fast electron. Then electrons initially in s-wave bound states are ejected by predominantly dipole interactions as outgoing p-waves with small admixtures of other partial waves. In the first Born approximation, where no net angular momentum $\boldsymbol{L}$ is transfered from b to a, the angular distribution exhibits minima in directions perpendicular to the momentum transfer. Because other partial waves are out of phase with the p-wave amplitude, the TDCS does not vanish exactly at any point.

The situation differs dramatically when angular momentum transfer is included, as in the Coulomb Born approximation. In that case the dipole transitions eject p-wave electrons with $m \neq 0$ so that the states are oriented. Then there is a vortex line, *i. e.* a line of zeros, along an axis perpendicular to the incident $\boldsymbol{k}_i$ and final $\boldsymbol{k}_b$ momentum of the scattered electron. The coordinate system is chosen so that this axis, parallel to $\boldsymbol{k}_b \times \boldsymbol{k}_i$, is the $y-$axis. The presence of other partial waves shifts the zero away from the $y-$axis to a non-zero value of $\boldsymbol{k}_b$ called the vortex location $\boldsymbol{k}_{b0}$.

In the Coulomb Born calculations of Botero and Macek [7] for K-shell ionization of hydrogen by fast electrons, zeros in the electron distribution were found but not recognized as vortices. A partial wave analysis shows that these zeros are due to orientation of the continuum electrons since the ionization amplitude had $p-$wave components with $L_y \neq 0$. Interference with s and d-waves shifts the zero away from the $y-$axis where it appears as a zero in the angular distribution of the ejected electron b.

Murray and Read [2] observed a similar minima in the angular distribution of electrons ejected from He by relatively slow electron impact in a geometry where $\boldsymbol{k}_a = \boldsymbol{k}_b$ and the momentum vectors were symmetrically placed relative to $\boldsymbol{k}_i$. This geometry is shown in Fig. 2, where the angle between the outgoing momentum vectors is 2Θ. Calculations by Berackdar and Briggs [8]

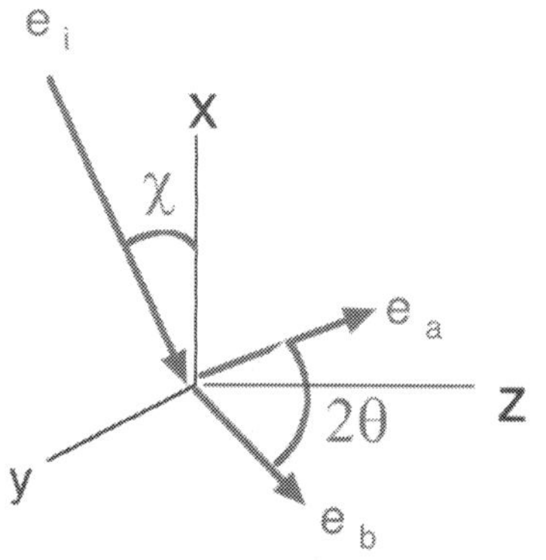

Figure 2. Symmetric (e,2e) geometry showing incident electron momentum inclined at an angle χ to an axis perpendicular to the plane of the final scattered and ejected electrons. The angle 2Θ is the angle between the outgoing electrons a and b in the final state.

using highly correlated final state wave functions, and essentially exactly by Colgan *et. al* [9] showed that the minima was related to an exact zero in the electron momentum distribution. We have repeated the calculations of Ref. [8] as the solid red curve in Fig. 3 and find a similar minima. Also shown are some experimental data points from Ref. [2] which show a minima in

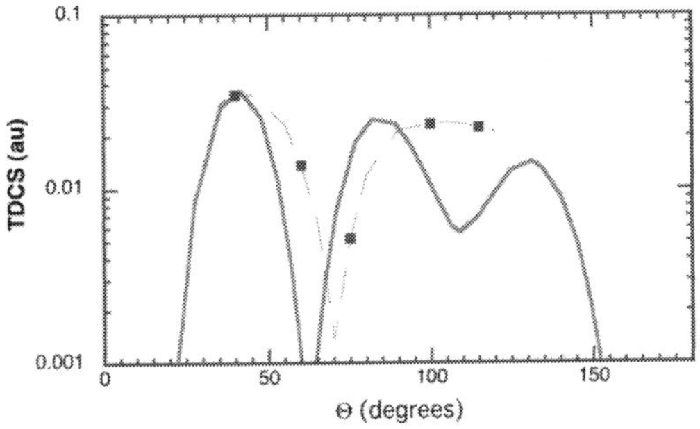

Figure 3. TDCS vs Θ for 67.4 eV electrons incident on helium at $\chi = 22.5^o$ ejecting 20 eV electrons. The solid (red) curve is the calculated cross section and the squares are experimental data from Ref. [2]. The dashed curve is a spline fit to the data.

the vicinity of the theoretical minima.

To analyze the minima for vortex structure we transform to a coordinate system with $\boldsymbol{k}_i$ and $\boldsymbol{b}_a$ fixed in plane and examine the direct ionization amplitude for vortex structure in the neighborhood of the computed minima. A contour plot of the TDCS for values of $\boldsymbol{k}_b$ near the minima seen in Fig. 3 is shown in Fig. 4.

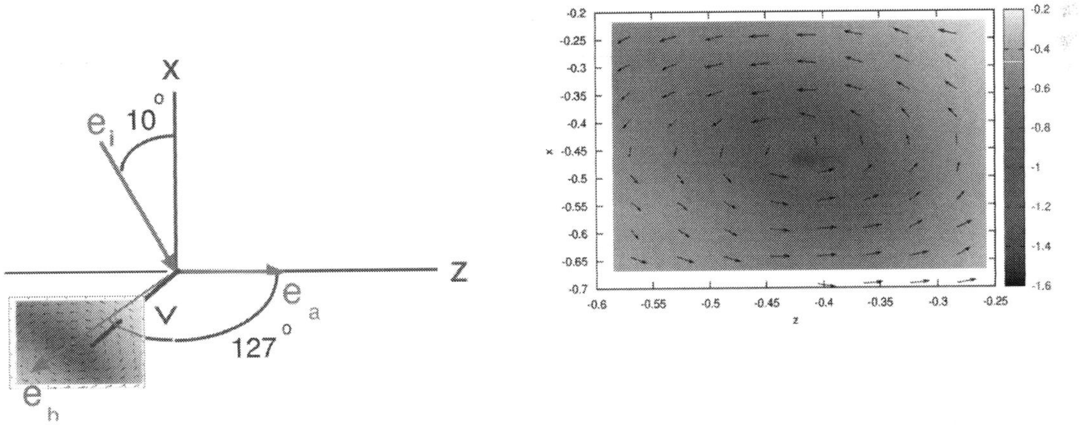

Figure 4. Contour plot of the direct ionzation amplitude in a plane parallel to the x-z plane. An arrow (red) points to the computed vortex position in a slice of the TDCS. The contour plot shows a vortex zero and the small black arrows point in the direction of the velocity current. An expanded version of the logarithmic contour plot is shown on the right.

Also shown on the contour plot in Fig. 4 are small black arrows pointing in the direction of the velocity current . It is apparent that there is a vortex near the position of the TDCS mimimum, corresponding to probability circulation in a counterclockwise direction. The direction is consistent with the transfer of angular momentum from the scattered to the ejected electron. Integration of the current around vortex gives a value of 2π consistent with a first order zero.

This analysis finds a vortex in the direct amplitude for fixed $\boldsymbol{k}_i$ and $\boldsymbol{k}_a$. Experiment does not observe this amplitude rather an incoherent mixture of singlet and triplet amplitudes are observed. An exact zero is unlikely for incoherent mixtures, however the zero was actually observed in the symmetric geometry where only the triplet amplitude contributes. The observed state is therefore a pure state and could exhibit an exact zero. Further analysis, not given here, shows that there is indeed a vortex in the triplet state momentum distribution, and that this zero gives rise to the cross section minimum seen in Fig. 3. It is apparent the role of vortices in atomic dynamics must be understood to form a complete picture of (e,2e) TDCS.

5. acknowledgements
This research is sponsored by the Office of Basic Energy Sciences, U.S. Department of Energy under grant number DE-FG02-02ER15283.

[1] Briggs J S 1989 *Comments At. Mol. Phys.* **23**, 155
[2] Murray A J and Read F H 1993 *J. Phys. B* **26** L359
[3] Lee T G, Ovchinnikov S Yu, Sternberg J B, Chupryna V, Schultz D R and Macek J H, 2007 *Phys. Rev. A* **76**, 050701R
[4] Macek J H, Sternberg J B, Ovchinnikov S Yu, Lee T G and Schultz D R, *Phys. Rev. Lett***102**, 143201 (2009).
[5] Bialynicki-Birula I, Bialynicka-Birula Z and Sliwa C 2000 *Phys. Rev. A* **61** 032110
[6] Fano U and Macek J H, 1973 *Rev. Mod. Phys.* **45** 553
[7] Botero j and Macek j H 1992 *Phys. Rev. A* **45** 15
[8] Berakdar J and Briggs J S 1994 *Phys. Rev. Lett.* **72** 3799;1994 J. Phys. B **27** 4271
[9] Colgan J and Pindzola M S 2006 *Phys. Rev. A* **74**; J. Colgan J, Pindzola M S, Childers G and Khakoo M A 2006*Phys. Rev. A* **73**, 042710;Colgan J, Private communication and Colgan J, *et al* 2009 *J. Phys. B* **42**, 171001

International Symposium on (e,2e) and the 15th ISPCEAC | IOP Publishing
Journal of Physics: Conference Series **212** (2010) 012009 | doi:10.1088/1742-6596/212/1/012009

Molecular-frame angular distribution of normal and resonant Auger electrons

D Rolles[1,2], G Prümper[3], H Fukuzawa[3], X-J Liu[3], J Harries[4], K Ueda[3], Z D Pešić[1], I Dumitriu[1] and N Berrah[1]

[1]Physics Department, Western Michigan University, Kalamazoo, Michigan 49008, USA
[2]Max Planck Advanced Study Group, Center for Free Electron Laser Science, 22761 Hamburg, Germany
[3]Institute of Multidisciplinary Research for Advanced Materials, Tohoku University, Sendai 980-8577, Japan
[4]Japan Synchrotron Radiation Research Institute, Sayo-gun, Hyogo 679-5198, Japan

E-mail: `Daniel.Rolles@asg.mpg.de`

Abstract. Molecular-frame angular distributions (MFADs) of resonant and non-resonant ("normal") Auger electrons were determined using angle-resolved electron-ion coincidence spectroscopy. Here, we concentrate on the MFADs of normal N_2 KVV Auger electrons. We conclude that their MFADs are independent of both photon energy and light polarization direction, thus confirming that the two-step model is a good approximation. All threshold and shape-resonance phenomena can be attributed to the absorption process and are irrelevant for the Auger decay in the molecular frame. Distinct differences in the MFADs are found as a function of the Auger final state.

1. Introduction

Molecular-frame angular distributions (MFADs) of photo- and Auger electrons offer an unprecedented level of detailed information on the photoionization dynamics of (small) molecules, which are hidden in conventional gas-phase electron spectroscopy due to the random orientation of the molecules that averages out most characteristic intensity modulation. In order to experimentally determine MFADs, the molecules have to be "fixed" in space. For dissociative photoionization, this can be realized by applying the angle-resolved photoelectron-photoion coincidence technique [1, 2, 3], which has developed rapidly with the advent of position-sensitive detectors [4, 5, 6, 7]. While most of these studies to date have focused on photoelectrons [8, 9, 10, 11, 12, 13, 14, 15, 16, 17], only few studies on Auger electrons have been performed. Apart from a measurement of the MFAD of Auger electrons in N_2 after electron impact [18], only two studies of Auger electron MFADs after photoexcitation of CO [19, 20] have been reported prior to our recent series of measurements. Since our recent results for resonant and non-resonant Auger decay in CO have already been described in detail [21, 22, 23], the present paper focusses on our results obtained for non-resonant Auger electrons in N_2.

2. Experimental Method

In contrast to most experiments that determine the photoelectron molecular-frame angular distributions, we did not record the emitted Auger electrons in a wide range of angles, but rather fixed the direction of the Auger electron detection while measuring the full three-dimensional ion angular distribution for two different orientations of the polarization vector. Assuming the validity of a strict two-step model, i.e. assuming that the Auger electron emission characteristics are independent of the initial ionization step, the coincident ion distribution should be a product of the angular selectivity of the ionization process and the molecular-frame angular distribution of the Auger electrons. The probability $I_{ei}(\overrightarrow{e}, \overrightarrow{n}, h\nu)$ to find an electron in the directions $\overrightarrow{e}$ in coincidence with an ion in the directions $\overrightarrow{n}$ at a given photon energy $h\nu$ is simply the probability $I(\theta_n^E, h\nu)$ to ionize a molecule with the axis aligned in the direction $\overrightarrow{n}$ at an angle θ_n^E to the polarization vector multiplied by the probability that this molecule emits an electron towards the electron detector, which is given by the molecular-frame angular distribution of Auger electrons, MFAD(θ). The angular selectivity, $I(\theta_n^E, h\nu)$, of the initial ionization process is described by the photon energy dependent molecular alignment parameter, β_m, and $I_{ei}(\overrightarrow{e}, \overrightarrow{n}, h\nu)$ can thus be expressed as

$$I_{ei}(\overrightarrow{e}, \overrightarrow{n}, h\nu) \propto (1 + \beta_{mol}(h\nu) \cdot \frac{3 \cdot \cos^2(\theta_n^E) - 1}{2}) \cdot MFAD(\theta). \quad (1)$$

Therefore, if β_m is known and the two-step model is assumed, a measurement of the angular distribution of ions coincident with Auger electrons emitted in a fixed direction is sufficient to determine MFAD(θ). At the same time, the validity of the strict two-step model can be tested within the same experiment by repeating the measurement with two different polarization directions of the light and with different photon energies, thus checking the consistency of the two corresponding molecular-frame angular distributions. Note that this method is slightly different from the one used in our work on the MFADs of resonant CO Auger electrons [21, 22], which was performed at the Advanced Light Source using a modified velocity map (VMI) ion imaging spectrometer [24] in combination with *two* electron time-of-flight analyzers [25] and which thus allowed measuring the MFADs without making further assumptions regarding the two-step model.

The present experiment was performed at the high-resolution photochemistry beamline 27SU at SPring-8 using the identical setup as for our measurement of the non-resonant Auger electron MFADs in CO [23]. CO was introduced as an effusive beam crossed by the synchrotron radiation. The pressure in the vacuum chamber was $2.0 \cdot 10^{-4}$ pa. The Auger electrons were measured in a hemispherical electron analyzer, for which a pass energy of 200 eV and an instrumental resolution of about 1 eV were used. The photon energy bandwidth was 120 meV. A pulsed-field velocity focussing ion time-of-flight (TOF) spectrometer [26] was mounted opposite to the electron spectrometer. From the exact time of flight of ions with the same mass together with the detector hit position, we reconstructed the initial ion momentum and thus the ion angular distribution.

For the coincidence experiment, all voltages of the electron spectrometer were set such only electrons in an energy interval from 354 to 371 eV were recorded. The ionizing soft x-rays were linearly polarized either horizontally along the axes of the electron spectrometer and the ion spectrometer or in the vertical direction perpendicular two both spectrometer axes. For each photon energy and each polarization direction, data was collected for about 10h. All figures shown in this paper show the data after the subtraction of the random coincidences as described in [27].

3. Results and Discussion

The N_2 normal Auger electrons were measured with high resolution at two photon energies, 420 eV (on the so-called "shape resonance") and far above threshold at 500 eV. By detecting the ions produced in coincidence with the Auger electrons, we were able to disentangle the channels that

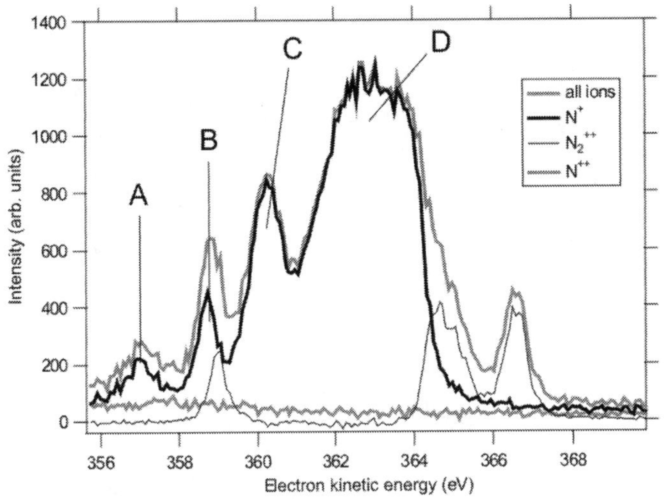

Figure 1. N_2 normal Auger spectrum at a photon energy of 420 eV measured along the polarization vector of the incoming synchrotron radiation. The spectra of the electrons detected in coincidence with N_2^{++}, N^+, and N^{++} ions are also shown.

yield N_2^{++}, N^+, and N^{++} fragments (see Fig. 1). The decay channels producing N^+ belong to four groups in the Auger spectrum labeled A through D, for which we determined the fragment ion angular distributions. For the ion angular distributions shown in Fig. 2, we analyzed N^+ ions within the kinetic energy range from 2 to 7eV. According to electron-optical simulations, the ion spectrometer should have a 4π collection efficiency for N^+ ions up to 7 eV, while faster ions may miss the detector for certain emission angles. The non-uniform efficiency of the ion detector was calibrated by making use of the symmetry of the ion distributions, which have to be cylindrically symmetric around the electron emission direction and the polarization vector when both directions coincide.

In order to discuss the observed ion angular distributions, let us first turn to the measurements at 500eV. While the angular distributions corresponding to each of the four Auger groups are noticeably different, they are (almost) identical for both polarization directions, as is expected for a photon energy far above threshold. Therefore, according to Eq. 1, the ion angular distribution in the laboratory frame is equivalent to the Auger electron angular distribution in the molecular frame.

If the two-step model underlying Eq. 1 is valid for all photon energies, the effect of the shape resonance on the measurements at 420eV can be described by multiplying the angular distributions measured at 500eV by a factor describing the selectivity of the molecular axis orientation in the core ionization process. Indeed, we find that this is possible for $\beta_m = 0.75$ at 420 eV and $\beta_m = -0.05$ at 500 eV. Therefore, in agreement with our findings for the normal Auger decay of CO [23], we can conclude that the two-step model is a very good approximation for the normal Auger decay of N_2. The first step, in which a core electron is emitted from the molecule, can have a high degree of selectivity for the molecular axis, especially in the shape-resonance region, which is described by the photon-energy dependent alignment parameter β_m. However, β_m does not depend on the Auger final state and the Auger decay can be considered as an independent second step since the core hole state does not contain any information on the direction of the electric vector or the photoelectron emission direction. The angular distribution of the Auger electrons thus depends only on the angle with respect to the molecular axis. In other words, the Auger electron MFADs do not depend on the photon energy or on the direction of the electric vector. Therefore, a measurement of the angular distributions of ions detected in coincidence with Auger electrons emitted in a fixed direction together with the known value of β_m is enough to determine the MFADs. In turn, knowing the photon energy dependence of β_m, one can predict the angular distribution of the coincident ions for any polarization direction of

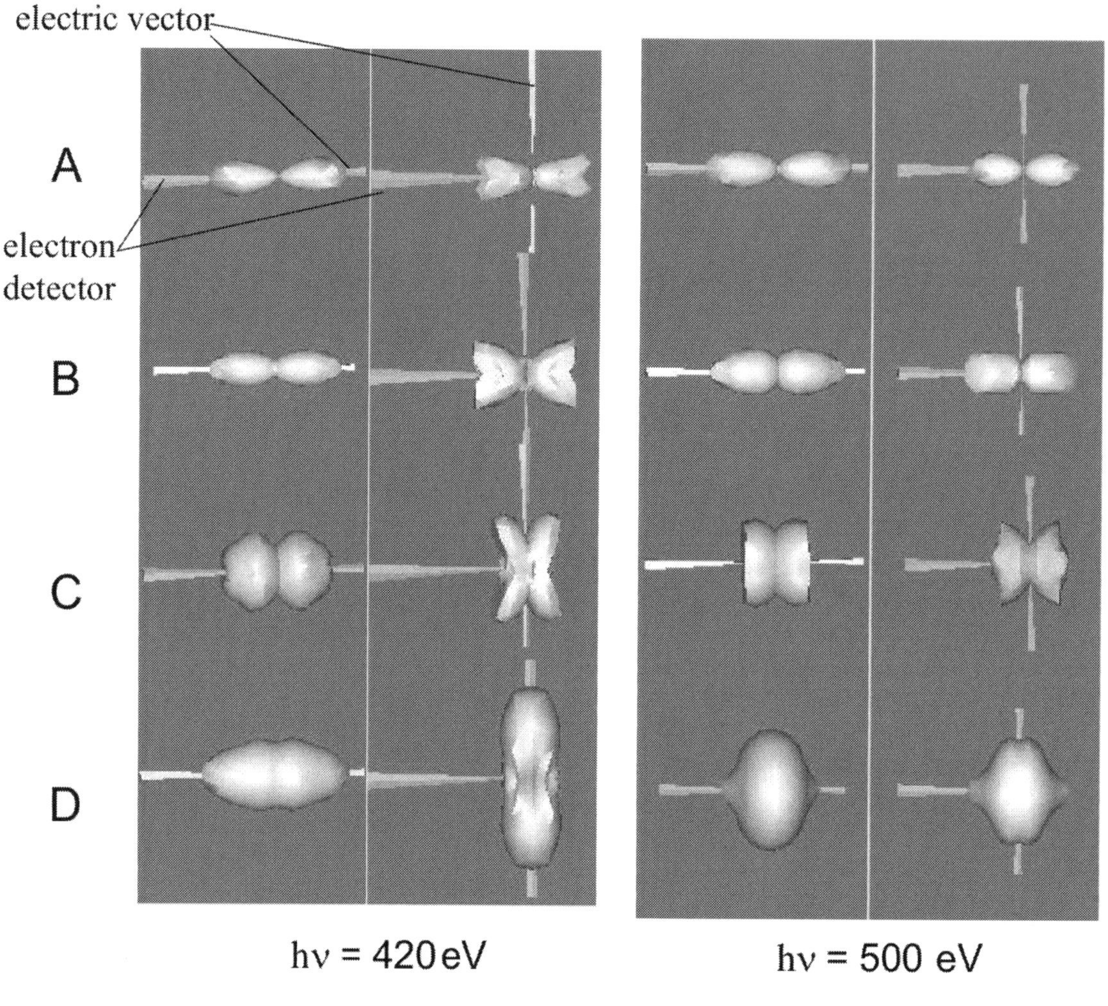

Figure 2. Angular distribution of N^+ ions in the plane containing the electric vector and the electron emission direction, measured in coincidence with an Auger electron detected along the direction marked by the horizontal white stick. The distributions for each of the four Auger groups A, B, C, and D are shown for two orientations of the electric vector, horizontal and vertical, and for two photon energies.

the light even in the photon energy region corresponding to the shape resonance.

Fig. 3 shows the molecular-frame angular distribution of N_2 normal Auger electrons as determined from the coincident ion angular distributions in Fig. 2. They differ distinctly for the four groups A, B, C, and D as defined in Fig. 1. For group A and B, the emission is directed primarily along the molecular axis, while it is more isotropic for group C and D. Although it is difficult to compare directly to the measurements of the Auger electron molecular-frame angular distribution by Schöffler *et al* [14], who did not resolve the Auger final state, we note a qualitative agreement to their angular distribution, which are dominated by the contribution from the most intense group D. The narrower and more pronounced maxima of the angular distribution measured by Schöffler *et al* could be attributed to a smaller effective acceptance angle for the orientation of the molecular axis in their experiment.

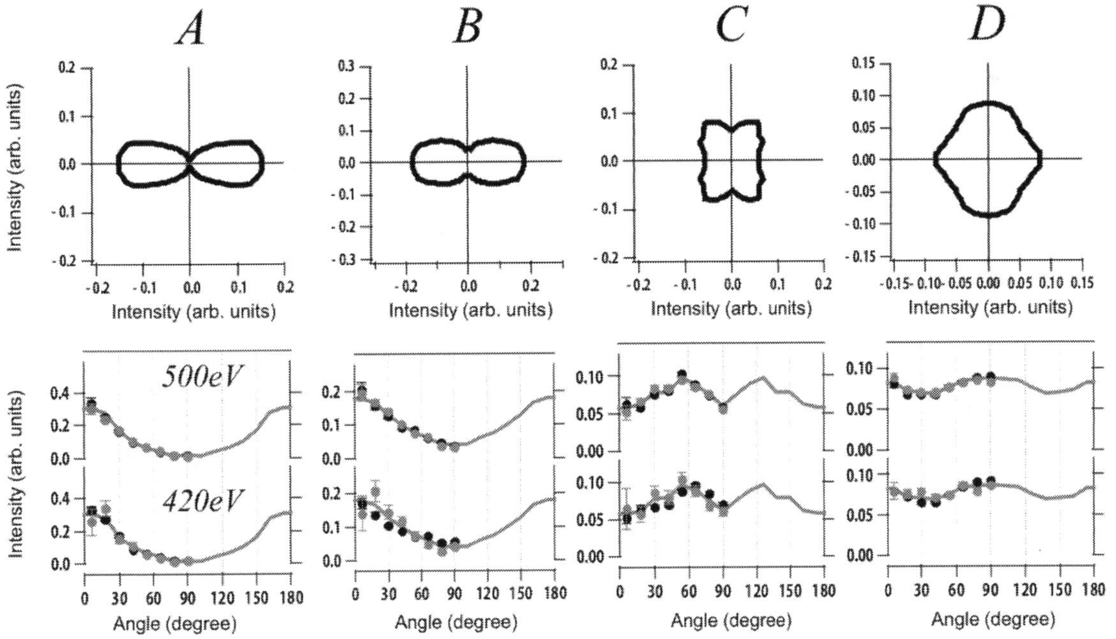

Figure 3. Molecular-frame angular distribution of N_2 normal Auger electrons in the four groups A, B, C, and D shown as polar plots in the plane containing the molecular axis (top row) and as cartesian plots as a function of electron emission angle with respect to the molecular axis (bottom row).

4. Conclusion

We have measured the angular distribution of N^+ fragments in coincidence with N_2 Auger electrons from different Auger groups and found significant differences for the different Auger groups. Our measurements at two photon energies and for two orientations of the polarization vector confirm the validity of the two-step model for normal N_2 Auger decay and thus allow extracting the molecular-frame Auger electron angular distributions, which are also significantly different for the different Auger final states.

Acknowledgments

The experiment was carried out with the approval of JASRI and was partly supported by the Japan Society of the Promotion of Science (JSPS) in the form of Grants-in-Aid for Scientific Research and by the BES, DOE, Chemical Sciences, Geosciences, Biosciences Division. D.R. acknowledges support from the Alexander von Humboldt foundation. The staff of SPring-8 are acknowledged for providing an excellent experimental facility.

References

[1] Golovin A V, Cherepkov N A and Kuznetsov V V 1991 *Opt. Spectrosc.* **71** 537
[2] Golovin A V, Cherepkov N A and Kuznetsov V V 1992 *Z. Phys. D* **24** 371
[3] Shigemasa E, Adachi J, Oura M and Yagishita A 1995 *Phys. Rev. Lett.* **74** 359
[4] Heiser F, Gessner O, Vielhaus J, Wieliczek K, Hentges R and Becker U 1997 *Phys. Rev. Lett.* **79** 2435
[5] Dörner R *et al* 1998 *Phys. Rev. Lett.* **81** 5776
[6] Downie P and Powis I 1999 *Phys. Rev. Lett.* **82** 2864
[7] Hikosaka Y and Eland J H D 2000 *J. Phys. B: At. Mol. Opt. Phys.* **33** 3137

[8] Motoki S, Adachi J, Ito K, Ishii K, Soejima K, Yagishita A, Semenov S K and Cherepkov N A 2002 *Phys. Rev. Lett.* 88 063003
[9] Gessner O, Hikosaka Y, Zimmermann B, Hempelmann A, Lucchese R R, Eland J H D, Guyon P-M and Becker U 2002 *Phys. Rev. Lett.* **88** 193002
[10] Martin F *et al* 2007 *Science* **315**, 629
[11] Saito N *et al* 2005 *J. Phys. B: At. Mol. Opt. Phys.* **38** L277
[12] Adachi J, Hosaka K, Teramoto T, Yamazaki M, Watanabe N, Takahashi M and Yagishita A 2007 *J. Phys. B: At. Mol. Opt. Phys.* **40** F285
[13] Fukuzawa H *et al* 2007 *Chem. Phys. Lett.* **436** 51
[14] Schöffler M *et al* 2008 *Science* **320** 929
[15] Rolles D *et al* 2005 *Nature* **437** 711
[16] Akoury D *et al* 2007 *Science* **318** 949
[17] Zimmermann B *et al* 2008 *Nature Physics* **4** 649
[18] Edwards A K, Zheng Q, Wood R M and Mangan M A 1997 *Phys. Rev. A* **55** 4269
[19] Guillemin R, Shigemasa E, Guen K L, Ceolin D, Miron C, Leclercq N, Morin P and Simon M 2001 *Phys. Rev. Lett.* **87** 203001
[20] Weber Th *et al* 2003 *Phys. Rev. Lett.* **90** 153003
[21] Rolles D, Prümper G, Fukuzawa H, Liu X-J, Pesic Z D, Fink R F, Grum-Grzhimailo A N, Dumitriu I, Berrah N and Ueda K 2008 *Phys. Rev. Lett.* **101** 263002
[22] Prümper G, Rolles D, Fukuzawa H, Liu X-J, Pesic Z D, Dumitriu I, Lucchese R R, Ueda K and Berrah N 2008 *J. Phys. B: At. Mol. Opt. Phys.* **41** 215101
[23] Prümper G, Fukuzawa H, Rolles D, Sakai K, Prince K C, Harries J R, Tamenori Y, Berrah N and Ueda K 2008 *Phys. Rev. Lett.* **101** 233202
[24] Rolles D, Pesic Z D, Perri M, Bilodeau R C, Ackerman G D, Rude B S, Kilcoyne A L D, Bozek J D and Berrah N 2007 *Nucl. Instr. Meth.* B **261** 170
[25] Berrah N, Langer B, Wills A A, Kukk E, Bozek J D, Farhata A and Gorczyca T W 1999 *J. Electron Spectrosc. Relat. Phenom.* **101-103** 1
[26] Prümper G, Lischke T, Fukuzawa H and Ueda K 2007 *Rev. Sci. Instr.* **78** 083104
[27] Prümper G and Ueda K 2007 *Nucl. Instr. Meth.* A **574** 350

Journal of Physics: Conference Series **212** (2010) 012010 doi:10.1088/1742-6596/212/1/012010

Photoemission dynamics in the molecular frame

A Yagishita[1], J Adachi[1] and M Yamazaki[1, 2]

[1] Photon Factory, Institute of Materials Structure Science, KEK
Tsukuba 305-0801, Japan

akira.yagishita@kek.jp

Abstract. Photoemission dynamics, based on the molecular frame photoelectron angular distribution, is reported. In the fast decay channel leading to Ne^{+} + Ne^{2+} of Ne_2 dimer, the molecular frame photoelectron measurements have been realized, and the core-hole localization observed. However, in the slow decay channel leading to Ne^{+} + Ne^{+}, such measurements have not been applicable as expected. The molecular frame photoemission study has been extended to a non-axially symmetric H_2O molecule, and its experimental result has been well explained by a simple scattering model which is widely used in surface science.

1. Introduction

In an axial-recoil dissociation process the two fragments fly off exactly anti-parallel to one another. If the dissociation is initiated by the ejection of an electron like inner-shell photoionization, and that electron is detected in coincidence with the fragment ion pair, then the electron is tied to the direction of the molecular axis at the moment of dissociation. Assuming that this whole process takes place on a time scale that is short compared with a rotational period, such coincidence measurements can register molecular-frame photoelectron angular distributions [1, 2]. Namely, the angle-resolved photoelectron-photoion coincidence enables one to select the process with a well defined orientation of molecules from the photoionization of an ensemble of randomly oriented molecules in a gas phase.

By broadening the scope of feasible and meaningful experiments in this manner, one may hope in the longer term to develop a broader understanding of molecular photoionization dynamics. However, note that up to now, such experiments were performed exclusively on linear molecules, [3-8] and references therein, because the coincidence rate of all fragment ions from bent molecules decreases radically compared with that of ions from back-to-back fragmentation of linear molecules.

In this paper, two recent topics of photoemission dynamics in the molecular frame are reported; decay channel dependence of the photoelectron angular distribution in core-level photoionization of Ne_2 dimer [9] and 3D mapping of photoemission from a single oriented H_2O molecule [10]. In the first topic, it is demonstrated that the photoelectron-photoion coincidence measurement is a good internal clock of the molecule. In the second topic, it is discussed that the molecular-frame photoelectron angular distribution is tied to a molecular structure.

2. Experimental

Referring the idea of Eppink and Parker [11] and Lebech *et al* [12], we have successfully applied the multi-coincidence velocity-map imaging technique to obtain the molecular-frame photoelectron

[2] Present address: IMRAM, Tohoku University, Sendai 980-8577, Japan

angular distribution (MF-PAD) [13]. The schematic of our experimental set-up is shown in figure 1. Linearly polarized soft x-rays from a planar undulator were monochromatized by a varied-space plane grating monochromator [14] and focused onto a supersonic molecular beam. Ions and electrons were guided by a uniform extraction electric field on opposite sides towards two area detectors with delay-line readout and detected in coincidence. The momenta of ions were determined from the flight times and the impact positions on the area detector. The projection of the electron momenta on a plane parallel to the electric field was determined from the positions. The polarization vector of the soft x-rays, being along the direction of the molecular beam, and the direction of the soft x-ray beam were in a plane parallel to the extraction electric field.

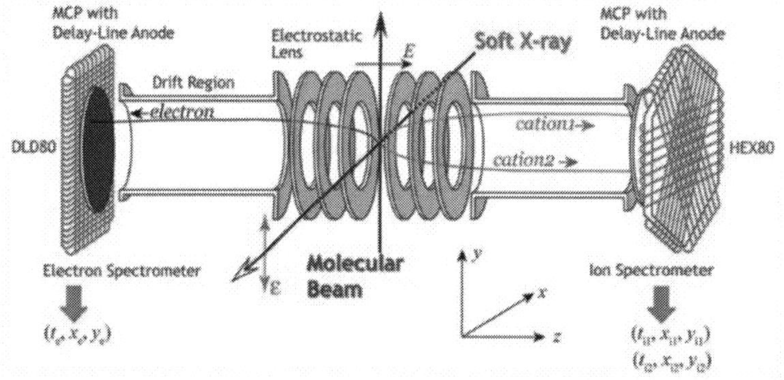

Figure 1. Schematic of multi-coincidence velocity-map imaging spectrometer

3. Results and discussion

3.1. Ne 1*s* photoemission from Ne_2 dimer

If one specific atom having the core-hole can be distinguished from the measurement, the measured parameters are described by the localized core-hole state, and if not, they are described by the symmetry-adapted core-hole state. The core-orbital photoionization process of Ne_2 dimer provides us an ideal showcase to resolve the elusive fundamental problem, because both removal of a 1s electron and successive Auger decay are essentially of atomic nature in the van der Waals molecules [15].

3.1.1. *Ne 1s photoemission for Ne^+-Ne^{2+} decay channel.* Figure 2 shows polar plots of Ne 1*s* photoemission relative to the dissociation axis of $Ne^+ + Ne^{2+}$ breakup channels, which are produced via Auger decay and successive interatomic Coulombic decay (ICD) [15-20] as shown in figure 3. In the calculations, we employed the time-dependent density functional theory (TDDFT). The calculations were conducted using localized description of the core-hole state. The MF-PAD for the localized core-hole is shown by thin curve in figure 2, where the photoelectron is ejected from the right Ne atom. As can be seen in the figure, the MF-PAD for the $Ne^+ + Ne^{2+}$ channel depends on the mutual angle between the polarization vector and dissociation axis. Surprisingly, broken inversion symmetry of the MF-PAD is seen clearly in the parallel geometry beyond the statistical errors of the data points, although the symmetry breaking is little recognized in the perpendicular geometry. The TDDFT calculations exhibit the lack of inversion symmetry as well, however, their patterns differ from the experimental ones in both the parallel and perpendicular geometries. The discrepancy is puzzling.

In weakly bound van der Waals dimers, both 1*s* photoemission and subsequent Auger decay with the lifetime of 2.4 fs take place at one site. On this basis, the decay process of core ionized Ne_2 can be described by a two-step model; in the first step the Auger decay occurs at the initial core-hole site leading to [Ne–Ne^{2+}], and in the second step a charge separation into two sites takes place during successive decays, see figure 3. The ICD in the second step produces the Ne^+ and Ne^{2+} ion pair. Since the Ne^+ fragment arises from the neutral atom of [Ne–Ne^{2+}] system due to the ICD, one can presume that the initial core-hole memory was carried by the resultant fragment Ne^{2+} ion. Moreover, the lifetime of the order of 100 fs of the ICD for [Ne–Ne^{2+}] is much faster than the rotational period of about 100 ps. This is the reason why we have succeeded in measuring the MF-PAD for the Ne^{+} + Ne^{2+} decay channel, which shows non-inversion-symmetric patterns reflecting symmetry lowering of the core ionized [Ne–Ne^{+}] system.

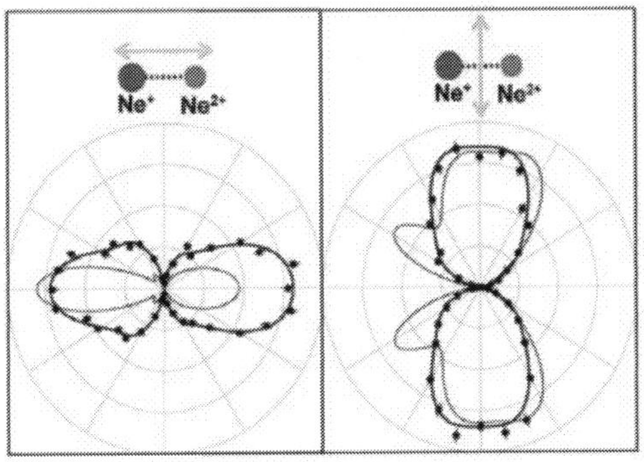

Figure 2. Polar plot of the Ne 1*s* MF-PAD from Ne_2 in coincidence with Ne^{+} + Ne^{2+} fragments at 10eV electron energy. The polarization vector (double headed arrow) and dissociation direction of the ion pair are indicated.

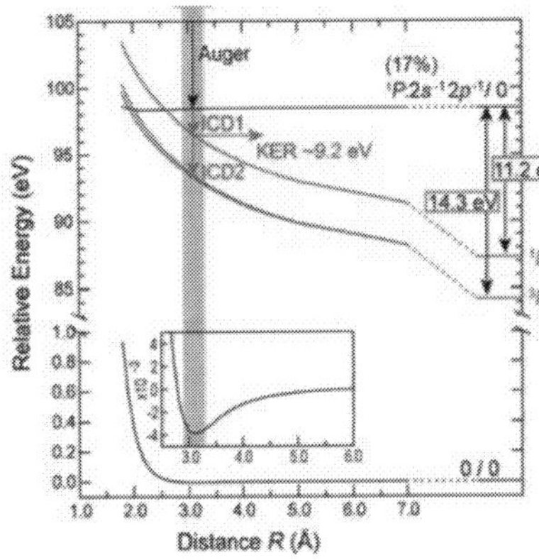

Figure 3. Calculated potential curves of Ne_2 dimer. Only the Auger final state, which can decay into the triply ionized states via ICD, is drawn.

3.1.2. *Ne 1s photoemission for* $Ne^{+}-Ne^{+}$ *decay channel.* Figure 4 shows polar plots of the Ne 1s photoemission relative to the dissociation axis Ne^{+} + Ne^{+} breakup channels which are produced via Auger decay and successive radiative decay as shown in figure 5. The TDDFT calculations were conducted using delocalized descriptions of the core-hole states. The MF-PAD for the delocalized core-hole state with σ_g symmetry is shown by thin curve in figure 4, where the photoelectron is ejected coherently from both left and right Ne atoms, and those with σ_u are also shown by thin curve. The superposition of the MF-PAD for σ_g and σ_u components is also shown with thin curve which is identical to the superposition of the MF-PAD for the localized core-holes on the left and right Ne atoms. The photoemission for the Ne^{+} + Ne^{+} channel depends only on the polarization vector. It is different from the calculated MF-PAD, but it is almost identical to the photoemission from randomly oriented dimers, although it is not shown here. This implies that the dissociation direction of two Ne^{+} ions has no correlation with the dimmer axis at the instant of the Ne 1*s* photoionization. Thus, the photoelectron angular distribution shown in figure 4 demonstrates an intriguing new phenomenon on the relaxation of the core-hole; the Ne^{+} + Ne^{2+} channel retains the memory of both the molecular orientation and the initial core-hole localization, while both are lost in the Ne^{+} + Ne^{+} channel.

The photoemission for the symmetric charge separation channel of Ne^{+} + Ne^{+} ought to be inversion symmetric, because the detection of such ions cannot distinguish the Ne atoms with their charge states. In fact, the experimental angular distribution is inversion symmetric. The present Ne 1*s* photoemission does not depend on the dimer axis at all, which results in a laboratory-frame photoelectron angular distribution (LF-PAD). In the two-step model, as shown in figure 5, a possible second step of fragmentation pathway to Ne^{+} + Ne^{+} from the Auger final [Ne + Ne^{2+} (^1D: $2p^{-2}$)] state is a radiative decay, because the Auger final state is energetically lower than the triply ionized state and cannot give off its excess energy by electron emission. During a lifetime (~ns) of the radiative decay leading to the symmetric charge separation, the dicationic dimer can rotate many times with a period of ~100 ps. Therefore, the memory of the molecular axis at the instant of photoionization is perfectly lost in the Ne 1*s* photoemission. As a consequence, the observed coincident photoemission is not "MF-PAD" but "LF-PAD". Actually, the observed PAD is well explained by the theoretical LF-PAD. That is, the

Journal of Physics: Conference Series **212** (2010) 012010 doi:10.1088/1742-6596/212/1/012010

experimental β values determined from the PAD are 1.72 ± 0.02 and 1.85 ± 0.03 for the parallel and perpendicular geometry, respectively, and the theoretical β is 1.878, which is also in good agreement with the experimental β values, 1.85 ± 0.03 determined from the LF-PAD. This good agreement certainly supports the idea that the coincident PAD for the Ne^+ + Ne^+ channel is reduced to the LF-PAD.

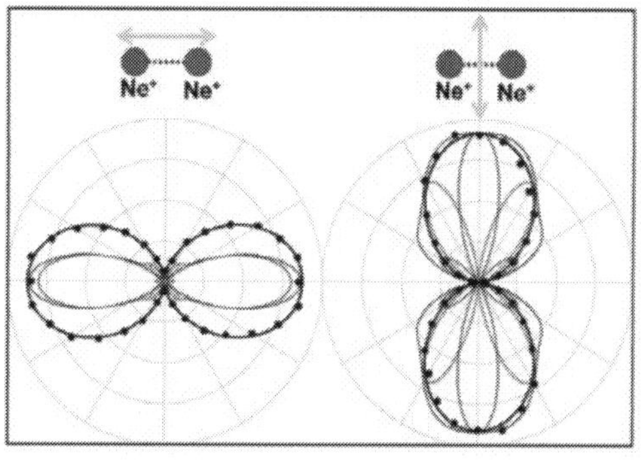

Figure 4. Polar plot of the Ne 1*s* LF-PAD from Ne_2 in coincidence with Ne^+ + Ne^+ fragments at 10eV electron energy. The polarization vector (double headed arrow) and dissociation direction of the ion pair are indicated.

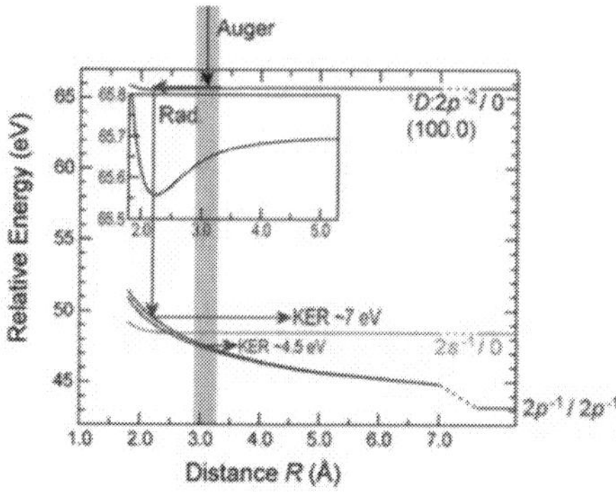

Figure 5. Calculated potential curves of Ne_2 dimer. Only the Auger final state, which can decay into the doubly ionized states via radiative decay, is drawn. The final state has the minimum at R = 2.2 Å

3.2. O 1*s* photoemission from a single oriented H_2O molecule

Figure 6 shows a vector correlation image of three particles, O^+, H^+, and H^+, produced by three-body break-up of H_2O^{3+} following double Auger decay after O 1*s* photoemission. The H^+ ion is emitted in a narrow angular range with respect to the O^+ ion and the momentum distributions of two H^+ ions are mirror symmetric to each other. It is also shown that the correlation angle between two H^+ momenta is slightly larger than the ground state bond angle ($\angle$HOH = 104.5°) because of the Coulomb repulsive force. These results verify that the Coulomb explosion leading to three-body break-up of H_2O^{3+} is immediate, maintaining the ground state shape of H_2O. Then, from the momentum vectors of the three fragment ions, the molecular orientation and intrinsic shape of H_2O in space at the instance of the O 1*s* photoionization can be determined.

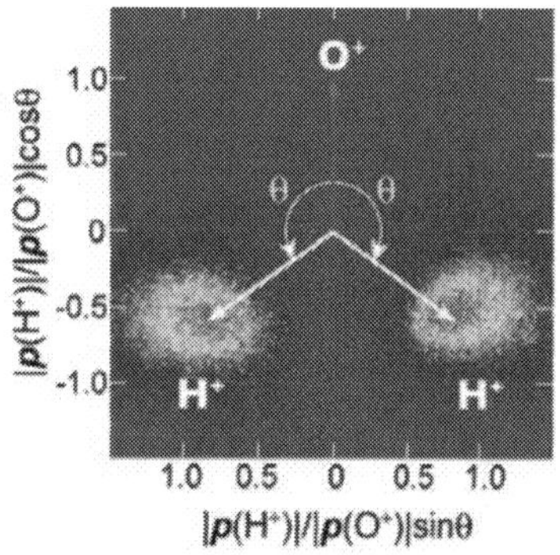

Figure 6. Momentum vector correlation among the three ions of O^+, H^+ and H^+ created in O 1*s* photoionization of H_2O.

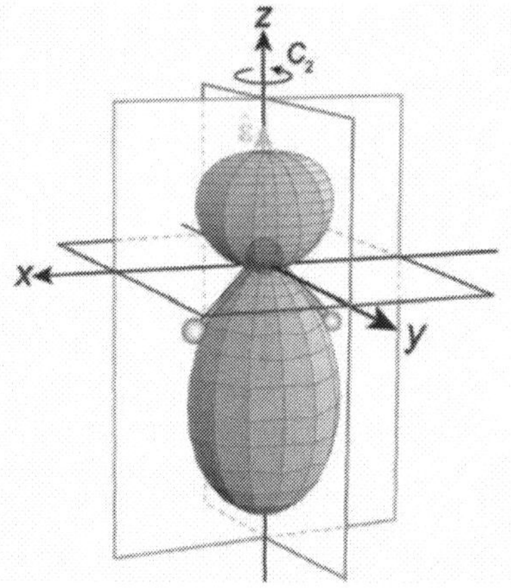

Figure 7. 3D angular distribution of O 1*s* photoelectrons from fixed-in-space H_2O. Fitted curves to the experimental data were used.

Journal of Physics: Conference Series **212** (2010) 012010 doi:10.1088/1742-6596/212/1/012010

By extending this approach to the quadruple coincidence of photoelectron–O^+–H^+–H^+, we have performed O 1*s* photoemission measurements from H_2O molecules fixed in space. The key result is shown in figure 7, which depicts a 3D mapping of MF-PAD produced by incident light polarized along the molecular *z*-axis, at the photoelectron energy of 14 eV. The distance from the origin to the surface corresponds to the probability of finding the photoelectrons emitted into the direction of (θ, ϕ), where θ is the polar angle measured from the C_2 axis, and ϕ is the azimuthal angle from the *x*-axis. The orientation of the MF-PAD can occur only through the interference of partial waves between odd- and even-ℓ [21–23]. This quantum interference effect results in the up-down asymmetry of the MF-PAD shown in figure 7, which is considerably deformed from the dipole shape of an atomic $1s \rightarrow kp$ transition. The photoionization dynamics seen in the MF-PAD is restricted by the symmetry of the photoelectron wavefunction as predicted by theories. Indeed, the a_1 symmetry (twofold symmetry about the C_2 axis) of the wavefunction created by the *z*-axis polarized photon is embodied in the 3-D MF-PAD.

Here, we apply a simple photoelectron diffraction model involving only a single scattering based on Huygens' approach, to get an intuitive understanding of the MF-PAD. From Huygens' point of view, the O atom acts an emitter of primary waves which are subsequently scattered against the two H atoms. In constructing the total wavefield as a coherent sum of the primary and secondary contributions we neglect the multiple scattering of the electron waves. The scattering scheme of the electron wave is depicted in figure 8. The MF-PAD can be expressed by superposing the primary wave and two scattered waves [24, 25]. In the simulation of the MF-PAD we used the ground state geometry of H_2O molecule, and introduced two parameters; the modulus and phase shift for the complex amplitude of e + H scattering. The fitting result of the simulation is shown in figure 9. As can be seen in the figure, the contribution of the two scattered waves from the H atoms, Young's double-slit interference, is negligible, but the contribution of the primary and scattered waves shows a considerable interference effect between them; strong negative values on the O atom side and weak on the opposite side. Due to the interference, the strong orientation of the MF-PAD is well explained.

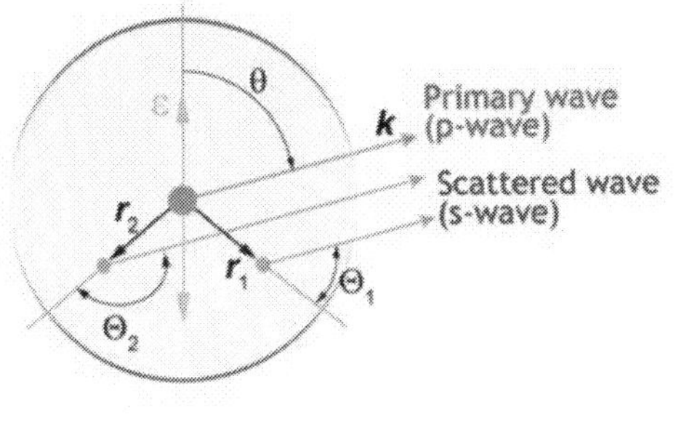

Figure 8. Scattering scheme of the electron wave. The origin of the position is set at the O atom. The positions of the two H atoms are indicated by $\mathbf{r}_1$ and $\mathbf{r}_2$.

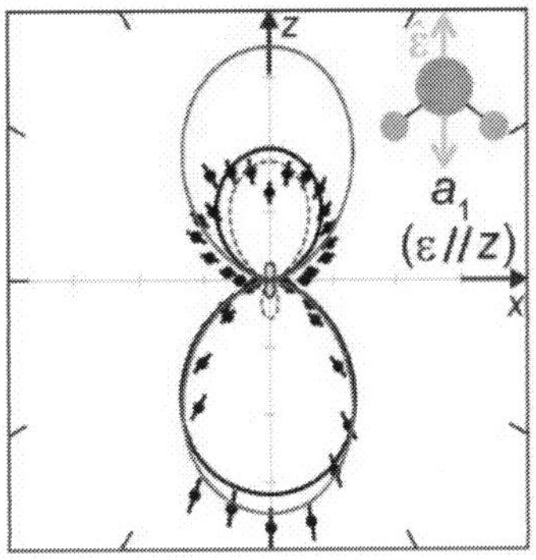

Figure 9. The experimental data on x-z plane in figure 7 and fitting result. The dipole shape is the primary wave contribution, the dotted curve is the interference between the primary and scattered waves, and tiny two loops along the z-axis are the double-slit interference. Solid lines are positive and dotted line negative.

4. Summary

We have demonstrated two new directions on photoemission dynamics in the molecular frame, based on the multi-coincidence velocity-map imaging measurement. One direction is the time scale discussion of core-hole decay owing to the high sensitivity of MF-PAD to the molecular symmetry. The other direction is the extension of MF-PAD to the non-axially symmetric molecule. The present results enable ones to compare the full 3D MF-PAD from a single oriented molecule with photoelectron diffraction patterns from the relevant adsorbed molecules on surface. It is quite interesting, because it might make the interaction between molecule and surface clear. Even more importantly, the successful application of the simple scattering model to the MF-PAD suggests that a new field of ultrafast molecular structure determination due to MF-PAD measurements will be opened, if possible to control its orientation in space, using a future x-ray laser.

Acknowledgements

The experiments have been performed under the approval of the Photon Factory Program Advisory Committee (Project No. 2005G099, 2005G108, 2007G056 and 2007G550). The present work was supported by a Grant-in-Aid for Scientific Research (B) 19340114. M. Y. is grateful for the support of the Japan Society for the Promotion of Science.

References

[1] Golovin A V, Cherepkov N A and Kuznetsov V V 1992 *Z. Phys. D* **24** 371
[2] Shigemasa E, Adachi J, Oura M and Yagishita A 1995 *Phys. Rev. Lett.* **74** 259
[3] Geßner O, Hikosaka Y, Zimmermann B, Hempelmann A, Lucchese R R, Eland J H D, Guyon P-M and Becker U 2002 *Phys. Rev. Lett.* **88** 193002
[4] Yagishita A, Hosaka K and Adachi J 2005 *J. Electron Spectrosc. Relat. Phenom.* **142** 295
[5] Weber T *et al.* 2004 Nature **431** 437
[6] Rolles D *et al.* 2005 Nature **437** 711
[7] Lebech M, Houver J C, Dowek D and Lucchese R R 2006 *Phys. Rev. Lett.* **96** 073001
[8] Adachi J, Hosaka K, Teramoto T, Yamazaki M, Watanabe N, Takahashi M and Yagishita A *2007 J. Phys. B: At. Mol. Opt. Phys.* **40** F285
[9] Yamazaki M, Adachi J, Kimura Y, Yagishita A, Stener M, Decleva P, Kosugi N, Iwayama H, Nagaya K and Yao M 2008 *Phys. Rev. Lett.* **101** 043004
[10] Yamazaki M, Adachi J, Teramoto T, Yagishita A, Stener M and Decleva P 2009 *J. Phys. B: At. Mol. Opt. Phys.* **42** 051001
[11] Eppink A T J B and Parker D H 1997 *Rev. Sci. Instrum.* **68** 3477
[12] Lebech M, Houver J C and Dowek D 2002 *Rev. Sci. Intrum.***73** 1866
[13] Hosaka K, Adachi J, Golovin A V, Takahashi M, Watanabe N and Yagishita A 2006 *Japan. J. Appl. Phys.* **45** 1841
[14] Watanabe M, Toyoshima A, Azuma Y, Hayaishi T, Yan Y and Yagishita A 1997 *Proc. SPIE Int. Soc. Opt. Eng.* **3150** 58
[15] Santra R and Cederbaum L S 2003 *Phys. Rev. Lett.* **90** 153401
[16] Cederbaum L S, Zobeley J and Tarantelli F 1997 *Phys. Rev. Lett.* **79** 4778
[17] Jahnke T *et al* 2004 *Phys. Rev. Lett.* **93** 163401
[18] Öhrwall G *et al* 2004 *Phys. Rev. Lett.* **93** 173401
[19] Morishita Y *et al* 2006 *Phys. Rev. Lett.* **96** 243402
[20] Jahnke T *et al* 2007 *Phys. Rev. Lett.* **99** 153401
[21] Chandra N 1987 *J. Phys.* B **20** 3417
[22] Reid K L and Powis I 1994 *J. Chem. Phys.* **100** 1066
[23] Kuznetsov V V, Cherepkov N A, Fecher G H, Schönhense G 2002 *J. Chem. Phys.* **117** 7180
[24] Fadley C S 1993 *Surf. Sci. Rep.* **19** 231
[25] Woodruff D P and Bradshaw A M 1994 *Rep. Prog. Phys.* **57** 1029

Intermediate-coupling phenomena in electron-impact excitation of Hg atoms

G. F. Hanne

Fachbereich Physik, Universität Münster, 48149 Münster, Germany

E-mail: hanne@uni-muenster.de

Abstract. In my talk the influence of the intermediate-coupling scheme of Hg on various electron scattering observables such as spin exchange cross sections, spin up-down asymmetries or spin-resolved circular polarization is discussed with special emphasis to recent experimental results from our group and recent theoretical data.

1. Introduction: intermediate coupling in Hg atoms

The study of spin-resolved electron-atom collisions reveals various scattering mechanisms that are masked when only spin-averaged observables are measured. The most significant spin effects in low-energy scattering generally result from electron exchange and the spin-orbit interaction, both in the target alone and in the projectile-target interaction.

As a benchmark target in atomic physics, Hg has been investigated with respect to intermediate-coupling phenomena in electron collisions since more than 40 years. The theoretical description of electron-impact excitation of heavy atoms such as Hg has managed to model more and more detailed structures of experimental data. However, along with this progress the complexity of the mathematical and numerical methods has increased as well. McConnell and Moiseiwitsch [1], used a Born-Ochkur approximation neglecting spin-orbit terms for the continuum electron in the interaction potential, but applied the intermediate-coupling scheme of the $6s6p$ states of Hg by

$$\begin{aligned}
\Psi(6s6p\,^3P_0) &= \Psi^0(6s6p\,^3P_0) \\
\Psi(6s6p\,^3P_1) &= \alpha\Psi^0(6s6p\,^3P_1) + \beta\Psi^0(6s6p\,^1P_1) \\
\Psi(6s6p\,^3P_2) &= \Psi^0(6s6p\,^3P_2) \\
\Psi(6s6p\,^1P_1) &= \alpha\Psi^0(6s6p\,^1P_1) - \beta\Psi^0(6s6p\,^3P_1)
\end{aligned} \tag{1}$$

where α = 0,987 and β = –0,171. The excitation energies are 4.71 eV, 4,89 eV, 5,46 eV and 6,70 eV for the states labelled with $^3P_0$, $^3P_1$, $^3P_2$ and $^1P_1$, respectively.

With this scheme the Born-Ochkur approximation lead to an astonishing quantitative agreement with the experimental data of the linear light polarization of the $6s6p \rightarrow 6s^2$ transitions excited by electron impact [2]. Since the Born-Ochkur calculations from the late 1960th more sophisticated approaches such as the relativistic distorted-wave Born approximation (RDWBA) or relativistic close-coupling approaches were developed and have improved the description of electron-Hg scattering considerably [3-8].

Journal of Physics: Conference Series **212** (2010) 012011 doi:10.1088/1742-6596/212/1/012011

2. Direct observation of exchange

It is interesting to note that the nature of a spin-exchange singlet-triplet transition was not quite clear in the 1960th. In their famous textbook "Electronic and Ionic Impact Phenomena" Massey and Burhop wrote [9]: "The multiplicity of the state may therefore change by ±2, but this may occur only if the spin direction of the outgoing electron is opposite to that of the incident electron – electron exchange must have taken place in the collision, for we are assuming that spin-orbit interaction is negligible, so the spin of the incident electron cannot be reversed by the impact. Exchange can also occur between electrons with the same spin but in that case it will not be associated with a change of the multiplicity." It is easy to see that this statement is wrong for a singlet-triplet transition by electron impact as can be seen from the following basic reactions which are obtained if spin conservation is assumed [10]:

$$\begin{aligned} &\mathrm{e}(\uparrow)+\mathrm{Hg}(S=0,M_s=0)\rightarrow\mathrm{Hg}^*(S=1,M_s=1)+\mathrm{e}(\downarrow) \quad :-\sqrt{2}g \\ &\mathrm{e}(\uparrow)+\mathrm{Hg}(S=0,M_s=0)\rightarrow\mathrm{Hg}^*(S=1,M_s=0)+\mathrm{e}(\uparrow) \quad : \ g \end{aligned} \tag{2}$$

The second reaction is a singlet-triplet transition with a change of the multiplicity but does not change the spin of the continuum electron. Only a phase change is required. As a result of a formal analysis of electron-impact excitation of triplet states [10,11], using the indicated amplitudes of the reactions (2), the spin polarization

$$P=\frac{N_\uparrow - N_\downarrow}{N_\uparrow + N_\downarrow} \tag{3}$$

of the incident electrons will be changed to $P' = P/3$ after the collision for a pure singlet-triplet excitation. Here $N_\uparrow$ and $N_\downarrow$ are the numbers of electrons with spin up and down with respect to a given quantization axis. It is true, however, that a singlet-singlet transition with electron exchange cannot result in a spin change at all, i. e. we will then have $P' = P$.

The cross section calculations of McConnell and Moiseiwitsch [1] using the intermediate coupling scheme (1) stimulated us to perform an experiment where spin exchange effects are directly observed by measuring P' and P. The idea was that one would observe $P' = -P/3$ if only the triplet parts of the wave functions contribute to the excitation of the triplet $6s6p$ levels, and $P' = P$ if the excitation is caused only by the singlet part in the wave function of the $6s6p\ ^3P_1$ state (1). We chose 4.9 eV as a mean excitation energy for the $6s6p\ ^3P$ excited states. The contribution of the singlet part to the excitation is expected to become more and more significant with increasing scattering energy, because the dominant excitation via exchange effects (triplet part) diminishes simultaneously. A scheme of the experiment and results [10] which confirm the anticipation about the energy behaviour of spin exchange are shown in Fig. 1.

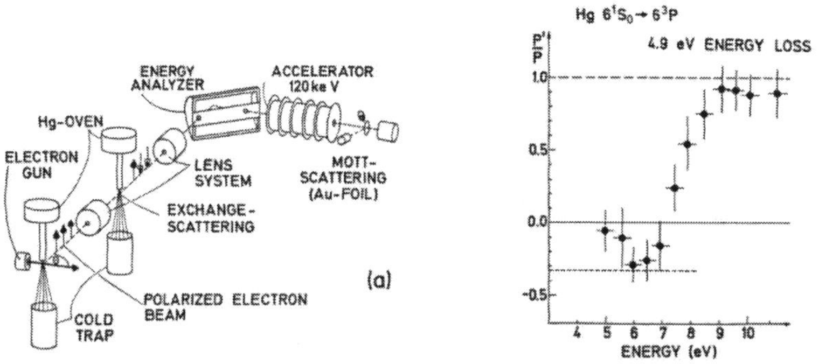

Fig. 1 Scheme of the experiment (left) and results (right) for electron-impact excitation of the $6s6p\ ^3P$ orbitals in Hg. Polarized electrons produced by Mott scattering from Hg atoms are fired onto another beam of Hg atoms and the polarizations P' (for scattering in the forward direction) and P are measured by means of a Mott detector.

Journal of Physics: Conference Series 212 (2010) 012011 doi:10.1088/1742-6596/212/1/012011

Spin-resolved fine-structure excitation

Another type of experiments are so-called S_A measurements where the observable S_A is given by the differential cross sections $\sigma(\uparrow)$ and $\sigma(\downarrow)$ for spin-up and spin-down electrons, respectively:

$$S_A = \frac{\sigma(\uparrow)-\sigma(\downarrow)}{\sigma(\uparrow)+\sigma(\downarrow)}. \qquad (4)$$

This spin-up-down asymmetry is interesting with respect to the fine-structure coupling in Hg [11] which causes an effect illustrated in Fig. 2. It is well known that in electron scattering the electrostatic Coulomb interaction leads to an orbital orientation $L_\perp$, i. e. a nonzero component of the expectation value of the orbital angular momentum operator perpendicular to the scattering plane (cf Fig. 2 left). Excitation of the 6*p* orbital in Hg may produce such an orbital orientation. The spin-orbit-coupling within the target will thus produce angular momentum configurations as illustrated in Fig. 2 (right) with an orbital angular momentum vector oriented perpendicular to the scattering plane.

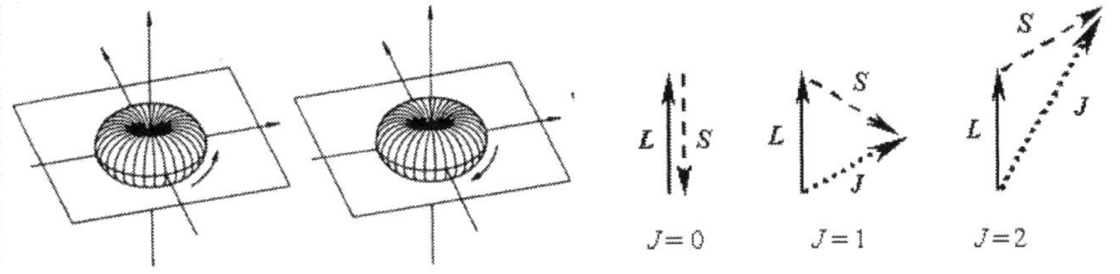

Fig. 2 Orbital orientation (left part): the differential cross section (in electron scattering to the left) for exciting a *p* orbital with positive orientation (left orbital) may be considerably different from that with negative orientation (right orbital) because of the different Coulomb interactions when continuum and target electrons move "parallel" or "antiparallel". Right part: the spin-orbit interaction within the target will couple the spin of a triplet state (for *LS* coupling with $L = 1$ and $S = 1$) with the oriented vector ***L*** to resulting fine-structure states with $J = 0$, $J = 1$ and $J = 2$.

If we use polarized electrons we can see from Fig. 2 (right part) that we must "offer" mainly spin-down electrons to excite a $^3P_0$ state and preferentially spin-up or spin-down electrons to excite a $^3P_1$ or a $^3P_2$ state, respectively. A simple angular momentum coupling procedure [11] yields indeed the following relation between the spin-up-down asymmetries for the different fine-structure states:

$$S_A(^3P_0) = S_A(^3P_1)/2 = -S_A(^3P_2)/2 \qquad (5)$$

In Hg we have a strong violation of *LS* coupling for the 6*s*6*p* $^3P_1$ state (1) and, therefore, we cannot expect the relation $S_A(^3P_1) = -S_A(^3P_2)$ to be valid. However, because of the coupling scheme shown in Eq. 1, one can show that sign sgn $S_A(6\ ^3P_1) = -\text{sgn}\ S_A(6\ ^1P_1)$ if the continuum spin-orbit interaction is almost negligible and exchange is the dominant spin effect [12]. Experimental and theoretical result confirm this model assumption. In Fig. 3 are shown experimental and theoretical results of the asymmetry functions S_A for the 6*s*6*p* $^3P_1$, 6*s*6*p* $^3P_2$ and 6*s*6*p* $^1P_1$ states at 20 eV. Clearly the relation $S_A(^3P_1) = -S_A(^3P_2)$ is violated, however, the results as well as the calculation (though not agreeing well with the experimental data) confirm the relation sgn $S_A(6\ ^3P_1) = -\text{sgn}\ S_A(6\ ^1P_1)$. This indicates that exchange in conjunction with the angular momentum coupling scheme is the dominant spin effect.

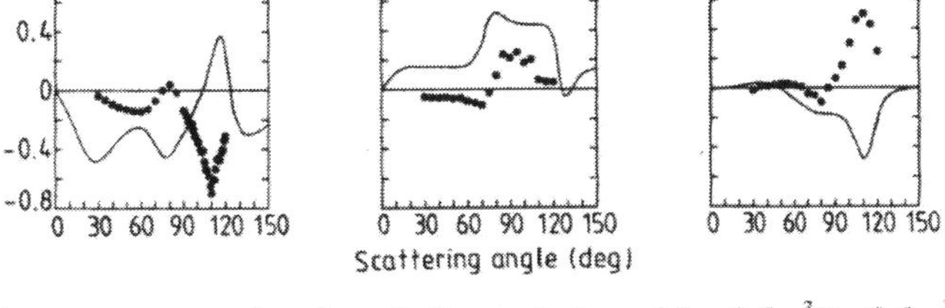

Fig. 3 Spin-up-down asymmetry functions S_A for excitation of the 6*s*6*p* $^3P_1$, 6*s*6*p* $^3P_2$ and 6*s*6*p* $^1P_1$ states (from left to right) in Hg at 20 eV. The results (dots) are taken from Borgmann et al [12] where the DWBA calculations (solid lines) are presented in Ref. [4].

4. Study of circular polarization

Another method to study electron-impact excitation processes and the role of intermediate coupling are measurements of the circular polarization of light emitted from Hg after excitation with polarized electrons. The scheme of such experiments is shown in Fig 4. Without electron detection we can measure the so-called integrated circular polarization of light emitted in the direction of the initial transverse electron polarization vector. In that case a nonzero circular polarization is mainly caused by electron exchange in conjunction with the spin-orbit coupling within the target. In so-called electron-photon coincidence studies the scattered electrons are detected in coincidence with the photons. In that case a scattering plane may be defined by the scattered electrons, and a circular polarization for photons emitted perpendicular to the scattering plane may be electron polarization dependent, but is already caused by the Coulomb interaction alone (orbital angular momentum orientation by electron impact, see Fig. 5).

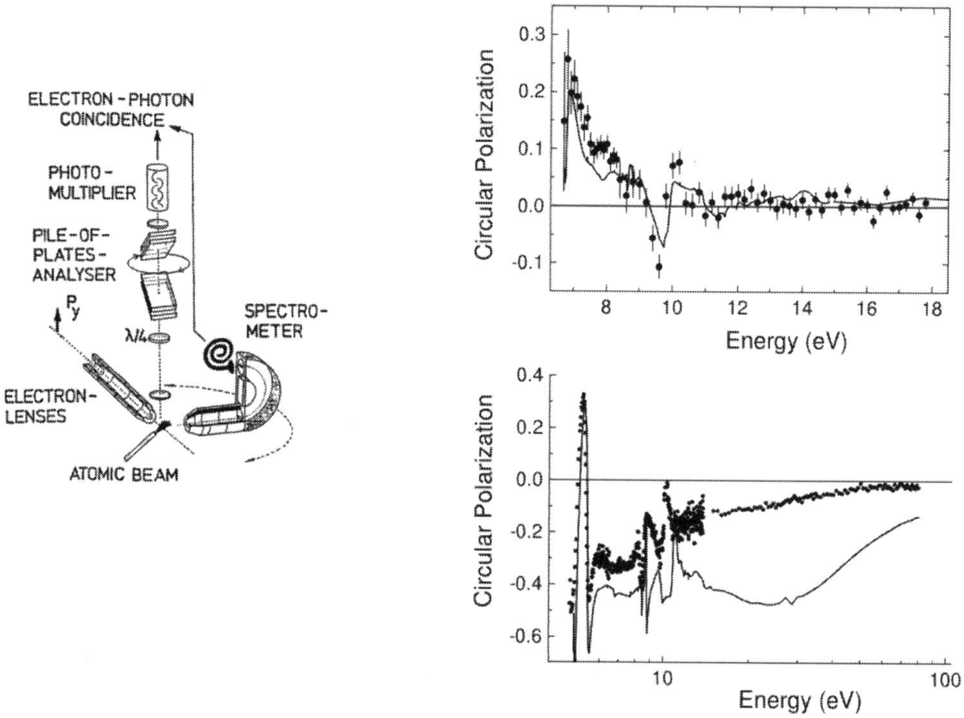

Fig. 4 Left: Scheme of an experiment where a target is excited by electron impact with polarized electrons and the circular polarization of photons emitted in the direction of the electron polarization is determined. In electron-photon-coincidence studies presented below, the electron polarization vector is perpendicular to the scattering plane. Right: Measured (dots) and calculated (lines) integrated circular polarization for the 185 nm line ($6s6p\ ^1P_1 \rightarrow 6s6p\ ^1S_0$, top) and the 254 nm line ($6s6p\ ^3P_1 \rightarrow 6s6p\ ^1S_0$, bottom) after excitation with polarized electrons [8,13,14]

Observable exchange processes are expected to contribute to electron-impact excitation of the $6s6p$ states of Hg with total angular momentum $J = 1$ because of the intermediate-coupling nature (1) of these states. They have excitation energies of 4.89 eV ($6s6p\ ^3P_1$) and 6.7 eV ($6s6p\ ^1P_1$) above the $6s^2\ ^1S_0$ ground state. Fig . 4 shows the results of the integrated circular polarization for the transitions $6s6p\ ^1P_1 \rightarrow 6s6p\ ^1S_0$ (185 nm) and $6s6p\ ^3P_1 \rightarrow 6s6p\ ^1S_0$ (254 nm) after excitation with polarized electrons. The experimental and theoretical results are from references [13,14] which we obtained recently. In the near-threshold regime, where the circular polarization is dramatically affected in the vicinity of resonances, i.e., temporary negative-ion states of the collision system, a fully relativistic 36-state Dirac

B-spline *R*-matrix (DBSR) approach with non-orthogonal orbital sets [8] provides very satisfactory agreement with the experimental data and represents a substantial improvement over an earlier 5-state semi-relativistic Breit-Pauli model (not shown here). At higher energies, the experimental data are in qualitative agreement with the expected reduction of exchange effects with increasing energy. Whereas the calculation and the experimental results for the 185 nm line show excellent agreement, there is a significant quantitative disagreement for the 254 nm line at energies above 12 eV where cascade effects are expected. And, unfortunately, cascade effects seem particularly strong for the spin-dependent light polarizations of the 254nm line, thereby making a straight-forward study of exchange effects by integrated circular polarization measurements virtually impossible. Note that the general sign of the circular polarization is different for the two lines. In the case of the 254 nm line the triplet part of the wave function dominates the sign, whereas for the 185 nm line interference between the singlet and the triplet part dominates which causes the sign change.

Finally, we present results of the angle-differential circular polarization of the $6s6p\ ^3P_1 \rightarrow 6s6p\ ^1S_0$ (254 nm) transition at a scattering energy of 25 eV from recent electron-photon coincidence studies, where cascading is eliminated. Clearly, the circular polarization is strongly electron polarization dependent with the astonishing result that even positive values at small scattering angles are possible (for spin-down electrons). This seems to contradict existing propensity rules, but it was shown [15] that this is an interference effect caused by the triplet-singlet mixture of the triplet state. The experimental results confirm the predicted *positive* circular polarization for scattering of *spin-down* electrons into small angles. The extraction of the *orbital* part of the orientation from the theoretical results, however, yields a consistently *negative* sign in both calculations, again confirming the validity of the propensity rules.

The results illustrate also that present calculations (which are the models mentioned above) are still a difficult task to describe specific observables such as P_3 for electron scattering from complex targets.

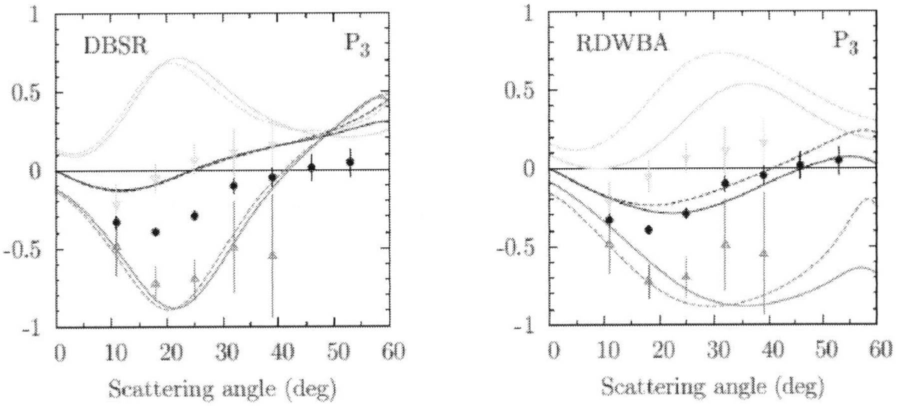

Fig. 5 Angle-differential, spin-resolved circular polarization (Stokes parameter P_3) of the 254 nm line ($6s6p\ ^3P_1 \rightarrow 6s6p\ ^1S_0$ transition) at a scattering energy of 25 eV. The experimental results (▲ spin-up, ▼ spin-down, • unpolarized electrons) are from Jüttemann [14]. DSBR theories: solid line DSBR-90, dashed line DSBR-36 [16]; RDWBA theories: solid line (MCGS), dashed line (SCGS) [17]:

Acknowledgements. Most of the presented experimental studies were supported by the DFG (Deutsche Forschungsgemeinschaft).

References

[1] McConnell J C and Moiseiwitsch B L *J. Phys. B* **1** (1968) 406
[2] Herting C and Hanne G F *J Phys B: At. Mol. Opt. Phys.* **35** (2002) L91
[3] N. S. Scott, P. G. Burke, and K. Bartschat, *J. Phys. B* **16**, (1983) L361
[4] Bartschat K and Madison D H *J. Phys. B: At. Mol. Phys.* **20** (1987) 1609

[5] Muktavat K, Srivastava R and Stauffer A D *J. Phys. B: At. Mol. Opt. Phys.* **36** (2003) 2341-2356
[6] Fursa D V, Bray I and Lister G *J. Phys. B: At. Mol. Opt. Phys.* **36** (2003) 4255-4271
[7] Fursa D V and Bray I *Phys. Rev. Lett.* **100** (2008) 113201
[8] Zatsarinny O and Bartschat K *Phys. Rev. A* **79** (2009) 042713
[9] Massey H S W and Burhop E H S *Electronic and Ionic Impact Phenomena 1* (2. Ed. Clarendon Press, 1969) 453
[10] Hanne G F and Kessler J *Phys. Rev. Lett.* **33** (1974) 341
[11] Hanne G F *J. Phys. B: Atom. Molec. Phys.* **9** (1976) 805
[12] Borgmann H, Goeke J, Hanne G F, Kessler J and Wolcke A *J. Phys. B: At Mol. Phys.* **20** (1987) 1619
[13] Jüttemann F, Hanne G F, Zatsarinny and Bartschat K *Phys. Rev. A* **79** (2009) 042712
[14] Jüttemann F *PhD thesis* (2009)
[15] Herting C, Hanne G F, Bartschat K, Grum-Grzhimailo A N, Muktavat K, Srivastava R and Stauffer A D *J. Phys. B: At. Mol. Opt. Phys.* **35** (2002) 4439
[16] Bartschat K *private communication* (2009)
[17] Srivastava R *private communication* (2009)

Electron correlations and spin effects activated by excitation of an inner 3d electron in zinc

D Cvejanović, S A Napier, J F Williams and L Pravica

ARC Centre for Antimatter-Matter Studies, Centre for Atomic, Molecular and Surface Physics, Physics Department, University of Western Australia, Perth 6009, Australia

jfw@physics.uwa.edu.au

Abstract. Excitation of the $4s4p^3P_{0,1,2}$ states of zinc was studied in the energy region of the lowest $3d^{-1}$ autoionizing state using spin polarized incident electrons. The energy dependence of angular differential excitation intensities was measured for incident electrons with spin up and down. Negative ion resonances with a $3d^94s^24p^2$ electron configuration show prominent structure in the angular differential cross section and also show significant spin dependence which is prominent especially at 90°.

1. Introduction

Experiments with incident spin polarized electrons have revealed subtle details of momentum coupling in a target atom and how this coupling may influence the scattering dynamics [1]. Our previous studies of inelastic scattering of polarized electrons from free zinc atoms explored the spin effects in the scattering process of ionization-with-excitation of open and closed 3d-shell states [2]. Also measurements of the three Stokes polarization parameters [3] of the $3d^{10}4s4d\,^1D_2 \rightarrow 3d^{10}4s4p\,^1P_1$ decay photons observed and disentangled the spin orbit and exchange effects in excitation via 3d core-excited negative ion resonances. Here we report on the difference in the angular differential cross sections for excitation of the $3d^{10}4s4p\ ^3P$ states for spin-up and spin-down incident electrons in the energy region where excitation is affected by 3d core excited negative ion resonances. A combination of different experimental techniques provides complementing information.

Asymmetries for electrons scattered left and right in elastic and inelastic collisions have been measured previously for a series of atoms [4,5,6]. In elastic scattering from atoms with a closed shell ground state electron configuration the polarization effects are caused only by the spin-orbit interaction of the scattered electron in the field of the target atom, that is, Mott scattering. In inelastic scattering, polarization effects may be caused also by the exchange interaction between the scattered and atomic electrons, the inter-atomic spin-orbit interaction and their combined effect [1]. Left/right asymmetries for zinc have been measured previously only for elastic scattering [4] and when experimental conditions were not adjusted to observe the effects of negative ion resonances. Here we study excitation via a negative ion resonance with the open 3d shell structure of a zinc atom for which there is a significant distortion of the atomic charge cloud such that the associated angular momentum effects can be observed [4,7]. In this way a clear distinction of the spin-orbit interaction via fine structure effect and via Mott scattering is possible.

2. Experimental

The apparatus used in the present study consists of a scattering chamber incorporating a cylindrical 127° scattered electron analyzer, an evaporative source of free zinc atoms and a source of spin polarized electrons. The scattering chamber and components used for detection and characterization of the collision products, scattered electrons and decay photons, are described elsewhere [8]. Briefly, electrons scattered by the beam of free zinc atoms are collected, energy analyzed and detected at a variable angle using a single channel electron multiplier. Two different modes of operation of the electron spectrometer were used, namely an energy loss mode to identify and isolate a particular scattering process and secondly an excitation function mode to observe the energy dependence of the angular differential cross section. In the latter mode, the incident electron energy and residual scattered electron energy were scanned simultaneously over a fine energy mesh to observe negative ion resonances.

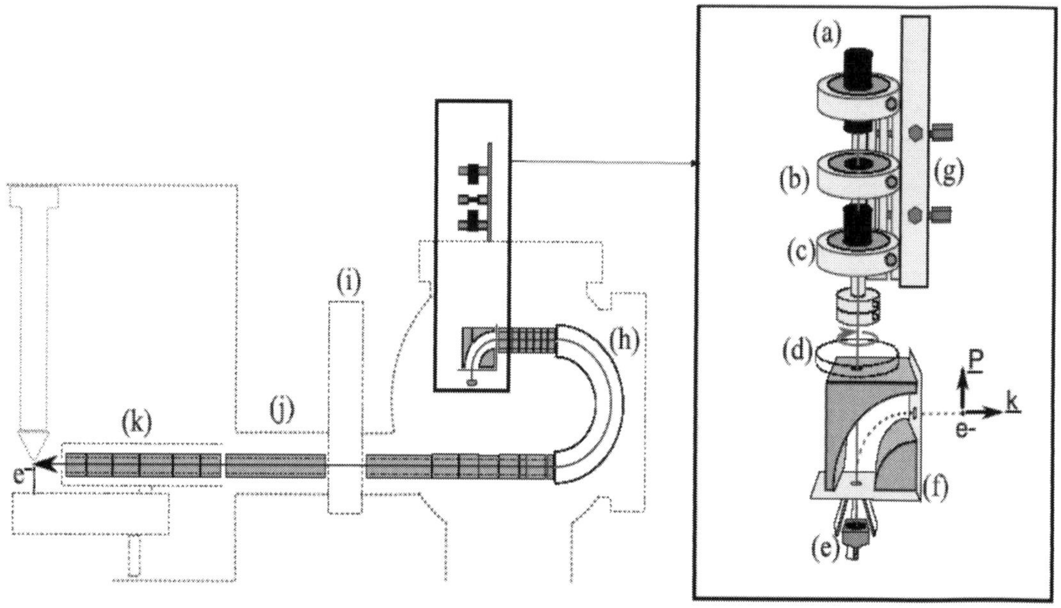

Figure 1. Representation of the spin polarized electron source and beam. The left-hand section shows the source and scattering chambers with dotted lines within which are (h) the electrostatic 180° energy selector; (i) the isolation valve between UHV source chamber and the scattering chamber; (j) the electron optical lenses for transfer and (k) focusing of electrons into the scattering region. The right-hand section shows the components used to obtain the transversely spin polarized electrons: (a) laser; (b) linear polarizer; (c) liquid crystal variable retarder; (d) circularly polarized laser light passing through the vacuum chamber window; (e) GaAs crystal; (f) electrostatic 90° deflector from which electrons initially longitudinally spin polarized emerge with spins **P** perpendicular to their momentum vector **k** as indicated by the vector diagram.

A new feature of the present experiment, different from the one previously used in our laboratory for the study of negative ion resonances in excitation of zinc [8], is the source of spin polarized electrons. While the general design principle of this source is the same as described by [9,10], here we employ a 180° electrostatic electron energy selector to reduce the energy spread in the incident electron beam. The need for a high efficiency in data collection and the relatively large natural width of the observed features justified the use of an energy resolution of 120 meV. The electron optics controlling the electron beam and the 180° energy selector are shown in the left-hand section of figure 1. The source of the spin polarized electrons is indicated on the right hand section of the figure.

The spin polarized electrons are obtained via photoemission from the surface of a GaAs crystal irradiated by circularly polarized 830 nm wavelength laser light. A spin polarization P_e=28% was used. The photo electrons are emitted with the spins parallel or anti-parallel to the incident laser light direction. This initially longitudinal spin polarization is changed to transverse by deflecting the electron beam through 90° so that in the interaction region the electron spin is perpendicular to the scattering plane. Spin orientation, from spin-up to spin-down, is changed using a liquid crystal variable retarder controlled by data acquisition software. For each incident electron energy the intensity of electrons scattered at a given angle for spin up, $I^{\uparrow}$, and spin down, $I^{\downarrow}$, was recorded over a period of a few seconds in repeated energy scans. The spin up-down asymmetry is determined from the relation

$$S_A=1/P_e\cdot(I^{\uparrow}- I^{\downarrow})/(I^{\uparrow}+ I^{\downarrow}) \quad (1)$$

The energy scale was calibrated by detecting the photons from the decay of the $4s4d\,^1D_2$ state while scanning the incident electron energy through the excitation threshold which for this state has a step-like shape [11]. The cross section is then fitted to theoretical BSRM data [12] as described previously [13] to obtain a calibration of the incident electron energy scale.

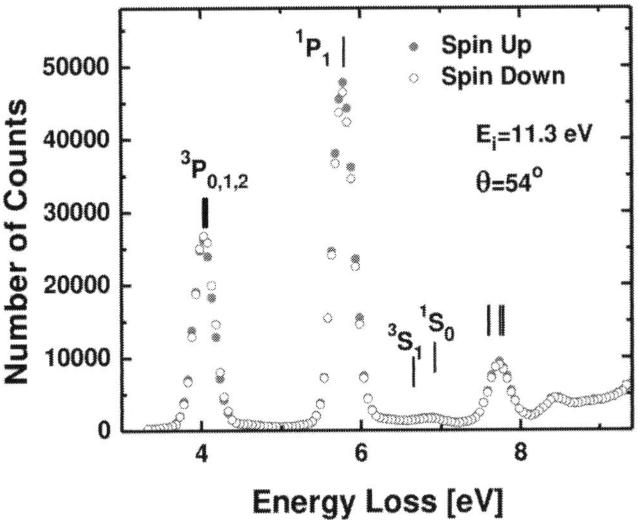

Figure 2. Electron energy loss spectrum in zinc using spin polarized incident electrons with spin up and spin down with respect to the scattering plane defined by the momentum vectors of the incident and detected scattered electrons.

3. Results and discussion

The electron energy loss spectrum, measured at an incident electron energy of 11.3 eV and scattering angle of 54° for spin up and spin down electrons, is shown in figure 2. The energy resolution and relative intensities of different excitation processes are well shown. The energy loss peaks, corresponding to the excitation of the $4p\,^3P_{0,1,2}$ and $4p\,^1P_1$ states studied here, are well isolated from each other and from other excitations and also from the elastic peak, (not shown in the spectrum). From this spectrum one can see the following features. At an electron impact energy of 11.3 eV and a scattering angle of 54°, a non-zero spin up-down asymmetry in excitation of both the $4s4p\,^3P_{0,1,2}$ and $^1P_1$ states is indicated by different intensities for spin-up and spin-down incident electrons. For these specific scattering conditions of the incident energy and scattering angle, the asymmetry is negative for the triplet state and positive for the singlet state. Also, the variation of the asymmetry is observed along the $^3P_{0,1,2}$ peak where there are larger values in the high energy wing. The fine structure components, the J=0 (at 4.006 eV), J=1 (at 4.030 eV) and J=2 (at 4.078 eV), of the lowest excited $4s4p\,^3P_{0,1,2}$ states are not resolved and the observed variation of asymmetry might be associated predominantly with the J=2 excitation. In contrast, the asymmetry associated with the $4s4p\,^1P_1$ excitation does not vary along the corresponding energy loss peak, as expected.

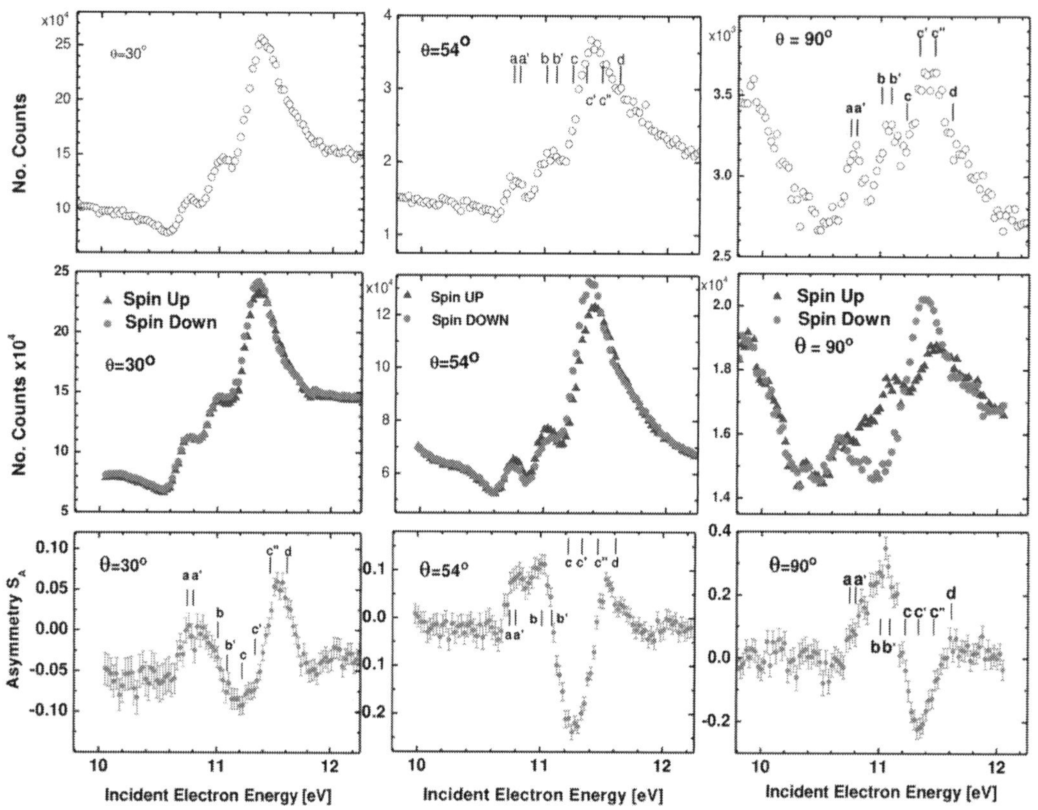

Figure 3. Scattered electron intensities following excitation of $4s4p^3P_{0,1,2}$ states using unpolarized electrons (top row) and spin polarized electrons with spin up (triangles) and spin down (closed circles) (middle row). The spin up-down asymmetry [calculated using expression (1)] is shown in the bottom row. The vertical bars and labels indicate the negative ion resonance values recommended by Napier et al 2009 [14].

The energy dependence of the angular differential scattered electron intensities and spin up-down asymmetries for excitation of the $4s4p^3P_{0,1,2}$ states are shown in figure 3 for scattering angles of 30°, 54° and 90°. The top spectrum at each angle was obtained using a source of unpolarized electrons [8] which was optimized for low energy electrons. The middle spectrum was measured with the new source (see above) of spin polarized electrons. The two sets of data correspond to the two orientations of spin up and down. As shown on the figure, the smoothly varying continuous part of the cross section in the experiments with unpolarized and spin polarized electron beams is very similar. That indicates a negligible change of transmission of electron optical parts in the energy region studied here. Also the appearance of structures caused by the scattering via negative ion resonances is in good general agreement for unpolarized and spin up and down incident electrons at the small scattering angles of 30°. However the cross section shape and resonance structure is remarkably different for spin-up and spin-down incident electrons for the larger scattering angle of 90°. While the cross section for spin up electrons at 90° still resembles the cross section for the scattering of spin unpolarized electrons, the spin-down data show one of the resonance groups, b b', missing and one resonance, most likely c, enhanced. As a consequence the spin up/down asymmetry increases progressively with increase of scattering angles from a variation inside the -0.10 to +0.07 range at 30°, to a -0.25 to +0.35 range at 90° respectively.

The resonance structure indicated by vertical bars in figure 3 and labeled a a', b b', c c' c'' and d, was investigated previously, and here is observed clearly in the intensities but less clearly in the asymmetry functions. The asymmetries at all scattering angles in figure 3 are dominated by a broad smooth variation with energy with a hint of a possible doublet structure for the first broad peak at 30°, and a clearly resolved doublet at 54° where the maxima align best with the resonances a' and b. No clear indication of individual effects of any of the resonances a to d is observed at 90°.

While the observed variation of the asymmetry function is related clearly to negative ion resonance phenomena and indicates the spin dependence in the creation and decay of these short-lived negative ions, an association with individual a to d resonances is less clear. The behavior of the spin up-down asymmetry at 90°, and to some extent at other angles, can be viewed as an unresolved contribution of several broad overlapping and possibly interfering resonances, but another explanation in terms of the existence of shape resonances as inferred from both the elastic scattering [14] and excitation [8,15] needs to be considered. Most of the resonances in the lowest autoionizing region, shown in figure 3 with labels a to d, have been assigned a $3d^94s^24p^2$ electronic configuration. The angular differential elastic scattering indicates that the labeled a a',b b', c c' resonances can be characterized by total angular momentum J=5/2 and 3/2. In the absence of any theoretical modeling the symmetries of corresponding states are discussed by comparison to the corresponding autoionizing states of a Gallium atom. In view of these results some of the properties of the asymmetry function can be discussed further using a simple picture [1,16]. Based on conservation of angular momentum, the collisional alignment of the excited atomic particle and the assumption that spin and orbital momentum have well defined values, the resonance scattering of electrons with spin-up and spin-down will be different and will excite preferentially negative ions with J=5/2 or J=3/2 if the scattered electrons are detected at a positive scattering angle. This observation will correspond to the orientation of orbital momentum and spin in the direction of the positive z-axis. If the effect of a resonance was to enhance the cross section, then this would lead to an absence or reduction of one of the J=5/2 or J=3/2 components and would be observable in the spin up/down asymmetry. For example, in the case of the 54° scattering in figure 3, the structures in the asymmetry values seem to be associated best with the a' and b resonance energies which were determined separately from electron transmission measurements [14]. Consequently their positive asymmetries indicate J=5/2 state assignments for both of them. This deduction is in agreement with the assignments for b but not for a' by Napier et al [14] who proposed J=5/2 and 3/2 respectively. The proposed a' assignment is also contrary to the situation in Ga where the J=3/2 state is lower in energy for the equivalent isoelectronic auto-ionising states. However, if angular dependent interference exists, either with non-resonant scattering or between the resonances themselves, the picture would be less straightforward to interpret and theoretical modeling would be essential. For a meaningful comparison with experimental data there is a need for theoretical models to investigate the effects of negative ion resonances and their associated spin effects in both the integral and angular differential cross sections.

The previous experimental studies using spin polarized electrons reported the angular behavior of asymmetries in elastic scattering but the widely spaced energies did not permit the detection of resonances. Measurement of spin dependent Stokes polarization parameters by detecting the $4d^1D_2 \rightarrow 4p^1P_1$ decay photons by Pravica *et al* [3] was the first to observe a spin dependence associated with the unresolved resonances b b' and also c' specific to this scattering channel. That study used predictions from Bartschat and Blum [15] to identify and disentangle the roles of the spin orbit and exchange interactions. Similar theoretical analyses for the spin asymmetries are not yet clear. The present data with overlapping unresolved resonances and possible interference effects requires a theoretical study specifically for zinc atoms.

4. Acknowledgements

This work was supported by Australian Research Council and the University of Western Australia.. SAN is supported by an Australian Postgraduate Award.

Journal of Physics: Conference Series **212** (2010) 012012 doi:10.1088/1742-6596/212/1/012012

References

[1] Hanne GF 1983 *Phys. Rep.* **95** 95.
[2] Pravica L, Williams JF, Cvejanović D and Napier SA 2007 *Phys. Rev. A* **75** 012721.
[3] Pravica L, Cvejanovic D, Williams JF and Napier SA 2007 *Phys. Rev. A* **75** 030701(R).
[4] Bartsch M, Geesmann H, Hanne GF and Kessler J 1992 *J. Phys. B: At. Mol. Opt. Phys.* **25** 1511.
[5] Dummler M, Bartsch M, Geesmann H, Hanne GH and Kessler J 1990 *J. Phys. B: At. Mol. Opt. Phys.* **23** 3407.
[6] Dummler M, Hanne GH and Kessler J 1995 *J. Phys. B: At. Mol. Opt. Phys.* **28** 2985.
[7] Haberland R and Fritsche L 1987 *J. Phys. B: At. Mol. Phys.* **20** 121.
[8] Napier SA, Cvejanovic D, Williams JF and Pravica L 2009, *To be published.*
[9] Hayes PA, Yu DH and Williams JF 1997 *Rev. Sci. Instrum.* **68** 1708.
[10] Hayes PA, Yu DH, Furst J, Donath M and Williams JF 1996 *J. Phys. B: At. Mol. Opt. Phys.* **29** 3989.
[11] Napier SA, Cvejanovic D, Williams JF, Pravica L, Fursa D, Bray I, Zatsarinny O and Bartschat K 2009 *Phys. Rev. A* **79** 042702.
[12] Zatsarinny O and Bartschat K 2005 *Phys. Rev. A* **71** 022716.
[13] Napier SA, Cvejanovic D, Williams JF and Pravica L 2007 *J. Phys. B: At. Mol. Opt. Phys.* **40** 1323.
[14] Napier SA, Cvejanovic D, Burrow PD, Williams JF, Michejda JA and Pravica L 2009 *To be published.*
[15] Napier SA, Cvejanovic D, Williams JF and Pravica L 2008 *Phys. Rev. A* **78** 032706.
[16] Kessler J 1985 *Polarized Electrons*, 2nd edn. (Berlin: Springer-Verlag).
[17] Bartschat K and Blum K 1982 *Z. Phys. A* **304** 85.

Experimental Investigation of Atomic Collisions in Time Scales Varying from Nanosecond to Microseconds

R.D. Glover, D.E. Laban, K.J. Matherson, W. Wallace and R.T. Sang

Centre for Quantum Dynamics, Griffith University, Nathan, Queensland 4111, Australia

R.Sang@griffith.edu.au

Abstract. We present the results from two experiments investigating collisions that differ in time scale by three orders of magnitude. The first experiment enables the determination of absolute total collision cross sections using a technique that measures a change in the loss rate of trapped atoms from a magneto optical trap (MOT). We also investigate light assisted collision processes between cold metastable neon atoms in the $^3P_2$ metastable state within the MOT. A catalysis laser is scanned in frequency across the $^3P_2$ – $^3D_3$ cooling transition and the ionization rate was observed. Ionization spectra are obtained which demonstrate a dependence on the magnetic sublevels of the transition that the catalysis laser is exciting.

1. Introduction

Cold and ultra-cold atomic collision processes have been the subject of intense scientific investigations over the past decade as a result of the advent of laser cooling and trapping techniques [1]. These techniques enable the trapping of atoms and create a source of dense atomic gases at mK to μK temperatures and have served as the source of many detailed collision experiments. At low temperatures such as this, atomic collision times between trapped atoms are relatively long and of the order of microseconds [2], compared to room temperature collisions where collision partners have velocities in the order of 10^3 s yielding collision times of the order of 10^{-12} s. Such long interaction times yield interesting long-range interactions that are negligible at short collision times. Long range interactions between trapped atoms can also be modified though the presence of light that is mediated by the dipole interaction, whereby ground and excited state atoms interact though a potential created by the light field and has a $1/r^3$ dependence where r is the separation distance between the interacting atoms. At low temperatures, the induced potential can exceed the kinetic energy of the colliding species even at large separation distances of $r > 100$ nm [2].

The experimental investigation of collisions of atoms that are in atomic metastable states has been limited. This is primarily a result of the difficulty in preparing experiments with large numbers of atoms in atomic beams with a significant fraction of the atoms in the metastable state of interest. The limited number of experiments is surprising given the number of technological devices such as fluorescent lights and lasers that rely on collision processes involving these states. The work presented in this paper describes a recently developed technique by us [3] to measure absolute total atom-atom collision cross sections between neon atoms in the $^3P_2$ metastable state (designated as Ne*) with a variety of collision partners. In these experiments collision cross sections are obtained for collisions between the cold trapped Ne* atoms and an introduced gas species at room temperature. Such collisions take place over times scales of the order of nanoseconds. The technique relies on the observation of the population dynamics of trapped metastable atoms in a magneto-optical trap (MOT). It has the significant advantage that the trap is a pure sample of atoms in this metastable state and does not require knowledge of the number density of the atoms in the trap. Accurate collision cross sections can be measured and the technique is capable of producing benchmark atom-atom collision cross sections [3].

In the last section of this paper we present the progress of the modification of cold collisions between trapped metastable atoms with the introduction of an additional laser. The additional laser can dramatically alter the ionization rate between colliding atoms in the MOT. This additional laser, called a catalysis laser, can produce an attractive or repulsive potential between colliding atoms that depends on the frequency tuning of the laser relative to an appropriate dipole allowed transition [1]. We present the progress on initial experiments that measure ionization rates as a function of catalysis laser frequency detuning and demonstrate a rich structure that is dependant on the magnetic sublevels of the optical transition being excited. This effect has not been observed in previous cold collision studies on rare-gas metastable atoms.

2. Atom-Atom Absolute Total Collision Cross sections Using a Magneto-Optical Trap

The rate of change of the number of trapped atoms is determined by the loading and loss rates from the trap and in the case of a low density MOT is given by [3]

$$\frac{dN_t}{dt} = R_L - N_t\Gamma_B - \frac{\beta N_t^2}{V_{eff}} \qquad (1)$$

where N_t is the number of trapped metastable neon atoms, R_L is the loading rate, Γ_B is the trap loss rate due to the collisions with the residual background gas atoms, β is the density-dependent binary collision loss rate between trapped atoms and V_{eff} the effective volume of the trap and is given by $V_{eff}=8\pi^{3/2}w$ where w is the full width at half maximum of the trap distribution. With the addition of a collision species introduced into the interaction region the trap population dynamics is modified to

$$\frac{dN_t}{dt} = R_L - N_t\left(\Gamma_B + \Gamma_C\right) - \frac{\beta N_t^2}{V_{eff}} \qquad (2)$$

where Γ_C is an additional trap loss rate due to collisions of the trapped atoms with the introduced collision species. The collisions cross-section between the trapped atoms and the collisions species is given by

$$\sigma = \frac{\Gamma_C}{n\bar{v}} \qquad (3)$$

where n is the density of the collision species and $\bar{v}$ is the relative collision velocity which is approximately given by the mean velocity of the collision species viz:

$$\bar{v} = \sqrt{\frac{8k_BT}{\pi m}} \qquad (4)$$

where T is the temperature and m the mass of the collision species. If the loading of the trap is suspended from steady state, the population in the trap can be found by solving (2) yielding:

$$N_t(t) = \frac{N_t(0)e^{-t(\Gamma_B+\Gamma_C)}}{1+\left[\beta N_t(0)/V_{eff}\left(\Gamma_B+\Gamma_C\right)\right]\left(1-e^{-t(\Gamma_B+\Gamma_C)}\right)} \quad (5)$$

Γ_C can be determined by fitting eqn (5) to the measured trap decay. The cross section is then found though plotting Γ_C as a function of n, the absolute density of the collision species. As seen in eqn (3) this is a linear relationship where the gradient is proportional to the cross section σ.

The apparatus for the experiment is discussed in detail in the references given in [3], but we provide a brief description here. A beam of neon gas supersonically expands though a hollow cathode discharge source that is cooled via a liquid nitrogen cold finger. Neon atoms emerge from the source in a variety of internal quantum states at a mean velocity of 500 m/s. A small percentage of this atomic beam has Ne atoms excited to the $^3P_2$ metastable state that has a lifetime of 14.7 s. Atoms in this metastable state are decelerated using a Zeeman slower in combination with a counter-propagating laser detuned to the red from the $^3P_2 - {}^3D_3$ transition at 640 nm. Atoms in the $^3D_3$ state can only decay to the $^3P_2$ state and hence there are no loss channels from the cooling transition. This optical transition is also used for the MOT trap beams. Following the Zeeman slower, the MOT is formed by the intersection of three pairs of counter propagating laser beams detuned to the red of the cooling transition in combination with a set of anti-Helmholtz coils which create the appropriate magnetic field gradient. A 12-bit CCD camera is used to monitor the trapped atom population via resonance fluorescence from the trapped atoms interacting with the MOT laser beams. All laser beams are produced by a Coherent 899 single-mode ring dye laser pumped by a Coherent Verdi V5 solid state pump laser.

The experiments involved the measurement of the change in the atom number (i.e. the measurement of a decay curve) with a collision gas added to the MOT vacuum chamber at different pressures. A calibrated residual gas analyzer was used for pressure measurements and was also used to verify that the partial pressure of the collision species in the vacuum system had come to equilibrium. Once this equilibrium was confirmed, the loading of the MOT was interrupted and the decay of the MOT was measured using the imaging system.

Collision Species	$\bar{v}$ (m/s)	E_{rel} (meV)	σ_M (Å^2)[a]	σ_R (Å^2)[b]	σ_{TI} (Å^2)[c]
He	1249.0	27.0	164.8±0.5±15.5	123	
Ne	556.3	16.2	499.1±4.1±47.1	143	
Ar	395.4	10.9	837.7±15.7±79.0	398	3.46-30.5
H_2	1760.0	29.4	230.1±2.8±21.7		2.6-23.4
O_2	441.8	12.5	1026.8±12.7±96.9		25.5
N_2	472.1	13.5	1259±16.5±118.8		10.4

Table 1 (a) the measurements for this work where E_{rel} is the relative collision energy. The errors quoted respectively are the statistical and systematic errors respectively. (b) Measurements are from reference [4] with E_{rel} ~115 meV and estimated uncertainties of the order of 25%. (c) The total ionization cross sections for collision with metastable Ne in a mixed composition of $^3P_2$ and $^3P_0$ states has been included for comparison purposes. In this case E_{rel} is between 33 and 47 meV [5].

The results in Table 1 show the measured total absolute cross sections for collisions between trapped Ne* atoms with six different collision species. The total cross section is a combination of elastic collisions plus, in some cases, ionization events. As a comparison to our measured cross sections, Table 1 also reports the measured cross sections of Rothe and Neynaber that were taken at higher relative energies using crossed atomic beam techniques [4]. Given that these experiments are at higher energy, it is expected that these cross sections are smaller than our cross sections. It should be noted that the contribution to the total ionization cross section is relatively small as shown by the work of Yencha [5]. In the case of Ne* - He collisions, the cross section is purely elastic since the Ne* atoms do not have enough energy to ionize the He atom. At low velocities it is expected that elastic, van der Waals type interactions dominate and the cross section can be approximated as

$$\sigma_{tot}(v_{rel}) \approx 8\left(\frac{C_6}{\hbar v_{rel}}\right)^{2/5} \quad (6)$$

To estimate the total cross section over the velocity range of this experiment, the rate of collisions leading to trap loss is calculated at each velocity using equation (6) and integrated from the capture velocity of the MOT over the Maxwellian velocity distribution. Using this method and taking C_6 to be 340 kcal mol^{-1} Å^6 [6], the total elastic cross section for Ne* in the $^3P_2$ state with thermal ground state He is calculated to be 168.9 Å^2. As can be seen from our data for this cross section our measured cross section is in good agreement with this value. Our data is unique due to the low energy of the collisions as well as the relatively low uncertainties. A full analysis of the contributions to errors are detailed in Matherson et. al. [3].

3. Cold Collision Experiments in the Presence of a Catalysis Laser

The collision processes discussed in the previous section investigated binary collisions that were dominated through long-range electrostatic dispersion forces. We have also commenced an experimental investigation on the effects of light assisted collision processes. In these experiments another retro-reflected laser beam is inserted into the MOT. Such a laser is termed a catalysis laser as it can alter the collisional interactions between atoms in the trap [7]. Consider two colliding Ne* separated by a distance r in the presence of such a catalysis laser. If the catalysis laser is frequency tuned to a dipole transition moment, then the two atoms will be influenced by the resonant dipole-dipole interaction potential given by [7]:

$$V(r) = \pm a\hbar\gamma\left(\frac{\lambda}{2\pi r}\right)^3 \quad (7)$$

where λ and γ are the wavelength and the natural linewidth of the atomic transition respectively and a is configuration constant of order 1. If the transition moment is large, such as in the case of a cooling transition, then the magnitude of this potential can be significantly greater than the kinetic energy of the trapped atoms even at relatively large separation distances. The effect of such a potential between an atom in the ground state and an atom in the excited state is to create an attractive potential between the approaching atoms, thereby enhancing the probability of collision if the laser is frequency detuned to the red of an appropriate transition. In the case of a blue detuned laser beam the induced interaction potential is repulsive and collision rates are reduced [1]. A MOT trapping rare-gas metastables can also undergo ionization in the binary collision process. There are two possible ionization channels that can occur; the first is Penning ionization whereby following a binary collision event, three collision products are produced, a neutral ground state atom, an ionized ground state atom and an electron. The second process is associative ionization in which an ionized neon molecule is produced as well as a free electron.

Ionization rates for atoms in metastable states can be greatly enhanced using a catalysis laser as shown by the work of Katori and Shimizu [7]. In this work Katori and Shimizu scanned the frequency of the catalysis laser over the cooling transition and observed the change in the ionization rate of trapped metastable Kr atoms. When the catalysis laser is frequency detuned to the red of the cooling transition, the ionization rate increased dramatically and peaked close to resonance. Past resonance, when the laser is frequency tuned to the blue side of the transition, the ionization rate decreased dramatically and was inhibited for a small frequency range due to optical shielding as a result of a repulsive potential induced by the catalysis laser [1].

The goal of this work was to investigate the enhancement of ionization processes in the Ne* MOT using such a catalysis laser. The apparatus described for the measurement of the absolute total collision cross sections was modified to facilitate the detection of charged products resulting from ionizing collisions in the MOT. Ionization processes from the trap were monitored by the use of a channel electron multiplier with a simple electrostatic discrimination system that assisted in the detection of Ne^+ ions. A catalysis laser beam was provided by an additional ring-dye laser (Spectra Physics 380D operating with kyton red laser dye pumped by a Verdi V5 laser) and was inserted into the MOT in a retro-reflecting configuration so as to minimize trap loss resulting from momentum transfer from the light beam due to photon absorption. The laser beam was circularly polarized in the same polarization configuration as the MOT trap beams and could be scanned over the cooling transition with approximately 1GHz frequency detuning range. The laser power was typically of the order 100 times the saturation intensity (~4mW/cm^2) of the cooling transition.

The resulting ionization spectrum is shown in the left of figure 1 and displays a complex spectrum unlike that of the work of Katori and Shimizu. The spectrum shows multiple ionization enhancement peaks when the catalysis laser is frequency detuned over the cooling transition. At a first glance, one might deduce that this spectrum is displaying these features due to the formation of molecular ions through the associative ionization process with the peak structure a result of resolving molecular vibrational energy levels. However, this is unlikely due to the probability of Penning ionization having an order of magnitude greater probability [5]. A further point to note is the enhancement in the ionization peak at approximately +50 MHz when the catalysis laser is frequency detuned to the blue of the cooling transition. This is completely unexpected as optical shielding should occur in this region of the spectrum and therefore reduce Ne^+ ion production.

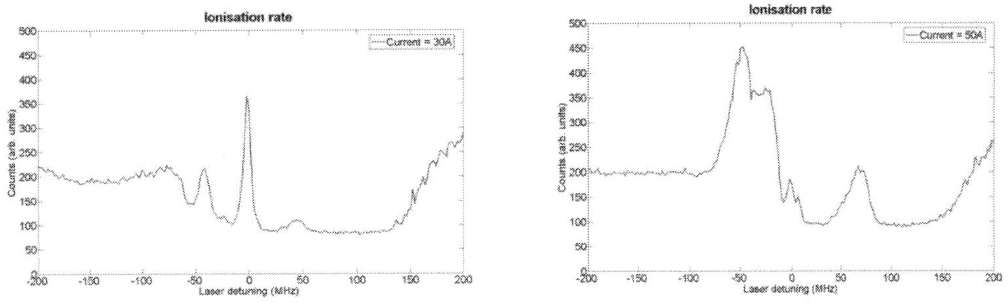

Figure 1: Ion count rate Vs catalysis laser frequency detuning relative to the $^3P_2$-$^3D_3$ cooling transition; left figure MOT trap gradient ~ 0.25 T/m, right figure MOT trap gradient ~0.4 T/m.

Further experimental investigation shows that the ionization spectrum is dependent on the magnetic field produced by the MOT coils. Increasing the current in the coils produced more peaks in the ionization spectrum as well as shifting the peaks as shown in the spectrum on the right of figure 1. The cooling transition that is used for the catalysis laser displays fine structure and the magnetic sublevels of the two participating atomic states in the MOT have their degeneracy lifted due to the Zeeman shift as a result of the magnetic field created by the MOT coils. In the case of the magnetic field gradient of 0.4 T/m the various transitions between magnetic sublevels that can be excited are given in table 2 at a distance of 1mm from the centre of the trap.

$^3P_2$	Δm_J = -1 (MHz)	Δm_J = 0 (MHz)	Δm_J = +1 (MHz)
2	-58.9	-11.8	35.3
1	-53.0	-5.9	41.2
0	-47.1	0	47.1
-1	-41.2	5.9	53.0
-2	-35.3	11.8	58.9

Table 2: Zeeman frequency shifts for optical transitions between magnetic sublevels for the $^3P_2 - {}^3D_3$ optical transition for σ^- light (Δm_J=-1), π light (Δm_J=0) and σ^+ light (Δm_J=+1) at a displacement of 1mm from the trap centre with magnetic field gradient of 0.4 T/m. The left hand column displays the magnetic sublevel of the lower energy magnetic sublevel.

The spectrum on the right of figure 1 shows that there are two peaks in the ionization spectrum at approximately ±50 MHz which are close to the two transitions for both, σ^- from the $^3P_2$ (m_J=+2, Δm_J=-1),) and σ^+ from the $^3P_2$ (m_J=-2, Δm_J=+1) transitions respectively. It is likely that the peaks in the ionization spectrum are from these transitions as well as contributions from the other transitions that are not frequency resolved. The general shape of each of the peaks is in good agreement with that of the work of Katori and Shimizu [7], in that each of the peaks in the spectrum show a gradual increase in the ionization rate on the red side of the transition with a fast drop off in the ionization rate when tuned to the blue of the transition. Our spectra also show that the background ionization rate is suppressed on the blue side of the transition compared to the background rate at far red detuning. This is consistent with optical shielding effects as demonstrated in previous experimental results [7]. We should point out that the work of Katori and Shimizu does not show a multi-peak structure like ours, however, the magnetic field gradient in that work is nearly an order of magnitude smaller than this work and it would be unlikely they could resolve the fine structure. This would also explain why their work shows a broader ionization peak due to the influence of all of the magnetic sublevel transitions. To confirm our findings we could use a smaller magnetic field gradient but it is not currently possible with our current experimental configuration. Experiments are now underway that will perform these measurements in a pulsed mode so that the ionization rates are measured without the influence of any magnetic field to confirm our expectation.

4. Acknowledgements

This research was supported by the Australian Research Council and Griffith University.

References

[1] J. Weiner, V.S. Bagnato, S. Zilio, and P. S. Julienne, *Rev. Mod. Phys.* **71**, 1 (1999).
[2] S.D. Gensemer and P.L. Gould, *Phys. Rev. Lett* **80**, 936 (1998).
[3] K.J. Matherson, R.D. Glover, D.E. Laban, and R.T. Sang, *Rev. Sci. Instrum.* **78**, 073102 (2007).
K.J. Matherson, R.D. Glover, D.E. Laban, and R.T. Sang, *Phys. Rev. A* **78**, 042712 (2008).
[4] E.W. Rothe and R.H. Neynaber, *J.Chem. Phys.* **42**, 3306 (1965).
[5] A.J. Yencha, Electron Spectroscopy: Theory and Technique, and Application, edited by C.R. Brundle and A.D. Baker (Academic Press, New York, 1984), Vol 5.
[6] T. Fukuyama and P. Siska, *Chem. Phys. Lett.* **39**, 418 (1976).
[7] H. Katori and F. Shimizu, *Phys. Rev. Lett.* **73**, 2555 (1994).

Frustration of direct photoionizations of rare gas clusters in intense extreme ultraviolet free-electron laser pulses

H. Iwayama [1,7], K. Nagaya [1,7], H. Murakami [1,7], M. Yao [1,7], H. Fukuzawa [2,7], X.J. Liu [2,7], G. Prümper [2,7], M. Okunishi [2], K. Shimada [2], K. Ueda [2,7], T. Harada [2], M. Toyoda [2], M. Yanagihara [2], M. Yamamoto [2], K. Motomura [2,7], N. Saito [3,7], A. Rudenko [4,7], J. Ullrich [4,5,7], L. Foucar [6,7], A. Czasch [6], R. Dörner [6], M. Nagasono [7], A. Higashiya [7], M. Yabashi [7], T. Ishikawa [7], H. Ohashi [7,8], H. Kimura [7,8], T. Togashi [7]

[1] Department of Physics, Kyoto University, Kyoto 606-8502, Japan

[2]Institute of Multidisciplinary Research for Advanced Materials, Tohoku University, Sendai 980-8577, Japan

[3] National Metrology Institute of Japan, AIST, Tsukuba 305-8568, Japan

[4] Max-Plank Advanced Study Group within CFEL, D-22607, Hamburg, Germany

[5] Max-Planck-Institut für Kernphysik, 69117 Heidelberg, Germany

[6] Institut für Kernphysik, Universität Frankfurt, D-60486 Frankfurt, Germany

[7] RIKEN, XFEL Project Head Office, Kouto 1-1-1, Sayo, Hyogo 679-5148, Japan

[8] Japan Synchrotron Radiation Research Institute, Kouto 1-1-1, Sayo, Hyogo 679-5198, Japan

E-mail: yao@scphys.kyoto-u.ac.jp, and ueda@tagen.tohoku.ac.jp

Abstract. We performed a momentum imaging measurements of Ar clusters irradiated by extreme ultraviolet free electron laser pulses (61nm 1.3×10^{11}W/cm^2) for various cluster sizes. Dominant fragment ions were Ar^+ and Ar_2^+ ions with significant kinetic energy (up to 40eV). This indicates that Ar clusters absorbed many photons, become highly ionized and finally dissociated into many fragments through Coulomb explosion. The size dependence of the ion kinetic energy distribution indicates that larger clusters significantly suffer from the frustration of direct photoelectron emissions and multiple ionizations.

1. Introduction

The interaction of clusters with intense femtosecond laser radiation has been of considerable interests.

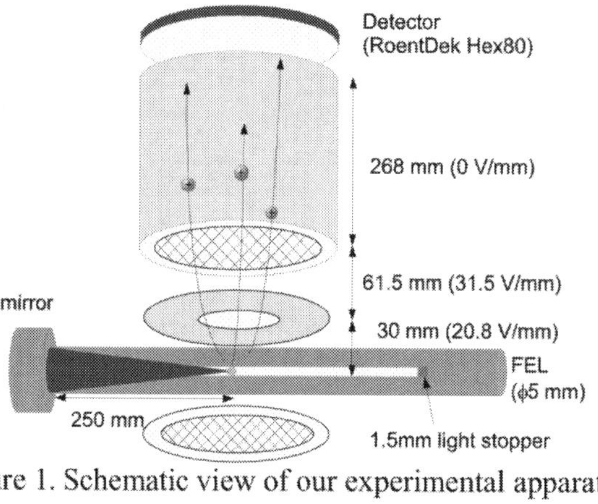

Figure 1. Schematic view of our experimental apparatus.

When the clusters are irradiated with strong visible or near-infrared laser pulses, it is known that high energy ions and electrons are generated from the clusters [1-3]. Many experimental and theoretical works revealed that a cluster absorbs a large amount of photon energy by the creation of transient nanoplasma and the inverse Bremsstrahlung absorption [4]. This efficient energy absorption results in Coulomb explosion of highly charged clusters, which dissociates into many high energy ions. The ion energy distributions give us valuable information on the energy absorption process of clusters in intense laser fields.

Recent progress in free electron lasers (FEL) based on self-amplified spontaneous emission (SASE) allows to utilize intense laser pulses in the short wavelength region. At the TESLA test facility (TTF) at Deutsches Elektronen-Synchrotron (DESY) , Wabnitz *et al.* investigated multiple ionization of Xe clusters by 98-nm FEL radiation, and found complete Coulomb explosion and ions with charge states up to 8+. This result shows extremely strong absorptions of the laser radiation by Xe clusters [5]. The efficient energy absorption of clusters is interesting topics even in the short wavelength laser.

Very recently, the SPring-8 Compact SASE Source (SCSS) test accelerator, has started operation in Japan [6]. It provides linearly polarized EUV-FEL pulses with ~30 μJ per pulse, ~100 fs pulse width, and 10–20 Hz repetition rate in the wavelength region of 51–61 nm. This energy regime is of particular interest because all atoms in any forms of matter can be ionized by just a single photon owing to huge photoionization cross sections.

Recently we reported experimental results of rare gas clusters irradiated by extreme ultraviolet free electron laser (EUV-FEL) pulses from the new FEL facility [7-10]. In this paper, we describe the details on experimental results of Ar clusters irradiated by 61nm 1.3×10^{11}W/cm^2 EUV-FEL pulses.

2. Experiment

The experiments were performed at the SPring-8 Compact SASE Source (SCSS) test accelerator in Japan [6]. In the present experiments, the wavelength was 61 nm, its polarization axis was in the horizontal plane and the laser pulse repetition rate was 10 Hz. The laser power was calculated to be $\sim1.3\times10^{11}$ W cm^{-2} with an estimated uncertainty of a factor of 2 on the basis of the laser pulse energy (~30μJ), pulse duration (~100 fs), effective focus size (~15μm), total reflectivity of three mirrors (~13%), the size of the skimmer and partial light stopper and a laser power reduction of 10% by the variable horizontal slits upstream the beam stopper.

Figure 1 shows a schematic view of our experimental apparatus. We performed momentum imaging measurements with a position sensitive delay-line detector (HEX80 RoentDek [11]). To focus the FEL beam onto the cluster beam of 1 mm in diameter, we used a concave mirror at normal incidence. This mirror was fabricated at Tohoku University, using a tungsten-vanadium coating on a superpolished quartz substrate with a focal length of 250 mm. The FEL beam was partially blocked by a 1.5-mm wide vertical beam stopper. We calibrated the position of this vertical beam stopper to avoid

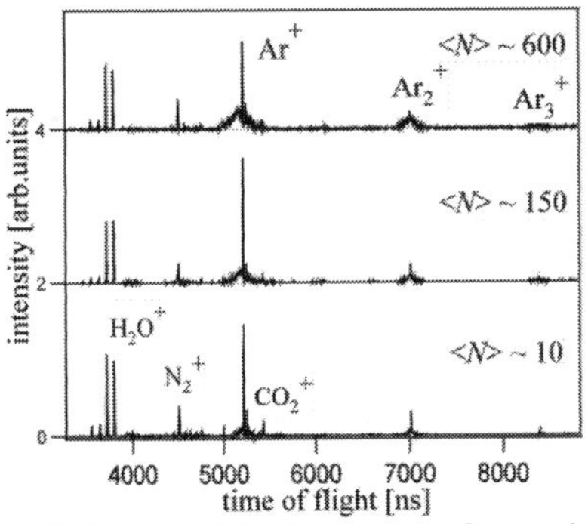

Figure 2. Time-of-flight spectra for various cluster size (<N>=10,150 and 600).The ions such as H_2O^+ and N_2^+ come from residual gas on the FEL beam.

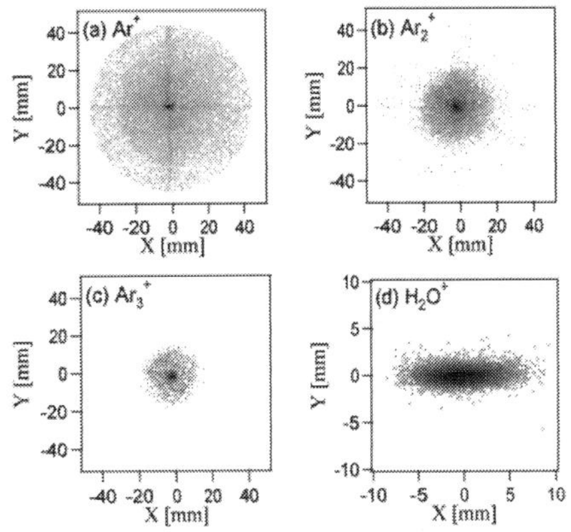

Figure 3. Ion image on the detector for (a) Ar^+, (b) Ar_2^+, and (c) Ar_3^+ ions for the cluster size <N>=150. The ion image of H_2O^+ ions is also shown in (d).

undesirable irradiation of unfocused FEL beam on the cluster beam. Argon clusters were produced by a supersonic expansion through a pin hole (φ = 30 μm) kept at 130 K. The stagnation pressures of Argon gas were 1.5, 3, 4.5, 6 and 8bar. We estimated the average cluster size <N> to be 10, 60, 150, 300 and 600 for each gas pressure using the Hagena's scaling law [12].

The kinetic energy of fragment ions was recorded by momentum imaging measurements with a position sensitive delay-line detector (HEX80 RoentDek). The three dimensional momentum vector for each ion was extracted from times of flight of ions and their positions on the detector [13]. We recorded the signal waveforms from the detector with two 4-channel digitizers (Acqiris DC282), which can substantially reduce dead times of signals compared to a conventional time-to-digital converter system [9].

3. Results

Time-of-flight spectra for <N> = 10, 150 and 600 are shown in Fig. 2. The spectra reveal that dominant fragment ions were Ar^+, Ar_2^+ and Ar_3^+ ions, and relative abundances of Ar^+ ions are roughly 80% for each cluster size. The peaks of Ar^+ ions have sharp and broad components. While the ions in the sharp peak come from atomic Argon in the cluster beam, the broad component corresponds to Ar fragment ions which have a considerable kinetic energy. These energetic fragments indicate that the parent clusters were multiply ionized and dissociated through Coulomb explosion after intense FEL irradiation. With increasing cluster size, the peaks of Ar fragment ions become broader. This indicates that the kinetic energy of fragment ions become larger with the increase of the cluster size.

Figure 3 shows the ion images on the position sensitive detector. The ion images of residual gas ions such as H_2O^+ are nearly lines because they were mainly ionized by incident and unfocus FEL pulses (see Fig. 1). Ion images of Ar_n^+ (n=1, 2 and 3) are circular and broad. This means that only focused FEL beam irradiated the cluster beam. These broad images show that fragment ions come from Coulomb explosion of highly ionized clusters. The ion image of Ar^+ ions seems to be larger than the detector (φ=86mm), and some Ar^+ ions should have missed the detector.

Using the time of flight and the detection position, we calculate the ion kinetic energy for each ion. Figure 4 represents the ion kinetic energy of Ar^+, Ar_2^+ and Ar_3^+ ions. The kinetic energies of Ar^+ ions are several tens eV. In Fig 4, only ions are selected within a cone of 45 deg relative to the spectrometer axis. In this cone, the spectrometer has a energy independent solid angle up t a maximum energy of 50eV. The kinetic energy distributions of Ar^+, Ar_2^+ and Ar_3^+ ions increase with the increase of the cluster size. This result directly shows that larger clusters absorb more photons and dissociate more violently than smaller ones.

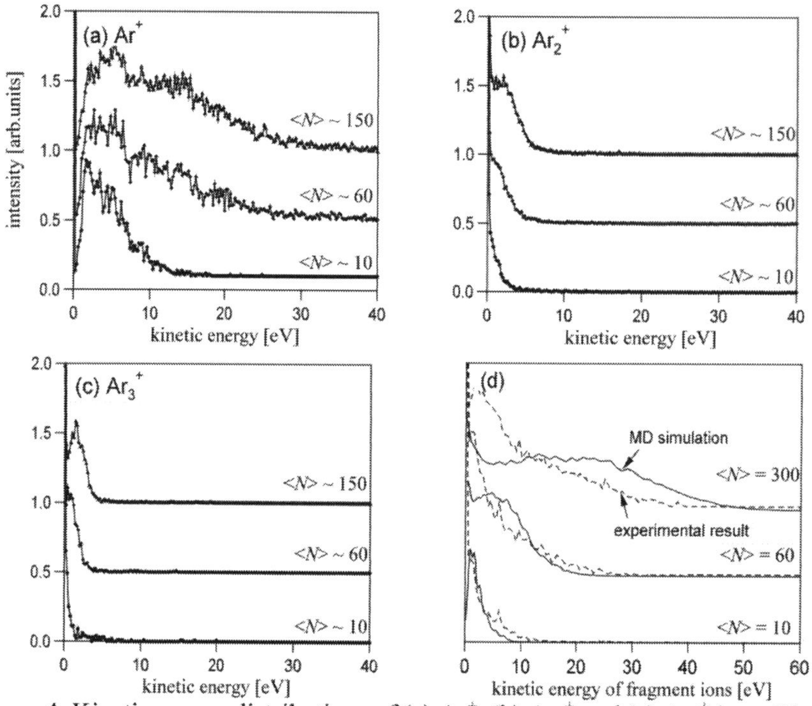

Figure 4. Kinetic energy distributions of (a) Ar^+, (b) Ar_2^+ and (c) Ar_3^+ ions. The kinetic energy distribution of fragment ions in (d) is the sum of Ar_n^+ (n=1, 2 and 3) ions. The simulated results are also shown in (d).

In our laser condition, the ponderomotive energy is 4×10^{-5} eV, and the Keldysh parameter is much higher than one [14]. This means that an optical field ionization does not play an important role. Instead the main photon absorption process is the sequence of single photon absorptions by individual Ar atoms in the cluster. The kinetic energies of fragment ions are estimated by using a classical molecular dynamics calculation. At first, we assume that all photoelectrons completely and instantaneously escape from a cluster. In this case, the charge state of cluster ions is equal to the number of absorbed photons. Here, the cluster size distribution is assumed as the log-normal function, and the distribution of absorbed photon numbers is evaluated by using the laser power of 1.3×10^{11} W/cm^2 and a Monte Carlo method for each calculation. Single charges are randomly distributed to atoms within a cluster.

Figure 4(d) shows simulated and experimental kinetic energy distributions. The experimental distribution is the sum of Ar_n^+ (n=1, 2 and 3) kinetic energy distributions based on their relative abundances. For small clusters, the simulation can reproduce the experimental distribution well. For large clusters, however, the size dependence of the kinetic energy for experimental results seems to be saturated, while kinetic energies of the simulation rapidly become higher with increasing the cluster size. We believe that this is because a considerable number of photoelectrons could not escape and were trapped in the strong Coulomb potential of the highly charged cluster. It is suggested that the larger clusters significantly suffer from the frustration of direct photoelectron emission and multiple ionization.

4. Summary

We have measured ion kinetic energy distributions of Argon clusters irradiated by EUV-FEL pulses for various cluster size. Main fragment ions were singly charged Ar^+, Ar_2^+ and Ar_3^+ ions. With increasing the cluster size, the kinetic energy of fragment ions becomes larger. We observed Ar^+ ions with up to 40 eV. Molecular dynamics simulation results suggest the clusters significantly suffer from the frustration of multiple ionizations.

5. Acknowledgements

We are grateful to SCSS Test Accelerator Operation Group at RIKEN for continuous support in the course of the studies. This study was supported by the X-ray Free Electron Laser Utilization Research Project of the Ministry of Education, Culture, Sports, Science and Technology of Japan (MEXT), by the Grant-in-Aid for the Global COE Program "The Next Generation of Physics, Spun from Universality and Emergence" from the MEXT, by Japan Society of Promoting Science, and by the MPG Advanced Study Group within CFEL.

References

[1] Ditmire T, Tisch J W G, Springate E, Mason M B, Hay N, Smith R A, Marangos J and Hutchinson M H R 1997 *Nature* 386 54
[2] Snyder E M, Buzza S A and Castleman A W Jr 1996 *Phys. Rev. Lett.* 77 3347
[3] Shao Y L, Dimire T, Tisch J W G, Springate E, Marangos J P and Hutchinison M H R 1996 *Phys. Rev. Lett.* 77 3343
[4] Saalmann U and Rost J-M 2003 *Phys. Rev. Lett.* 91 223401
[5] Wabnitz H *et al* 2002 *Nature* 420 482
[6] Shintake T *et al* 2008 *Nature Photonics* 2 555
[7] Fukuzawa H *et al* 2009 *Phys. Rev.* A 79 031201R
[8] Iwayama H *et al* 2009 *J. Phys.* B 42 134019
[9] Motomura K *et al* 2009 *Nucl. Instr. and Meth.* A 606 770
[10] Liu X J *et al* 2009 *Rev. Sci. Instrum.* 80 053105
[11] Jagutzki O *et al* 2002 *IEEE Trans. Nucl. Sci.* 49 2477
[12] Karnbach R, Joppien M, Stapelfeldt J, Wörmer J and Möller T 1993 *Rev. Sci. Instrum.* 64 2838
[13] Ullrich J, Moshammer R, Dorn A, Dörner R, Schmidt L Ph H and Schmidt-Böcking H 2003 *Rep. Prog. Phys.* 66 1463
[14] Augst S, Strickland D, Meyerhofer D D, Chin S L and Eberly J H 1989 *Phys. Rev. Lett.* 63 2212

Confinement and electron correlation effects in photoionization of atoms in endohedral anions: $Ne@C_{60}^{z-}$

V K Dolmatov, G T Craven and D Keating

Department of Physics and Earth Science, University of North Alabama, Florence, AL 35632, USA

E-mail: vkdolmatov@una.edu

Abstract. Trends in resonances, termed confinement resonances, in photoionization of atoms A in endohedral fullerene anions $A@C_{60}^{z-}$ are theoretically studied and exemplified by the photoionization of Ne in $Ne@C_{60}^{z-}$. Remarkably, above a particular nl ionization threshold of Ne in *neutral* $Ne@C_{60}$ ($I_{nl}^{z=0}$), confinement resonances in corresponding partial photoionization cross sections σ_{nl} of Ne in any charged $Ne@C_{60}^{z-}$ are not affected by a variation in the charge z of the carbon cage, as a general phenomenon. At lower photon energies, $\omega < I_{nl}^{z=0}$, the corresponding photoionization cross sections of *charged* $Ne@C_{60}^{z-}$ (i.e., those with $z \neq 0$) develop additional, strong, z-dependent resonances, termed Coulomb confinement resonances, as a general occurrence. Furthermore, near the innermost 1s ionization threshold, the 2p photoionization cross section σ_{2p} of the outermost 2p subshell of thus confined Ne is found to inherit the confinement resonance structure of the 1s photoionization spectrum, via interchannel coupling. As a result, new confinement resonances emerge in the 2p photoionization cross section of the confined Ne atom at photoelectron energies which exceed the 2p threshold by about a thousand eV, i.e., far above where conventional wisdom said they would exist. Thus, the general possibility for confinement resonances to resurrect in photoionization spectra of encapsulated atoms far above thresholds is revealed, as an interesting novel general phenomenon.

1. Introduction

Endohedral fullerenes $A@C_{60}$, where the atom A is encapsulated inside the hollow interior of the carbon cage C_{60}, are of the highest interest and importance both to the basic and applied sciences and technologies, as new modern building blocks of materials and devices with unique properties. Therefore, they have attracted much attention of many investigators in recent years. In particular, the photoionization, as a basic phenomenon in nature, of atoms A encapsulated in fullerenes $A@C_{60}$ has become a topical research subject both for theorists for some years now (see a review paper [1] as well some latest works [2, 3, 4] on the subject and references therein) and, since only very recently, experimentalists [5, 6]. Among the performed photoionization studies of thus confined atoms, only works [1, 7] have provided the initial understanding of how the photoionization spectrum of an atom could be modified by the environment of a negatively charged carbon cage C_{60}^{z-}. The understanding was exemplified by trends in photoionization of the innermost 1s subshell of Ne in $Ne@C_{60}^{z-}$ with various z's. However, how these and other possible trends might show up, as general phenomena, in spectra of intermediate and outer subshells of atoms A in $A@C_{60}^{z-}$ has remained unstudied. The present paper expands further the investigation started in [7] to the intermediate 2s and outermost 2p subshells of the confined Ne with the aim to reveal which new aspects of these spectra of endohedral anions are most interesting, as general phenomena.

2. Brief description of theoretical concepts

Following previous works, see, e.g., [1, 2, 7] and references therein, the neutral C_{60} cage is modelled by a short-range attractive spherical potential $V_c(r)$ of inner radius $r_0 = 5.8$ a.u., depth $U_0 = -8.2$ eV and finite thickness $\Delta = 1.9$ a.u.

$$V_c(r) = \begin{cases} U_0, & \text{if } r_0 \le r \le r_0 + \Delta \\ 0 & \text{otherwise.} \end{cases} \quad (1)$$

A neutral endohedral fullerene $A@C_{60}$ is formed by placing the atom A at the center of the cage. For small sized, compact atoms A there is no charge transfer to the cage, so that the confined atom A retains the general structure of the free atom A. Alternatively [1, 7], an endohedral anion $A@C_{60}^{z-}$ is modelled by the sum total of the potential V_c and the Coulomb potential $V_z(r)$ of an excessive negative charge on the cage C_{60}. Assuming that the charge z is uniformly distributed over the entire outer surface of C_{60},

$$V_z(r) = \begin{cases} \frac{z}{r_0+\Delta}, & \text{if } 0 \le r \le r_0 + \Delta \\ \frac{z}{r} & \text{otherwise.} \end{cases} \quad (2)$$

Next, the sum total of these two potentials is added to nonrelativistic Hartree-Fock (HF) equations for a free atom. Solutions of these new HF equations, i.e., electronic energies and wavefunctions of the confined A atom, are used in well-known expressions for nl photoionization amplitudes, angle-differential and/or angle-integrated nl photoionization cross sections, *etc.*, for free atoms; see [8] for the latter. To account for interchannel coupling in the photoionization of a confined atom, the *random phase approximation with exchange* (RPAE) [8] is utilized to meet the aim. This is because RPAE, which uses a HF approximation as the zero-order approximation, has proven to be a very reliable methodology over the years. Accordingly, for the sake of "theoretical" consistency, HF values of ionization thresholds of free and confined Ne atoms are used in the present study.

3. Results and discussion

3.1. Conventional confinement resonances: irrelevance of a charged state of C_{60}^{z-}

RPAE calculated photoionization cross sections σ_{2p} of Ne in variously z-charged $Ne@C_{60}^{z-}$ are displayed in figure 1. Corresponding photoionization cross sections σ_{2s} and σ_{1s} are depicted in figure 2. In these calculations, interchannel coupling in the atom was accounted for at an intra-shell approximation level, as the first step in the study. One can see that, above the 2p threshold $I_{2p}^{z=0} \approx 23$ eV of Ne in *neutral* $Ne@C_{60}^{z=0}$ (i.e., to the right of a vertical line marked with $z = 0$), all σ_{2p}'s oscillate about the 2p photoionization cross section of free Ne, as do σ_{2s} above $I_{2s}^{z=0} \approx 53$ eV and σ_{1s} above $I_{1s}^{z=0} \approx 892$ eV, respectively. The oscillations are due to the interference between the outgoing nl photoelectron wave and those scattered off the confining potential of C_{60}^{z-}. When the constructive interference occurs, there emerge maxima, i.e., resonances in σ_{nl}'s, termed confinement resonance [1, 2, 4, 7]. Furthermore, one can see that confinement resonances in *all* σ_{nl}'s are nearly z independent above related $I_{nl}^{z=0}$ thresholds (previously [7], the same was noted in the 1s spectrum of Ne in $Ne@C_{60}^{z-}$). This is because the energy ϵ of an outgoing nl photoelectron in any charged $Ne@C_{60}^{z-}$ exceeds by far the Coulomb potential barrier of the charged carbon cage, at $\omega \ge I_{nl}^{z=0}$. This is clearly seen in figure 3 where direct (Hartree) parts of the potentials "seen" by the ϵp and ϵs photoelectrons (due

Journal of Physics: Conference Series **212** (2010) 012015 doi:10.1088/1742-6596/212/1/012015

to the 2p $\to$ ϵp, ϵs transitions) are displayed. Hence, the presence of the Coulomb potential barrier is inconsequential for the outgoing nl photoelectrons at photon energies $\omega \geq I_{nl}^{z=0}$. Correspondingly, at $\omega \geq I_{nl}^{z=0}$, confinement resonances in any nl photoionization spectrum of the encapsulated atom will chiefly be governed by details of the confining potential well V_c, equation (1), as in neutral Ne@C$_{60}$. Thus, the resonances will nearly be z-independent. We term such confinement resonances as *conventional confinement resonances.*

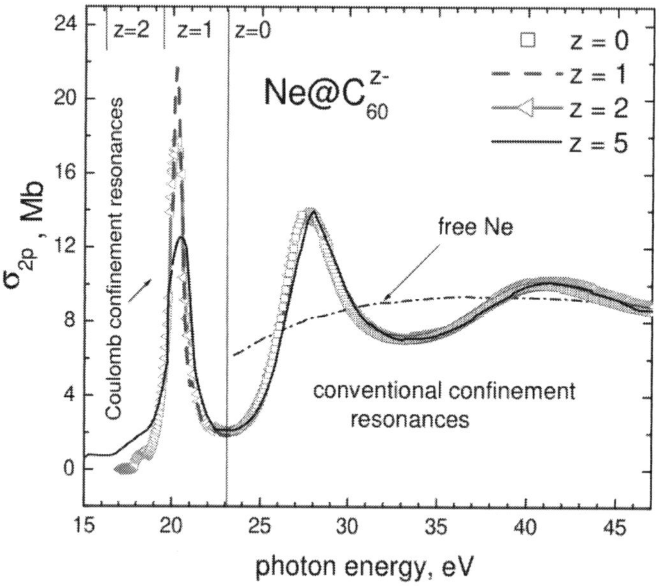

Figure 1. RPAE calculated data, at an intra-shell interchannel coupling approximation level, for the σ_{2p} photoionization cross section of Ne in Ne@C$_{60}^{z-}$ for z = 0, 1, 2, 5 as well as for free Ne, as marked. Vertical lines with marks $z = 0$, $z = 1$ and $z = 2$, show positions of the Ne 2p thresholds $I_{2p}^{z=0} \approx 23.1$, $I_{2p}^{z=1} \approx 19.6$, $I_{2p}^{z=2} \approx 16.1$ and $I_{2p}^{z=5} \approx 5.5$ eV in corresponding Ne@C$_{60}^{z-}$.

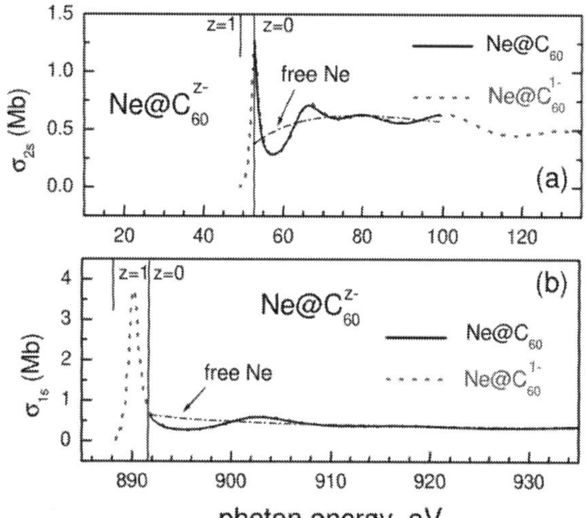

Figure 2. (a): RPAE calculated data, at an intra-shell interchannel coupling approximation level, for the σ_{2s} photoionization cross section of Ne in Ne@C$_{60}^{z-}$ for z = 0 and 1 as well as for free Ne, as marked. Vertical lines with marks $z = 0$ and $z = 1$ show positions of Ne $I_{2s}^{z=0} \approx 52.5$ eV and $I_{2s}^{z=1} \approx 49$ eV. (b): The same as in (a) but for the Ne 1s photoionization ($I_{1s}^{z=0} \approx 892$, $I_{1s}^{z=1} \approx 888$ eV).

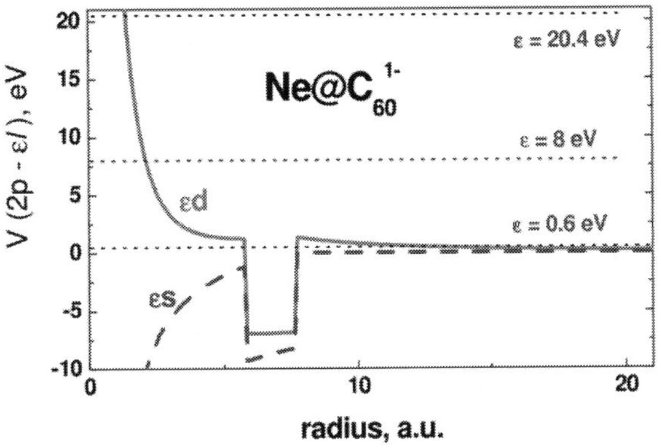

Figure 3. A direct (Hartree) part of the potential V(2p $\to \epsilon l$) "seen" by ϵs and ϵp photoelectrons in the 2p $\to \epsilon$s and 2p $\to \epsilon$d photoionization transitions in the Ne@C_{60}^{1-} atom, as marked. Horizontal lines show the photoelectron energies $\epsilon \approx 0.6$, 8 and 20.4 eV corresponding to maxima of the resonances in the Ne@C_{60}^{1-} photoionization cross section σ_{2p} (see figure 1) at about 20, 28 and 42 eV of the photon energy, respectively.

To conclude, the above results reveal, and the given explanation proves, that *conventional* confinement resonances in photoionization cross sections of inner, intermediate and outer subshells of an atom A in A@C_{60}^{z-} appear to be almost z-independent. This is an interesting general phenomenon. Another important observation (see figures 1 and 2) is that conventional confinement resonances vanish quite rapidly with increasing photon (or photoelectron) energy. This is in line with a theory of scattering of particles off a potential well/barrier. Indeed, starting at a sufficiently high energy of the outgoing photoelectron, the coefficient of reflection of the latter off a finite potential well/barrier decreases with increasing energy of the electron. As a result, the interference effect between the outgoing and scattered off the potential well/barrier photoelectron waves weakens, with increasing energy of the electron, and so are the associated conventional confinement resonances. In further, we will term this reasoning as "conventional thinking".

3.2. Coulomb confinement resonances

Above, trends in the Ne nl photoionization cross sections of Ne@C_{60}^{z-} were considered at photon energies ω beyond the corresponding $I_{nl}^{z=0}$ thresholds of Ne in neutral Ne@C_{60}. We now turn the attention to lower photon energy ($\omega < I_{nl}^{z=0}$) parts of figures 1 and 2 (to the left of the $z = 0$ marked line) where we consider photoionization of charged Ne@C_60^{z-} ($z \neq 0$). There, the additional resonance in each of σ_{2p}, σ_{2s} and σ_{1s} of Ne@C_{60}^{z-} is seen to emerge. It owes its existence to the Coulomb potential V_z, equation (2), of the charged carbon cage. The V_z potential brings up the Coulomb potential barrier at the outer surface of C_{60}. This engenders reflection of the low-energy continuum photoelectron wave from the Coulomb barrier causing additional resonances one of which is depicted in figures 1 and 2 at photon energies under discussion. Originally, the emergence of this kind of a resonance was noted in the Ne 1s photoionization of Ne@C_{60}^{z-} [7], where it was named *Coulomb confinement resonance*, in view of its association with the Coulomb potential barrier of the charged carbon cage. The present paper establishes that the phenomenon emerges in the photoionization of intermediate and outer subshells of the encapsulated atom as well, as a general occurrence. Coulomb confinement resonances appear to be z-dependent, as is clearly exemplified by depicted in figure 1 σ_{2p}'s. This is because, in this instance, the energy ϵ of the outgoing photoelectron is near or, generally, below the top of the Coulomb potential barrier (see, as illustration, the energy line $\epsilon = 0.6$ eV in figure 3). This makes the photoionization process to be sensitive to details of the latter, and, hence, to a charged state of the carbon cage as well.

To conclude, the established co-existence of z-dependent Coulomb and z-independent conventional confinement resonances in photoionization spectra of endohedral *anions* is an exclusive feature of these systems.

3.3. Correlation confinement resonances: the resurrection of confinement resonances far above thresholds

In the above, the discussion was related to RPAE calculated data for photoionization cross sections $\sigma_{nl}(\omega)$ of the encapsulated Ne atom which were obtained at an intra-shell interchannel coupling approximation level. However [1, 2, 9], the effect of *inter-shell* interchannel coupling in the encapsulated atom may result in the emergence of new confinement resonances, termed *correlation confinement resonances*. These resonances were previously interpreted as resonances which are induced in an outer-shell photoionization spectrum of the encapsulated atom by *conventional confinement resonances* in inner-shell photoionization transitions in the atom, via interchannel coupling. The earlier finding was illustrated by RPAE [1, 9] and recently seconded by relativistic RRPA [2] calculated data for the Xe 5s photoionization of Xe@C_{60} where interchannel coupling between the 5s and 4d transitions was accounted for. However, the same effect may occur via interchannel coupling with Coulomb confinement resonances as well. It may even be bigger in this case since Coulomb confinement resonances dominate over conventional confinement resonances; see figure 2 for the most illustrative supporting evidence. Furthermore, the effect of interchannel coupling may show up strongly in an outer-shell photoionization spectrum at photoelectron energies which are thousands eV above the threshold, when interchannel coupling involves very deep inner-shell transitions. The effect is going to be strong, because at such big differences in ionization thresholds of the inner and outer subshells, photoionization transitions from the former will be strong whereas those from the latter will be weak, at photon energies above the inner shell threshold. As a result, both inner-shell Coulomb and conventional confinement resonances may be effectively "funneled" through a thousands-eV-distance to the outer-shell spectrum. However, to which extent the "funneled" confinement resonances may indeed perturb the outer-shell spectrum is not clear. To clarify this point, we performed RPAE calculations both of the 2p photoionization cross section σ_{2p} and dipole photoelectron angular-asymmetry parameter β_{2p} for Ne in Ne@C_{60}^{1-}, above the Ne 1s ionization threshold. This time, inter-shell interchannel coupling between the 1s, 2s and 2p transitions was included in the calculations. Thus obtained RPAE calculated data for σ_{2p} and β_{2p} for the encaged Ne are depicted in figure 4 along with data for free Ne.

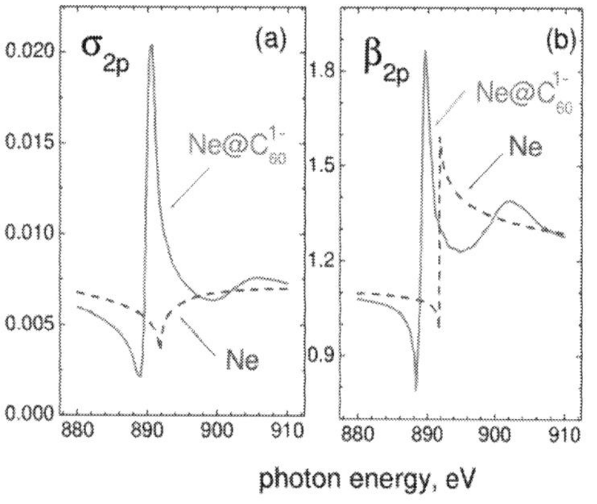

Figure 4. (a): RPAE calculated data for $\sigma_{2p}(\omega)$ (Mb) of Ne in Ne@C_{60}^{1-} and free Ne near their 1s thresholds, respectively. RPAE calculations included interchannel coupling between transitions from the 1s, 2s and 2p subshells. (b): The same as in (a) but for the dipole photoelectron angular-asymmetry parameter $\beta_{2p}(\omega)$.

One can see that both σ_{2p} and $\beta_{2p}(\omega)$ for the encapsulated Ne atom possess a strong sharp resonance at about 890 eV which is followed by a lower but broader resonance at about 905 eV. As a result, these "encapsulated" σ_{2p} and $\beta_{2p}(\omega)$ differ considerably from the free Ne σ_{2p} and $\beta_{2p}(\omega)$, far above threshold. Thus, the confinement matters in this case, so that the two prominent resonances in "encapsulated" σ_{2p} and $\beta_{2p}(\omega)$ *are* confinement resonances.

The striking novelty of the above finding is that the resonances emerge far-far above where conventional thinking said they would exist. Indeed, when considering confinement resonances, one normally thinks in terms of conventional confinement resonances which occur due to the interference between the directly outgoing and reflected off the confining potential photoelectron waves. However, in line with "conventional thinking", as was discussed above, conventional confinement resonances fade away relative rapidly with increasing energy and do vanish far above threshold. The discovered emergence, or better say resurrection of confinement resonances in the 2p photoionization spectrum of Ne@C_{60}^{z-} far above threshold implies that, in contrast to "conventional thinking", a few eV deep/high confining potential well/barrier may, once again, be felt by a far-above-potential-barrier-electron. The effect may as well be called *reemerging confinement effect* for a high-energy scattering electron, as a general phenomenon. This general phenomenon may result in the emergence of far above threshold confinement resonances in the *nl* photoionization spectrum of a confined atom, as in the above discussed particular example of the Ne 2p photoionization of Ne@C_{60}^{1-}. The latter effect may rightly be termed as *resurrection of confinement resonances* effect. Both the reemerging confinement and resurrection of confinement resonances effects owe their existence to inter-shell interchannel coupling in the encapsulated multielectron atom. Indeed, a trial calculation for Ne@C_{60}^{1-} showed that removal of the Ne 1s transition (and, thus, associated with it Coulomb and conventional confinement resonances) from RPAE calculations of the Ne 2p photoionization leaves no traces of the two resonances in "encapsulated" σ_{2p} and $\beta_{2p}(\omega)$. As a result, the 2p photoionization spectra of the confined and free Ne atoms become virtually identical far above threshold, as they previously have been thought to remain nearly identical at all high energies, on the basis of "conventional thinking". Hence, the resurrected confinement resonances in "encapsulated" σ_{2p} and $\beta_{2p}(\omega)$, far above threshold, are due to interchannel coupling with the conventional and Coulomb confinement resonances in the 1s spectrum. The latter are "funneled" to the 2p spectrum via interchannel coupling. This appears to perturb the outer-shell spectrum of the confined atom dramatically.

Clearly, there is nothing particularly special about the Ne@C_{60}^{z-} system. Therefore, both the reemerging confinement effect and the resurrection of confinement resonances effect are expected to appear in, and be qualitatively similar for, spectra of other endohedrally confined atoms as well. In other words, the two discovered effects step in as novel general features of spectra of endohedrally confined atoms $A@C_{60}^{z-}$ the existence and significance of which has been convincingly proven in the performed study.

In conclusion, neither Coulomb or conventional confinement resonances, not to mention the just discovered resurrected confinement resonances far above threshold, have been experimentally observed yet for technical reasons. We hope that the data presented herein will prompt experimentallists to look into the matter, thereby promoting such developments.

Acknowledgments
This work was supported by the NSF Grant No. PHY-0652704.

References
[1] Dolmatov V K 2009 *Photoionization of atoms encaged in spherical fullerenes Theory of Confined Quantum Sytems*: Part Two (Adv. Quantum Chem. vol 58) ed John R. Sabin and Erkki Brändas (USA: Academic Press) chapter 2 pp 13–68
[2] Govil R, Siji A J and Deshmukh P C 2009 *J. Phys.* B **42** 065004
[3] Chakraborty H S, Madjet M E, Renger T, Rost J-M and Manson S T 2009 *Phys. Rev.* A **79** 061201(R)
[4] Amusia M Ya, Baltenkov A S and Chernysheva L V 2008 *JETP Lett.* **87** 200
[5] Müller A, Schippers S, Phaneuf R A, Habibi M, Esteves D, Wang J C, Kilcoyne A L D, Aguilar A, Yang S and Dunsch L 2007 *J. Phys.: Conf. Series* **88** 012038
[6] Müller A, Schippers S, Habibi M, Esteves D, Wang J C, Phaneuf R A, Kilcoyne A L D, Aguilar A and Dunsch L 2008 *Phys. Rev. Lett.* **101** 133001
[7] Dolmatov V K and Manson S T 2006 *Phys. Rev.* A **73** 013201
[8] Amusia M Ya 1990 *Atomic Photoeffect* (New York: Plenum)
[9] Dolmatov V K and Manson S T 2008 *J. Phys.* B **41** 165001

Combining laser and electron interactions – current experiments and future possibilities in excitation and ionization studies

Andrew James Murray
Photon Science Institute, School of Physics and Astronomy, University of Manchester, Manchester M13 9PL, UK.

E-mail: Andrew.Murray@manchester.ac.uk

Abstract. In this brief review a number of different experimental techniques that combine the advantages of laser and electron interactions with atoms and molecules will be described. These processes range from super-elastic electron scattering from laser excited atoms so as to ascertain the electron collision parameters over all scattering geometries, through new electron interactions with laser cooled and trapped atoms, to the process of electron impact ionization of laser-aligned targets. Examples from work carried out in Manchester are given, and new types of measurements that may be possible in the future are described.

1. Introduction

Studying electron interactions with atomic and molecular targets has a long and distinguished history from the time Franck and Hertz first studied excitation of mercury to confirm the quantum nature of the atom, as predicted by Bohr. Since then scattering experiments have markedly improved in sophistication, and the corresponding quantum models describing these complex interactions now accurately predict much of the data. Studies include the processes of excitation, where the target is left in an excited state following the collision, through to interactions where the target is ionized. The most sophisticated measurements adopt coincidence techniques to correlate the reaction products, allowing the most accurate comparison to quantum models. In the case of excitation the scattered electron is measured in coincidence with an emitted photon to determine the collision parameters [1], whereas for ionization, (e,2e) studies determine the probability of ionization as a function of the scattered and ejected electron momenta [2].
The interaction of laser radiation with atomic and molecular targets usually adopts a tuneable laser source that can resonantly excite transitions within the target. High power pulsed lasers may also directly ionize the target, however these type of reactions are not considered here. For resonant excitation either a pulsed or continuous wave (CW) laser may be used, the limited wavelength range accessible from these lasers usually confining the number of targets that can be directly excited to only a few. By contrast, combining electron and laser interactions opens up a far wider range of targets for study, and further allows the unique properties of the laser and electron interaction to be exploited simultaneously. It is these types of experiments that are discussed here, using the examples of super-elastic scattering from a laser prepared target, electron impact ionization from laser cooled atoms, and ionization from laser aligned targets in CW and pulsed laser beams.

2. Super-elastic electron scattering from laser prepared atoms.

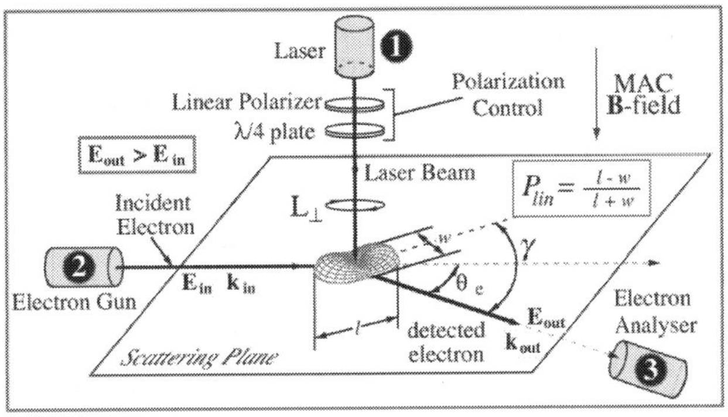

Figure 1. The super-elastic geometry showing the direction of incident laser and electron beams. The electron detector measures electrons which have gained energy from the simultaneous laser and electron interaction.

In this type of study a high resolution polarized single mode laser (1) coherently and resonantly excites an atomic ensemble from the ground state to an excited state, as shown in figure 1. The excited targets are then de-excited using an incident electron beam (2), so that the reaction products carry away from the interaction both the excess energy given to the atom by the laser beam, and the excited sub-state information that is imparted to the atom by the coherent excitation process. This information is determined at a given electron scattering angle as a de-excitation probability (3), which depends on the polarization and direction of the exciting laser beam. The super-elastic scattering process can hence be considered as a time reversal of the electron-photon coincidence technique. By contrast, the super-elastic process accumulates data many thousands of times faster, since the laser is always input in the same direction and the probability of excitation is high.

The data from super-elastic scattering measurements is usually given as a set of 'natural frame' parameters [1] that precisely define the electron excited atom. These are the angular momentum transferred to the atom during the collision $L_\perp$, the alignment parameter P_{lin} describing the 'length' to 'width' of the charge cloud and γ, which describes the alignment angle as in figure 1. These parameters relate directly to the scattering amplitudes for excitation that are predicted theoretically.

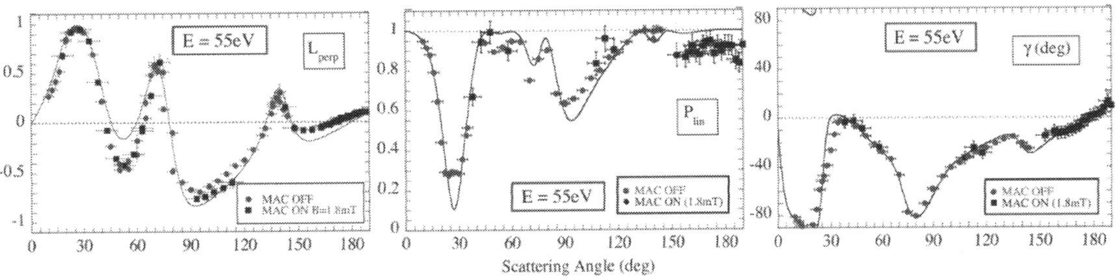

Figure 2. The collision parameters $L_\perp, P_{lin}$ & γ for excitation of calcium at 55eV incident energy, as a function of scattering angle. These data were taken using a MAC device inside the super-elastic spectrometer.

Figure 2 shows an example of these measurements for a calcium target taken at an equivalent coincidence incident energy of 55eV, compared to the theoretical prediction of Stauffer and colleagues [3] who used a relativistic distorted wave calculation (r-DWA) to derive the scattering amplitudes. Theory and data are in impressive agreement at this energy. It is also noted that the data have been taken at scattering angles up to and beyond 180 degrees, which was possible using a Magnetic Angle Changer (MAC) in the spectrometer [4-6]. These experiments were the first to adopt this new technique in these laser-based studies, and are the first to fully ascertain the collision parameters over all geometries. Agreement between theory and experiment becomes less favourable as the incident energy is lowered, and at energies below ~20eV the r-DWA calculation no longer performs well [7]. At these energies it then becomes necessary to adopt different theoretical methods [8], and it is in this low energy regime where current experiments are being conducted.

3. Collisions with cold atoms – the AC-MOT.

One of the key advances in laser-atom interactions over the last 20 years has been in the field of atom cooling and trapping, which adopts near resonant CW laser fields to interact with atoms in a magnetic field gradient [9]. The direct-current Magneto-Optical Trap (DC-MOT) invented by Chu and co-workers [10] has found wide use in areas ranging from metrology through to fundamental studies of Bose Einstein Condensation (BEC). Although DC-MOT's have been used in selected high energy ion collision studies (eg MOTRIMS experiments [11]), they have not been adopted in electron collision studies. For this type of collision the main limitation of the DC-MOT arises due to the **B**-fields from the DC-MOT, including induced fields generated by eddy currents in the spectrometer which occur when the DC-MOT current is switched on and off.
A new type of atom trap has recently been invented in Manchester which uses an *alternating current* to produce the required **B**-field, and which simultaneously switches the polarization state of the trapping lasers at the same rate [12]. This AC-MOT has proven to be extremely effective, and will allow many new types of collision studies to be conducted in the near future. The AC-MOT has the same trapping performance as a conventional DC-MOT, but the generated **B**-fields can be switched off more than 300 times faster than for the conventional MOT. This breakthrough means many different collision studies can now be routinely carried out with cold targets, so as to exploit the low Doppler profile and low momentum that these targets possess.
As an example of the enhanced speed of the AC-MOT, experiments were conducted which measured current from a low energy electron beam provided by an electron gun, where the electron beam passed through the centre of the atom trap. The trap was initially switched on with a field gradient in the trapping region of 10G/cm (as required to trap potassium atoms in the experiment). This field resulted in a large deflection of the electron beam, so that no current was observed on a Faraday cup located on the opposite side of the trapping region to the gun. The current in the MOT coils was then switched off in ~20μs, and the Faraday cup current measured as a function of time.
Figure 3 compares the results for the DC-MOT with those for the AC-MOT. At the electron energies used here (~20eV), **B**-fields greater than ~5mG were found to deflect the electron beam out of the Faraday cup. These measurements therefore provide a very sensitive test of the influence of the magnetic fields generated by the different types of trap when used for electron collision studies.

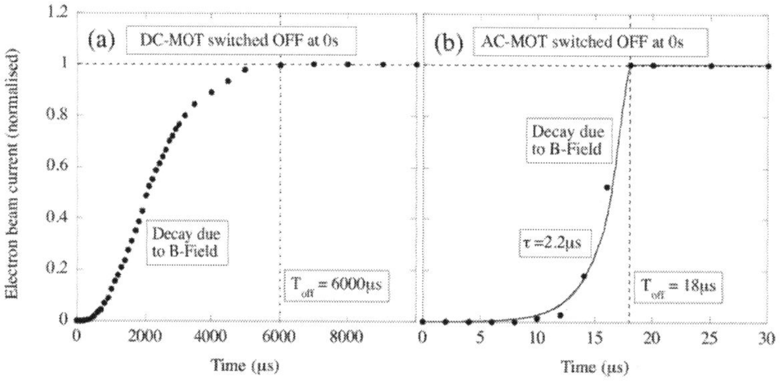

Figure 3. Faraday cup current measured as a function of time for (a) the DC-MOT and (b) the AC-MOT, where the current through the MOT coils was switched off at t = 0sec.

Figure 3(a) shows that experiments with trapped atoms in a conventional DC-MOT would need to wait ~6000μs before the **B**-field due to induced eddy currents reduce to <5mG, at which time electron beam collision experiments could be performed. This severely constrains the experiments, as many trapped atoms will have diffused out of the trapping region in this time, and the pulsed repetition rate of these experiments must be kept low (~80Hz). By contrast, figure 3(b) shows for the AC-MOT, the **B**-fields decay to <5mG within 20μs. In this case, eddy currents generated in the spectrometer are effectively eliminated by adopting an equal number of positive and negative cycles in the AC-MOT drive current. In a time of 20μs the trapped atoms are effectively frozen in space, and so diffusion from the trapping region is negligible. Further, since the AC-current can be driven at greater than 6kHz, pulsed collision experiments from these trapped atoms can be carried out many times faster.

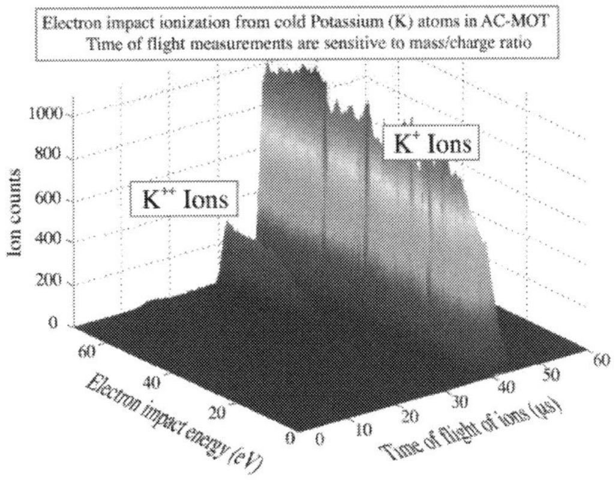

Figure 4. Time of flight spectrum for ionization of cold trapped potassium using electron impact.

Figure 4 shows an example of collision data from cold potassium atoms (T~300μK) in the AC-MOT, where a time of flight detector was installed to measure the total ionization cross section as a function of incident electron energy. These are the first comprehensive results from electron collisions with cold atoms. The spectrum shows the cross section for K^+ and K^{++} formation, the incident electron energy ranging from 2eV to 70eV. In these experiments the laser field was on while the collision occurred, and so there is an additional contribution to the cross section from excited potassium.

Adoption of the AC-MOT will allow many new experiments to be conducted. As an example, since the Doppler profile of the trapped atoms is effectively eliminated, it becomes possible to carry out super-elastic scattering experiments from targets with high principle quantum numbers. This will allow collision theories to be tested for excitation to highly excited states, with a precision impossible to obtain using conventional methods. Further, the well defined momentum of trapped atoms will allow target deflection techniques that measure cross sections to be more widely adopted. The AC-MOT provides much higher precision than is possible using conventional COLTRIMS and other types of deflection techniques, since the uncertainty in the initial momentum of the target that limits the resolution in these experiments is markedly reduced.

4. Ionization from laser-aligned targets.

The final topic discussed here is that of electron impact ionization, as measured by the (e,2e) technique. These experiments determine the angular correlation between the electron scattered from an ionized target and the electron that is ejected. Since the electrons can emerge from the interaction at any angle, the ionization probability may be measured over the full scattering geometry for a range of incident energies and outgoing electron momenta. As such, experiments usually measure a subset of this five-dimensional space. Often a coplanar geometry is chosen (where the incident, scattered and ejected electron momenta all lie in a plane), however it is also possible to measure the ionization probability in non-coplanar geometries. The (e,2e) spectrometer in Manchester allows measurements from a coplanar geometry through to the perpendicular plane (where the incident electron momentum is orthogonal to the scattering plane), and so provides a stringent test of current ionization theory.

Recent experiments in Manchester have been conducted over a wide range of scattering geometries where the target is either a diatomic molecule (such as H_2 or N_2) [13,14] or tri-atomic (such as H_2O or CO_2) [15,16]. These experiments paralleled significant theoretical developments which consider the collision using a non-spherical basis, or which adopt time dependent close coupling techniques. These theoretical methods predict the ionization probability as a function of the molecular alignment with respect to the scattering plane, or can average the results over all possible target orientations to compare to conventional experiments that do not select the alignment angle. These theories have recently proven to be very successful when compared to new experimental data being produced [17].

Of key significance in these calculations is the marked change in the cross section predicted for different molecular alignment angles. A few experiments have been conducted to test these theories, by adopting an energy scheme that ensures the molecular ion is left in a repulsive state following the interaction [18]. In this case Coulomb repulsion between the nuclei drives the molecule apart, and it is assumed the final ion direction defines the alignment direction of the target prior to interaction. These experiments are very difficult, and have a low yield and low energy resolution due to the steep gradient of the repulsive molecular energy curve. As such the data uncertainties are high.

Experimental techniques exist to partially align a polar molecule (eg using a hexapole field prior to the interaction region [19]), however these techniques are not applicable to molecules such as H_2, for which current collision theories are proving successful. It is hence worthwhile considering whether other techniques are possible to control the alignment of these targets.

A technique that has proven successful in recent experiments adopts a non-resonant high power pulsed laser beam to adiabatically align supersonically cooled molecules along the direction of the laser polarization [20]. In these experiments the laser field induces a dipole moment in the molecule, the molecule then re-orienting so as to minimise the interaction energy in the field. By supersonically cooling the target ensemble prior to this interaction, the rotational energy of the molecules is minimised, allowing the coupling of the laser field to the molecule to be maximized. These experiments clearly demonstrate that the molecules are aligned in the field, as predicted.

Unfortunately this alignment only occurs when the laser illuminates the target, and for all experiments carried out so far the temporal width of the laser pulse is only ~5ns, with a repetition rate of 10Hz. The molecules are hence only aligned for 50ns in each second, and so carrying out electron scattering coincidence experiments from these laser aligned targets is currently unrealistic.

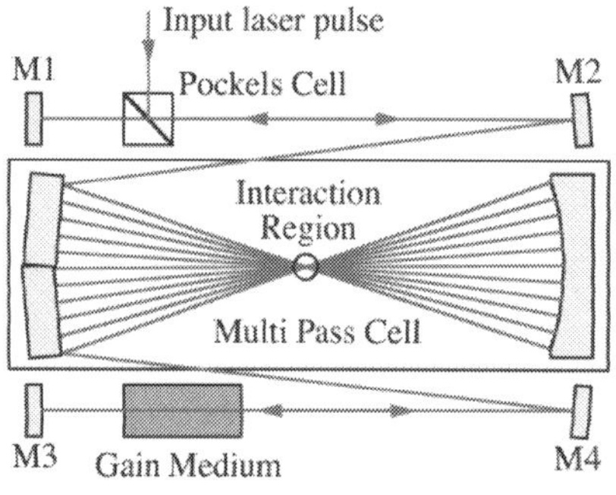

Figure 5. Design of a multi-pass high resolution laser cavity to extend the laser-molecule interaction time.

A new method is suggested here, as shown in figure 5. In this case an incident high power laser pulse is optically switched into a very high reflectivity cavity using a Pockels cell. A multi-pass cell (MPC) incorporated within the cavity ensures the laser pulse passes through the interaction region many times before the laser beam leaves the MPC, after which it is passed to a regenerative laser amplifier. This amplifier is constructed to add gain to the laser pulse so as to balance losses in the cavity. The regenerated laser pulse is then reflected back into the MPC (and hence through the molecules to align them), and returns to the Pockels cell which has been switched to ensure the laser pulse remains inside the cavity. By choosing the gain of the laser medium to balance cavity losses, it will then be possible to extend the time of alignment of the molecules by several orders of magnitude. Collision based experiments from laser-aligned molecules should then be possible.

A new pulsed (e,2e) spectrometer has been constructed in Manchester to commence preliminary studies in this area. The new spectrometer uses a pulsed supersonic jet to deliver rotationally cold molecules to the interaction region, and angular correlation measurements have now commenced from these targets. The experiments will initially look for differences between cold and hot molecules, by comparing (e,2e) results from the supersonic jet and from an effusive beam at room temperature. An optical multi-pass alignment cavity will then be designed and tested, to determine the feasibility of the proposed technique. Parallel experiments will also be conducted from laser aligned excited atoms in the existing (e,2e) spectrometer, to ascertain the effects of ionization from these targets. This will allow the new theoretical developments to be further refined and directly tested.

Acknowledgements.
This work could not be carried out without input from many people. These include Christian Kaiser, Matthew Harvey, Martyn Hussey, Alan Venables, Dave Coleman (UK), Bill MacGillivray (Australia), and Don Madison, Ola Al-Hagan and James Colgan (USA).

5. References

[1] Andersen N, Gallagher J W and Hertel I 1988 *Phys. Rep.* **165** 1
[2] McCarthy I E and Weigold E 1991 *Rep. Prog. Phys.* **54** 789
[3] Chauhan R K, Srivastava R and Stauffer A D 2005 *J. Phys. B.* **38** 2385
[4] Hussey M J, Murray A J, MacGillivray W R and King G C 2007 *Phys. Rev. Lett.* **99** 133202
[5] Murray A J, MacGillivray W R and Hussey M J 2008 *Phys. Rev. A* **77** 013409
[6] Hussey M J, Murray A J, MacGillivray W R and King G C 2007 *J. Phys B* **41** 055202
[7] Murray A J and Cvejanovic D 2003 *J. Phys B* **36** 4889
[8] Zatsarinny O, Barschat K, Bandurina L and Gedeon S 2007 *J. Phys. B* **40** 4023
[9] Metcalf H J and van der Straten P 1999 *Laser cooling and trapping* (Springer: New York)
[10] Raab E L, Prentiss M, Cable A, Chu S and Pritchard D E 1987 *Phys. Rev. Lett.* **59** 2631
[11] DePaola B D, Morgenstern R and Andersen N 2008 *Adv. At. Mol. Opt. Phys.* **55** 139
[12] Harvey M and Murray A J 2008 *Phys. Rev. Lett.* **101** 173201
[13] Murray A J 2005 *J. Phys. B* **38** 1999
[14] Murray A J, Hussey M J, Gao J and Madison D H 2006 *J. Phys. B* **39** 3945
[15] Kaiser C, Spieker D, Gao J, Hussey M J, Murray A J and Madison D H 2007 *J. Phys. B* **40** 2563
[16] Hussey M J and Murray A J 2005 *J. Phys. B* **38** 2965
[17] Colgan J, Pidzola M S, Robicheaux F, Kaiser C, Murray A J and Madison D H 2008 *Phys. Rev. Lett.* **101** 233201
[18] Takahashi M, Watanabe N, Khajuria Y, Udagawa Y and Eland J H D 2005 *Phys. Rev. Lett.* **94** 213202
[19] Baugh D A, Kim D Y, Cho A, Pipes L C, Petteway J C and Fuglesang C D 1994 *Chem. Phys. Lett.* **219** 207
[20] Stapelfeldt H and Seideman T 2003 *Rev. Mod. Phys.* **75** 543

International Symposium on (e,2e) and the 15th ISPCEAC
Journal of Physics: Conference Series **212** (2010) 012017
IOP Publishing
doi:10.1088/1742-6596/212/1/012017

Spin-resolved electron-impact ionisation of atoms

I Bray[1], **D V Fursa**[1], **A S Kadyrov**[1], **A S Kheifets**[2], **T Lepage**[1], **A T Stelbovics**[1]

[1]ARC Centre for Antimatter-Matter Studies, Curtin University of Technology, GPO Box U1987 Perth, Western Australia 6845, Australia
[2]Research School of Physics, Australian National University, Canberra ACT 0200, Australia

E-mail: `I.Bray@curtin.edu.au`

Abstract. We shall consider various aspects of electron-impact ionisation of atoms. Firstly, the recent progress in the formulation of the problem will be outlined. Then application of the convergent close-coupling (CCC) method to the calculation of the total ionisation spin asymmetries will be described for a range of neutral through to highly charged targets. We shall also consider the importance of electron spin in the problem of single and double photoionisation of lithium. Lastly, application of the CCC method to 64.6 eV e-He ionisation fully differential cross sections for in- and out-of-plane geometries will be considered.

1. Introduction

There has been much progress during the last decade in the theoretical treatment of electron-impact ionisation of atoms and the closely related problem of double photoionisation. The existence of readily accessible super-computer facilities has allowed the application to the problem of numerically intensive techniques that essentially solve the Schrödinger equation with minimal approximation. These are either based on the exterior complex scaling (ECS) approach [1–4] or the close-coupling method, time-independent [5–9] or time-dependent [10, 11]. Whereas total ionisation cross sections can be obtained from unitary theories utilising just the excitation amplitudes [12], when it comes to differential cross sections an ionisation amplitude is required. How that can be rigorously defined, even when the total wavefunction is available, has been problematic until very recently [13, 14]. Despite this an excellent agreement with experiment for fully differential cross sections has been obtained for quite some time, see for example [2, 8, 9]. So we have a very unusual situation where accurate agreement with experiment for fully differential ionisation cross sections has been obtained before it was clear how to go about calculating the underlying ionisation amplitudes. We begin by considering this question.

2. Surface-integral formulation of scattering theory

Few-body break-up problems involving long-range Coulomb potentials have been of interest for many decades. Yet even the two-body problem has not had an adequate mathematical formulation. Methods involving screening require renormalization due to singularities in the resultant on-shell amplitudes, see Kadyrov *et al* [14] and references therein for a detailed discussion. While the resolution of these problems is rather detailed and complicated, here we give just the underlying ideas.

Starting with a two-body scattering problem, for Hamiltonian $H = H_0 + V$ we write the Schrödinger equation

$$(\epsilon - H)\psi^{\pm}_{\boldsymbol{k}}(\boldsymbol{r}) = 0. \tag{1}$$

If V is short-ranged then

$$\psi^{+}_{\boldsymbol{k}}(\boldsymbol{r}) \overset{r\to\infty}{\sim} e^{i\boldsymbol{k}\cdot\boldsymbol{r}} + f(\widehat{\boldsymbol{k}}\cdot\widehat{\boldsymbol{r}})e^{ikr}/r, \tag{2}$$

or if V is long-ranged, i.e. Coulomb, then

$$\begin{aligned}\psi^{+}_{\boldsymbol{k}}(\boldsymbol{r}) \overset{r\to\infty}{\sim} & e^{i\boldsymbol{k}\cdot\boldsymbol{r}+i\gamma\ln(kr-\boldsymbol{k}\cdot\boldsymbol{r})}[1+O(1/r)] \\ & +f(\widehat{\boldsymbol{k}}\cdot\widehat{\boldsymbol{r}})\frac{e^{ikr-i\gamma\ln(2kr)}}{r}[1+O(1/r)],\end{aligned} \tag{3}$$

where $f(\widehat{\boldsymbol{k}}\cdot\widehat{\boldsymbol{r}})$ is the required scattering amplitude. Utilising surface-integral evaluation at some large r_0 one can show [14], for both short- and long-range potentials,

$$\langle\phi_{\boldsymbol{k}'}|\overleftarrow{H} - \epsilon|\psi^{+}_{\boldsymbol{k}}\rangle_{r_0} = t(\boldsymbol{k}',\boldsymbol{k}), \tag{4}$$

which yields the required amplitude, and where $\phi_{\boldsymbol{k}'}$ is the analytically known asymptotic form of the incident-channel wavefunction.

The main idea is that in the surface-integral form, in which the potential of interaction disappears, at large r_0, we are able to perform the integration with all quantities analytically known. The expressions given reduce to the usual ones in the case of short-range potentials.

The situation for the three-body problem relies on the same ideas, except that the asymptotic forms of three-body wavefunctions are more complicated. In the end, we have a definition for the amplitude and a practical method for its calculation for both the excitation and ionisation processes. The post form of the amplitude, Eq. (4) explains the success of the ECS and CCC methods in calculating ionization amplitudes. In the former, the asymptotic form of the total wavefunction is approximated by a product of two Coulomb waves, whereas in the CCC method the slow electron is represented by a Coulomb wave and the fast one by a plane wave. Further details may be obtained from Kadyrov *et al* [14].

3. Electron-impact total ionisation cross sections and spin asymmetries

Baum *et al* [15] measured, for Li, Na and K, the electron-impact total ionisation spin asymmetries A,

$$A = \frac{\sigma_0 - \sigma_1}{\sigma_0 + 3\sigma_1}, \quad \sigma = \frac{1}{4}(\sigma_0 + 3\sigma_1), \tag{5}$$

where σ_S are the ionization cross sections, dependent on the total spin S, and σ is the total spin-averaged ionisation cross section. They noted that the K asymmetries were quite different to the other two and speculated on the cause. Using the CCC method [16] we have calculated the asymmetries, and present the results so as to explain the origin of the unexpected behaviour.

In figure 1 we give the results for Li. We see good agreement between CCC and experiment for the spin asymmetries from near threshold to the higher energies. The general shape is typical of the asymmetries, at threshold the asymmetry is around 0.5 indicating that the singlet cross section is around five times larger than the triplet one, and then they gradually diminish to zero ($\sigma_0 = \sigma_1$) with increasing energy. Looking at the individual σ_S we see the underlying behaviour of the cross sections that define the spin asymmetries. The cross sections rise from zero at threshold to some maximum and then slowly diminish with increasing energy. The shape of the asymmetries is determined by the relative rise from the threshold and the position of the maxima of the σ_S.

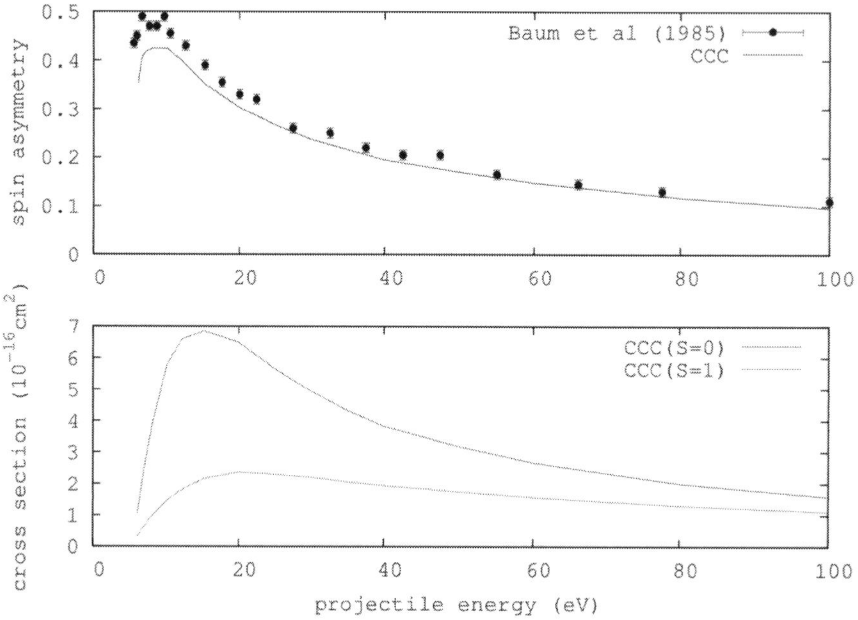

Figure 1. The e-Li total ionisation spin asymmetries and cross sections. The measurements are due to Baum *et al* [15]. The CCC calculations are due to Bray [17].

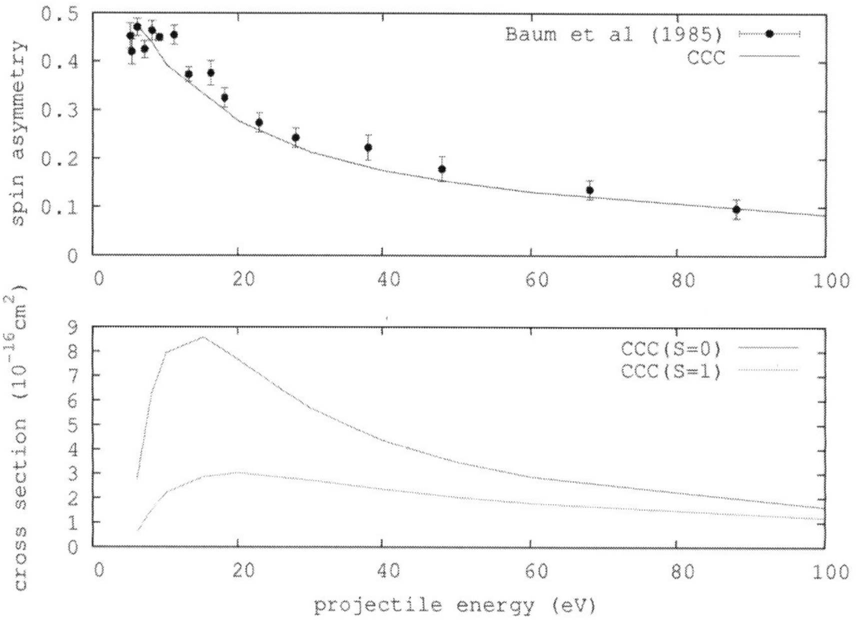

Figure 2. The e-Na total ionisation spin asymmetries and cross sections. The measurements are due to Baum *et al* [15]. The CCC calculations are due to Bray [12].

The Na case is considered in figure 2. This is quite similar to Li, for both the spin asymmetries and the magnitude of the cross sections σ_S.

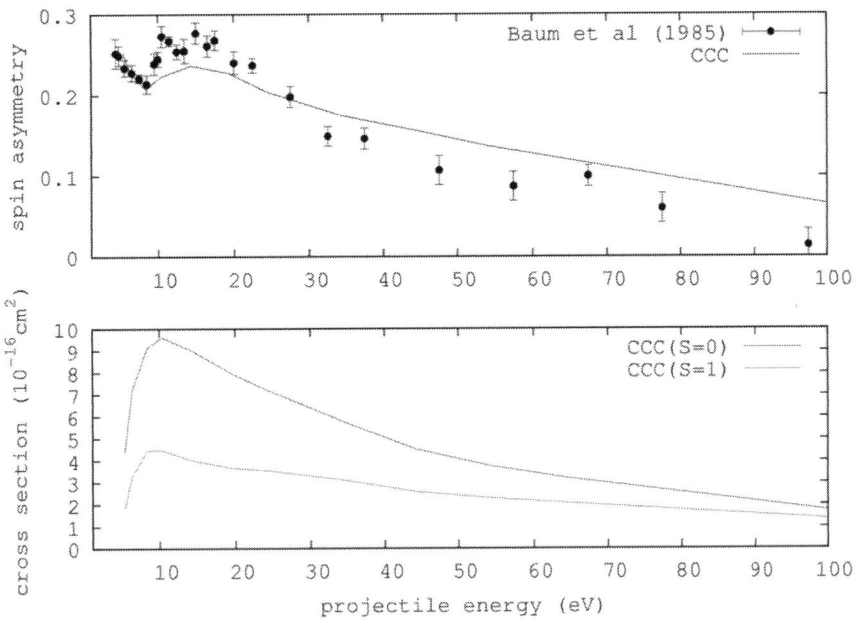

Figure 3. The e-K total ionisation spin asymmetries and cross sections. The measurements are due to Baum *et al* [15]. The present calculations are denoted by CCC.

However, the K case, given in figure 3, shows that the shape of the spin asymmetries has changed. Now there is a clear minimum around 9 eV, and yet there is little noticeable change in the qualitative behaviour in the corresponding σ_S. All that has happened is that the triplet cross section has increased a little in relative importance near threshold, and thereby reduced the spin asymmetry to around 0.25. The minimum observed is just due to the delicate interplay between the way the two cross sections rise, and is simply an indication of the sensitivity of ratio quantities such as spin asymmetries to the underlying absolute cross sections.

In figure 4 we consider the question of how do the spin asymmetries change if the target is a highly charged ion, specifically a Na-like ion Ar^{7+}. This question is interesting because we know that for highly charge ions the Coulomb-Born approximation yields reasonably accurate total ionization cross sections, but due to neglect of exchange yields identically zero spin asymmetries. We see that the CCC-calculated spin asymmetries have the same qualitative, and almost quantitative, behaviour as for Na, though now the ionisation threshold is much higher. Even though the Born approximation assumes that $\sigma_0 = \sigma_1$, which is clearly in substantial error over the entire energy range considered, the agreement for the total ionization cross section is remarkably good, particularly at energies above 400 eV. How can this be?

Agreement of Born plus exchange calculations with CCC suggests that coupling is negligible, and the cross sections can be obtained form the Born direct (D) and exchange (E) amplitudes. It then follows that these two amplitudes must satisfy $E^*E \approx \mathrm{Re}(D^*E)$ over the energy range where the Born approximation is sufficiently accurate. Why this should be the case is currently under investigation.

4. Single and double photoionisation of lithium

We report the extension of the CCC formalism for single photon double ionisation (DPI) of two-electron targets (He and its isoelectronic sequence of ions [18], valence and K-shell DPI of alkaline-earth atoms [19, 20]) to three-electron targets (Li presently). In the CCC calculations,

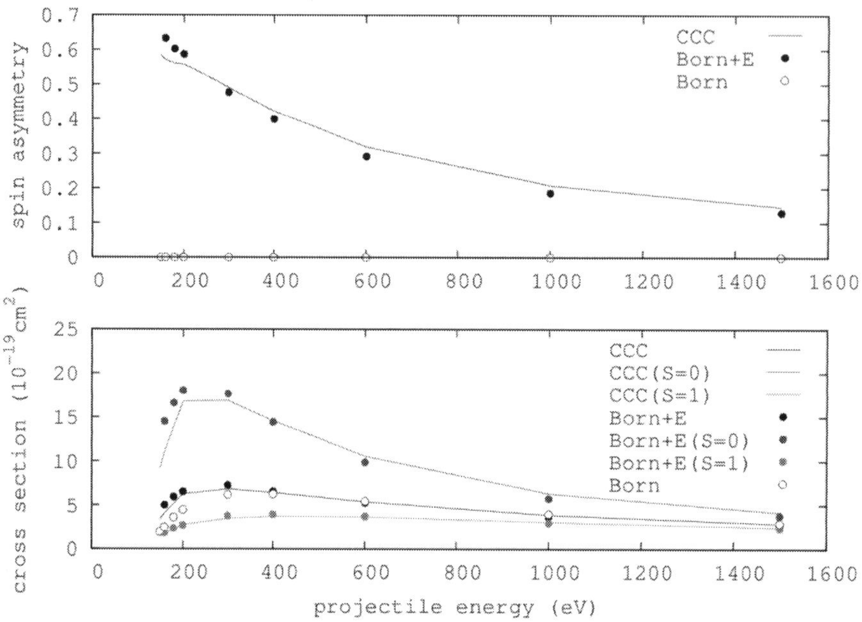

Figure 4. The e-Ar^{7+} total ionisation spin asymmetries and cross sections. The present convergent close-coupling calculations are denoted by CCC. The Coulomb-Born with Exchange approximation is labeled by Born+E. The pure Coulomb Born approximation is labeled by Born.

we treat DPI as a two step process. The first step is the full absorption of the photon energy by one electron. The second is electron impact ionisation of the singly charged ion which results in promotion of the remaining electron into the continuum. DPI calculations on two-electron targets requires CCC amplitudes of electron scattering on a singly charged hydrogenic target [16]. Similarly, DPI calculation on Li requires CCC amplitudes of electron scattering on a He-like Li^+ ion [6].

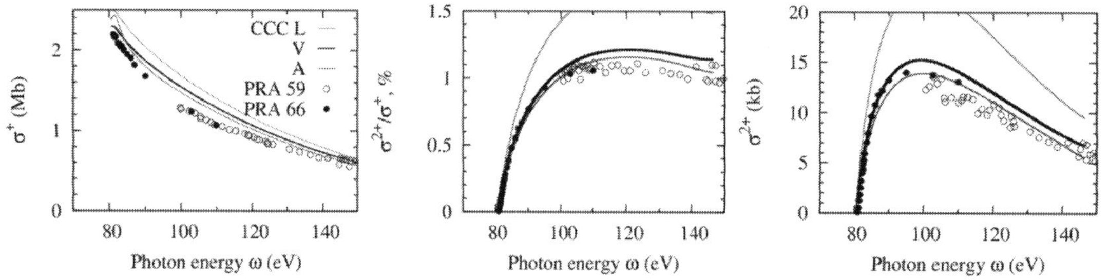

Figure 5. Photoionisation cross-sections of Li: single photoionisation σ^+ (left), double-to-single ratio σ^{2+}/σ^+ (center), DPI cross-section σ^{2+} (right). Experiments, labeled as PRA59 and PRA66, are from Refs. [21] and [22], respectively.

The total integrated cross-sections of single and double photoionisation of Li in the three gauges of the electromagnetic interaction (Length - L, Velocity- V and Acceleration - A) are presented in Figure 5 in comparison with experimental data. The L-gauge is unreliable for DPI calculations because of insufficient accuracy of the multi-configuration Hartree-Fock ground state. However, the two other gauges are in good agreement with experiment. This is the first successful attempt to match these experimental data with an *ab initio* non-perturbative calculation. Earlier time-dependent close-coupling calculations [23] were restricted to higher photon energy range above the second DPI threshold at 152 eV.

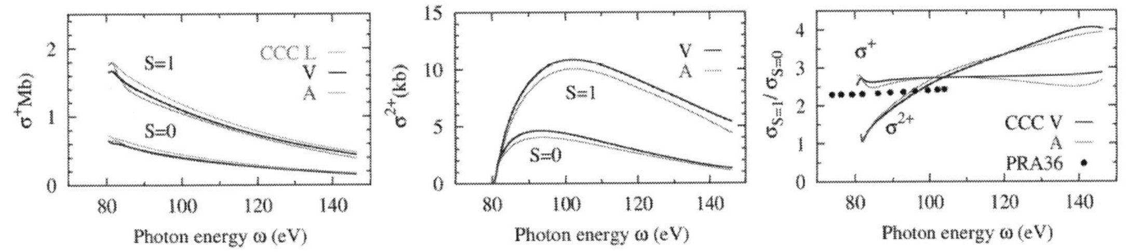

Figure 6. Spin-resolved photoionisation cross-sections of Li: single photoionisation σ^+ (left), DPI cross-section σ^{2+} (center), spin ratio $\sigma_{S=1}/\sigma_{S=0}$ (right). Earlier calculation of the spin ratio of single photoionisation reported in Ref. [24] is labeled as PRA36.

The first stage of DPI of Li can proceed via the singlet or triplet states $1s2s\,^{1,3}S$ of Li^+. It was argued in Ref. [22] that the single photoionisation of the triplet state of Li^+ should be the dominant channel. This conclusion was based on scaling of the DPI of Li versus DPI of He with the respective ionisation potentials of the singly charged ion (^1S or ^3S in the case of Li^+). Our spin-resolved results shown in Figure 6 demonstrate this propensity explicitly. Predominant population of the singly ionised triplet state, with the spin ratio $\sigma^+_{S=1}/\sigma^+_{S=0}$ close to 3, is translated into the dominant triplet DPI channel. The spin ratio $\sigma^{2+}_{S=1}/\sigma^{2+}_{S=0}$ shows more noticeable variation with photon energy, decreasing near the threshold.

5. 64.6 eV electron-impact ionisation of helium

The CCC method has been very successful in reproducing the angle-differential measurements of electron-impact ionisation of helium [6, 9]. Following the work of Schulz *et al* [25] there has been an increased focus on out-of-plane geometries [26, 27]. The corresponding cross sections are generally a good deal smaller than the in-plane ones and we would expect that these would be correspondingly more difficult to measure and calculate.

In figure 7 we present a CCC calculation of 64.6 eV electron-impact ionisation of helium in the coplanar symmetric and perpendicular-plane geometries corresponding to the experiment of Murray and Read [28]. The data are available for four energy sharings of the 40 eV excess energy. At the equal-energy-sharing there are also the in-plane data taken in Kaiserslautern and reported by Bray *et al* [29]. Looking at the left panels, corresponding to the in-plane geometries, we see that there is a systematic discrepancy of CCC with the data [28] in the angular range of 30° to 60°, where the cross sections are the largest. However, in the perpendicular plane case, agreement is excellent over the entire angular range, though the magnitude of the cross sections is much smaller. Note that the maximum at 90° in the perpendicular geometry corresponds to almost a minimum in the in-plane case. This common point was used to normalise experiment to theory, separately for each energy-sharing. We are at a loss to explain these discrepancies. We note the considerably better agreement with the Kaiserslautern in-plane data, available only for equal energy-sharing, but how it is possible at the same time to obtain such excellent agreement for the very small out-of-plane cross sections is not a question we can answer. Performing CCC calculations of various basis sizes shows greatest variations in the smaller cross sections, as might be expected.

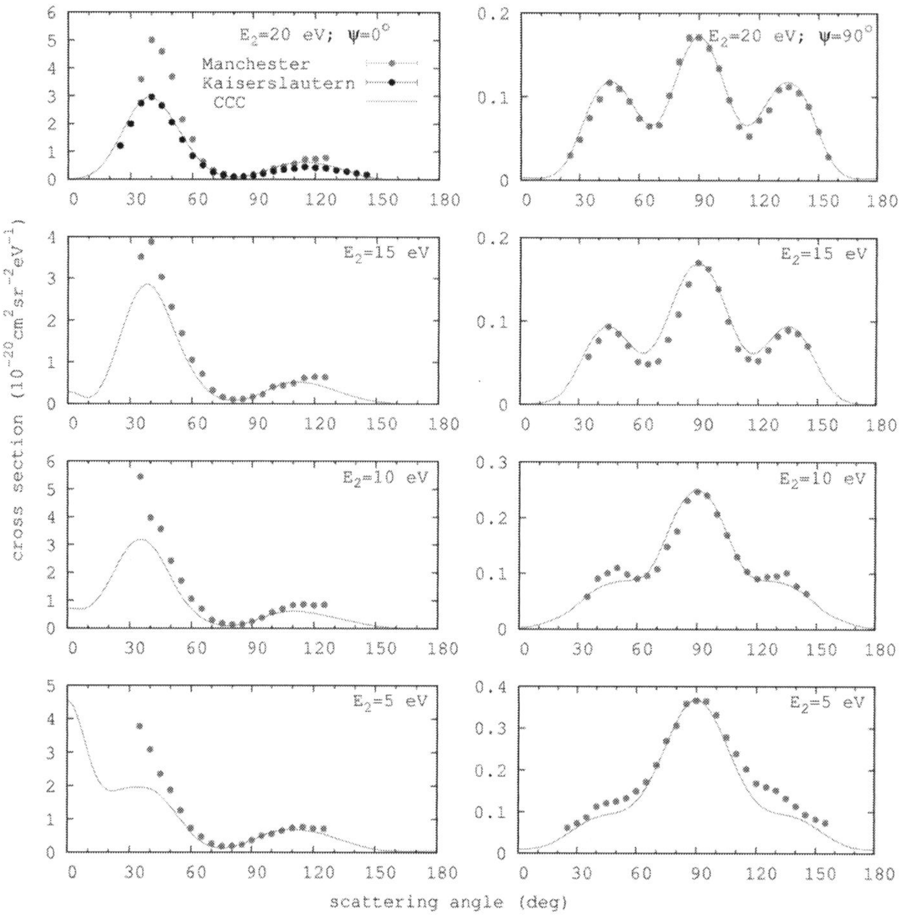

Figure 7. The 64.6 eV e-He fully differential ionisation cross sections for in-plane ($\psi = 0°$) and out-of-plane ($\psi = 90°$) incident electrons with outgoing electrons of energies E_2 and $40 - E_2$ measured in the symmetric ($\theta_1 = -\theta_2$) geometry. The present convergent close-coupling results are denoted by CCC. The Manchester and Kaiserslautern measurements are from refs.[28] and [29], respectively.

Acknowledgments

This work was supported by the Australian Research Council. We are grateful for access to the Australian National Computing Infrastructure Facility and its Western Australian node iVEC.

References

[1] Rescigno T N, Baertschy M, Isaacs W A and McCurdy C W 1999 *Science* **286** 2474–2479
[2] Baertschy M, Rescigno T N and McCurdy C W 2001 *Phys. Rev. A* **64** 022709
[3] Bartlett P L 2006 *J. Phys. B* **39** R379–R424
[4] Bartlett P L and Stelbovics A T 2004 *Phys. Rev. Lett.* **93** 233201
[5] Bray I and Stelbovics A T 1993 *Phys. Rev. Lett.* **70** 746–749
[6] Bray I and Fursa D V 1996 *Phys. Rev. A* **54** 2991–3004
[7] Badnell N R, Pindzola M S, Bray I and Griffin D C 1998 *J. Phys. B* **31** 911–924
[8] Bray I 2002 *Phys. Rev. Lett.* **89** 273201

[9] Stelbovics A T, Bray I, Fursa D V and Bartschat K 2005 *Phys. Rev. A* **71** 052716(13)

[10] Colgan J and Pindzola M S 2006 *Phys. Rev. A* **74** 012713 (pages 5)

[11] Colgan J, Foster M, Pindzola M S, Bray I, Stelbovics A T and Fursa D V 2009 *J. Phys. B* **42** 145002 (6pp)

[12] Bray I 1994 *Phys. Rev. Lett.* **73** 1088–1091

[13] Kadyrov A S, Bray I, Mukhamedzhanov A M and Stelbovics A T 2008 *Phys. Rev. Lett.* **101** 230405 (pages 4)

[14] Kadyrov A S, Bray I, Mukhamedzhanov A M and Stelbovics A T 2009 *Ann. Phys.* **324** 1516–1546

[15] Baum G, Roede M, Raith W and Schröder W 1985 *J. Phys. B* **18** 531–538

[16] Bray I 1994 *Phys. Rev. A* **49** 1066–1082

[17] Bray I 1995 *J. Phys. B* **28** L247–L254

[18] Kheifets A S and Bray I 1998 *Phys. Rev. A* **58** 4501–4511

[19] Kheifets A S and Bray I 2007 *Phys. Rev. A* **75** 042703 (pages 11)

[20] Kheifets A S, Bray I and Hoszowska J 2009 *Phys. Rev. A* **79** 042504 (pages 9)

[21] Huang M T, Wehlitz R, Azuma Y, Pibida L, Sellin I A, Cooper J W, Koide M, Ishijima H and Nagata T 1999 *Phys. Rev. A* **59** 3397–3401

[22] Wehlitz R, Bluett J B and Whitfield S B 2002 *Phys. Rev. A* **66** 012701

[23] Colgan J, Pindzola M S and Robicheaux F 2004 *Phys. Rev. Lett.* **93** 053201

[24] Ferrett T A, Lindle D W, Heimann P A, Brewer W D, Becker U, Kerkhoff H G and Shirley D A 1987 *Phys. Rev. A* **36** 3172–3183

[25] Schulz M, Moshammer R, Fischer D, Kollmus H, Madison D H, Jones S and Ullrich J 2003 *Nature* **422** 48

[26] Dürr M, Dimopoulou C, Dorn A, Najjari B, Bray I, Fursa D V, Chen Z, Madison D H, Bartschat K and Ullrich J 2006 *J. Phys. B* **39** 4097–4111

[27] Durr M, Dimopoulou C, Najjari B, Dorn A, Bartschat K, Bray I, Fursa D V, Chen Z, Madison D H and Ullrich J 2008 *Phys. Rev. A* **77** 032717 (pages 8)

[28] Murray A J and Read F H 1993 *J. Phys. B* **26** L359–L365

[29] Bray I, Fursa D V, Röder J and Ehrhardt H 1997 *J. Phys. B* **30** L101–L108

Journal of Physics: Conference Series **212** (2010) 012018 doi:10.1088/1742-6596/212/1/012018

Two-color photoionization of atoms in low and high radiation fields

M Meyer
LIXAM/CNRS, Centre Universitaire Paris-Sud, Bâtiment 350, F-91405 Orsay Cedex, France

E-mail: michael.meyer@u-psud.fr

Abstract. Atomic photoionization was studied by means of electron spectroscopy using two photon sources of different wavelengths for the excitation. In one part of the experiments, alkaline atoms were excited by an optical laser of high spectral resolution prior to their photoionization by short wavelength radiation from the synchrotron radiation source BESSY. These low field studies concentrated on measurements of the linear alignment dichroism in the angular distribution (LADAD) and of the linear magnetic dichroism (LMDAD) in the *2p* ionization of laser-polarized sodium atoms. In another type of experiments, the high intensity of a femtosecond optical laser was used for the investigation of dressed atoms, i.e. of the photoionization dynamics in strong external fields. In particular, the above threshold ionization (ATI) of rare gases was studied at the free electron laser facility FLASH in Hamburg, Germany. Measurements of the linear dichroism, here in the direct two-photon ionization of the *1s*-shell in He, were performed in order to determine the partial photoionization cross sections related to the emission of electrons with *s*- and *d*- symmetry, respectively.

1. Introduction

Two-color pump-probe experiments have demonstrated in the last decades their particular importance for the understanding of photoionization process and underlying dynamics [1, 2]. In many of these studies, an optical laser of high spectral resolution is used to change in a controlled way, via resonant excitation of an outer electron, the initial state for the photoionization, which is induced for example by synchrotron radiation (SR). The atoms are exposed to relatively low radiation fields created by the continuous-wave optical laser and the SR. Therefore, excitation and ionization processes can be treated independently. In addition to high-resolution spectroscopic studies (e.g. [3, 4]), special advantage is taken from the well-defined and easily changeable polarization of the laser. Due to the dipole selection rules for excitation with linearly and circularly polarized light, which induces only transitions of the type $\Delta m = 0$ and $\Delta m = \pm 1$, respectively, it is possible to prepare aligned or oriented atoms, i.e. to obtain an non-isotropic population of the sub-levels m_J in the electronic state with total angular momentum J. Subsequent resonant photoexcitation or direct photoionization with polarized SR has provided many valuable and very detailed information on the photoionization dynamics in polarized excited (e.g. [5, 6]) as well as ground state atoms polarized by optical pumping (e.g. [7, 8]). In general, the corresponding measurements of the linear (circular) dichroism provide ideal research grounds to access the dynamics of the photoionization process (e.g. [9] and references therein). As example for this type of studies, recent high-resolution experiments on laser-polarized sodium atoms and the corresponding theoretical treatment are described below.

In contrast to these studies, the interaction of an intense femtosecond optical laser with the atom enables the investigation of dressed atoms, i.e. of the photoionization dynamics in strong external fields [10-12]. When applying an optical field of 10^{10}-10^{13} W/cm^2 to the atom during the ionization by a XUV radiation pulse, the electron spectrum is strongly affected and the simultaneous action of both pulses has to be considered for the adequate description of the process (e.g. [13]). These experiments have attracted more and more interest in the last years due to the performances and availability of free electron laser (FEL) sources for the VUV and XUV wavelength regime, such as FLASH in Hamburg (Germany) [14]. In most cases, the experimental spectra of the two-color two-photon ionization process are well described by theoretical models based on solving the time-dependent Schrödinger equation or applying the so-called "soft-photon" approximation [15]. A typical experimental study undertaken at FLASH will be given for illustration.

2. Experimental

The experiments on laser-polarized atoms combining laser and SR were performed at the SR source BESSY II in Berlin (Germany). Details of the experiments are presented elsewhere [16]. In brief, the counter-propagating laser and synchrotron radiation intersect a beam of sodium vapor, which is produced by a radiatively heated oven, in the source volume of a high-resolution electron energy analyzer (Scienta SES-2002) mounted at the magic angle (54°44') with respect to the linear SR polarization vector. For the present study, electron spectra were recorded at a photon energy of hν(SR) = 54 eV with an total spectral resolution of 27 meV. For the excitation of the sodium atoms, two continuous-wave single-mode ring dye lasers (Rhodamine 6G) were used to excite efficiently both hyperfine components of the Na$3p^6 3s$ $^2S_{1/2}$ ground state when pumping the transitions to the Na*$3p^6 3p$ $^2P_{1/2}$ or the Na*$3p^6 3p$ $^2P_{3/2}$ excited state. Up to 40% of the sodium atoms could be brought to the excited state in this way. Under these conditions the atoms were exposed to a field of not more than about 10^4 to 10^5 W/cm^2 when the laser was operating at maximum average power of 1W and was focused to a diameter of about 50 μm in the interaction volume.

The experiments in the high field regime were performed using the two-color pump-probe set-up at FLASH [17]. Both the FEL and the optical laser beam were introduced into the experimental chamber in a collinear geometry and intersected an effusive gas jet within the acceptance volume of a magnetic bottle electron spectrometer (MBES). In order to ensure perfect time overlap between the pulses and to eliminate effects arising from the inherent time jitter of the FEL of about 500 fs [18], the experiments were performed using a picosecond optical laser synchronized to the 10-20 fs FEL pulses [14]. The relative orientation between the linear polarization vectors (purity better than 99%) was changed by means of a rotatable half-wave plate in the optical path. The FEL was operated in single-bunch mode at 13.7 nm (90.5 eV) with 5 Hz repetition rate and mean pulse energy of about 20 μJ. The optical field was produced by an intense (up to 2 mJ, 800 nm, 4 ps) Ti:Sapphire laser. The beam diameters of the FEL and the optical laser in the interaction volume were measured to be about 30 and 50 μm, respectively, which produces optical dressing fields of more than 10^{11} W/cm^2. The energy resolution in the electron spectra was mainly determined by the bandwidth of the FEL (0.5-1%) and the MBES (at best 1-2% of the electron kinetic energies) [17].

3. Results and discussion

3.1. Photoionization of laser-excited atoms

In many studies of laser-excited atoms sodium was chosen as primary example in order to highlight fundamental aspects of the photoionization process or to demonstrate the feasibility of a particular type of investigation [1]. Recently, it was therefore also used to investigate the effect of light polarization for the intensity distribution between the fine-structure states of the Na^+2p^5nl final ionic configuration [16, 19]. A series of high-resolution photoelectron spectra were recorded on laser-prepared sodium in the kinetic energy region of the $Na^+2p^5 3s$ $^{1,3}P_J$ and $Na^+2p^5 3p$ $^{1,3}S,P,D_J$ multiplets for ionization from the ground and the excited state, respectively. Photoionization from the $Na^*2p^6 3p$ $^2P_{3/2}$

excited state (Figure 1) results in a complex structure of 10 fine-structure components, which are, with the exception of the very close lying (1.5 meV) $^3P_0$ and $^3P_1$ lines, all resolved in the experiment. Assignment and energy positions of the states are given with respect to tabulated reference data [20]. The displayed spectrum was recorded with a relative angle $\theta = 0°$ between the linear polarization of the SR and the laser, i.e. parallel orientation of both vectors. The overall intensity distribution is in agreement with earlier experiments [3] showing in particular the strong intensity of the $^3D_2$ line upon photoionization from the $^2P_{3/2}$ excited state. This line is completely absent upon ionization from the $^2P_{1/2}$ state. In the lower part of the figure, the linear alignment dichroism in the angular distribution (LADAD) is displayed. The spectrum has been obtained by subtraction two spectra recorded at θ = 90° and θ = 0°. It is obvious that the individual lines behave quite differently, showing different amplitudes for the relative change of the intensity as well as different signs of the LADAD. For the interpretation of the results we have compared the experimental spectra with theoretical results obtained in the multi-configurational Dirac-Fock approximation as well as in the generalized geometrical model (GGM) [3, 19]. Different from the standard treatment of photoionization based on a geometrical model [21], the GGM takes into account intermediate coupling in the initial atomic and final ionic states, but neglects configuration mixing and energy dependence of the electron wavefunction within the energy range of the multiplet. The LADAD spectrum obtained from the GGM is included in the figure showing that almost all variations are perfectly reproduced. The main achievement of the GGM with respect to the geometrical model relies in the inclusion of intermediate coupling. When pure LSJ coupling is applied the agreement between and theory is less satisfactory [19], in particular in the energy region of the $^3P_{0,1,2}$ and $^1P_1$ lines, where the LSJ coupling gives very different signs and intensity ratios for the LADAD of the individual lines.

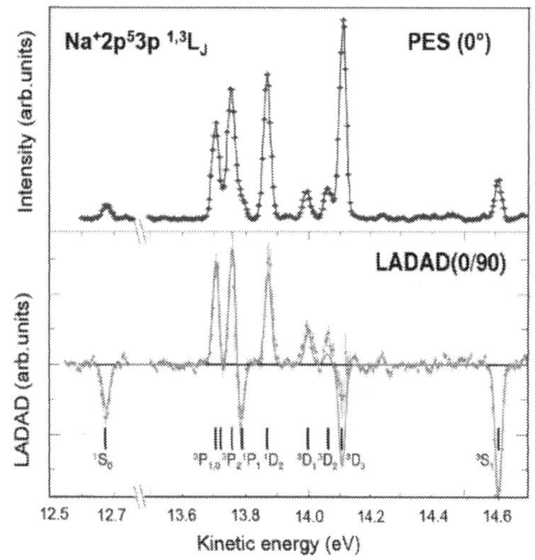

Figure 1. (Top) Photoelectron spectrum of sodium laser-aligned in the Na* $2p^63p$ $^2P_{3/2}$ excited state, recorded at hv(SR)=54eV and with parallel orientation between the linear polarization vectors of SR and laser. (Bottom) Experimental results (red dots) and theoretical simulation within the GGM (green solid line) for the LADAD.

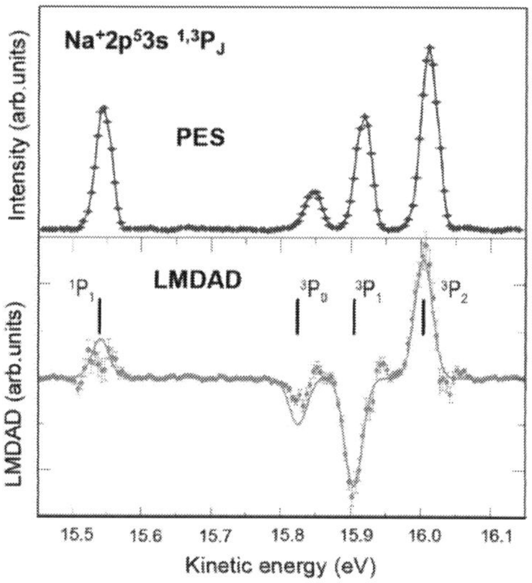

Figure 2. (Top) Photoelectron spectrum of sodium laser-oriented in the Na $2p^63s$ $^2S_{1/2}$ ground state, recorded using linearly polarized SR at hv(SR)=54eV and circularly polarized laser light. (Bottom) Experimental results (red dots) and theoretical simulation within the GGM (green solid line) for the LMDAD.

In a second part of the experiments, circularly polarized laser light was used enabling us to orient the excited state as well as the ground state. When probing the orientation of the Na $2p^63s$ $^2S_{1/2}$ ground with linearly polarized synchrotron radiation, the individual components of the Na^+2p^53s multiplet (Figure 2) show a strong variation with respect to the helicity of the circular laser polarization. The resulting spectra of the linear magnetic dichroism in the angular distribution (LMDAD), i.e. the difference between two spectra recorded with left-handed and right-handed circularly polarized laser, is given in the lower part. As for the LADAD, the experimental results are well described by the GGM. In this case, the LSJ coupling scheme fails also to describe the dichroism correctly. Especially for the $^1P_1$ line, the LSJ model does not produce any dichroism, although experiment and GGM show a clear effect. In the GGM, the dichroism of the $^1P_1$ line has its origin in the mixing with $^3P_1$.

In general, the dichroism measurements can serve to extract the complete information on the photoionization process, i.e. to determine the partial photoionozation amplitudes and the relative phases of the outgoing waves. In the present case of 2p photoionization of atomic Na, the relative importance of the emission of a continuum electron with s or d symmetry can be extracted. Theory predicts a value of $z = 3.4$ for the ratio between the d and s channel amplitudes and a phase difference of Δ between 1.2 and 1.4. The experimental results show the same tendencies, i.e. a ratio $z = 3.2 \pm 2.6$ and a positive sign of cos Δ, but the quite large error bar underlines the high sensitivity of the experiment to small fluctuations as well as the extremely high precision necessary in future studies in order to improve the present results.

3.2. Above Threshold Ionization of rare gas atoms

For the experiments in the high field regime, the two-color Above Threshold Ionization (ATI) of rare gases was investigated. ATI is the direct observation of a process where two photons interact simultaneously with the atom. When during the ionization process by the XUV photon an additional Visible/NIR field is present in the interaction volume, the outgoing electron is also exposed to this strong dressing field and can absorb or emit a second (or more) photon (Figure 3). In the photoelectron spectrum, this effect is observed through the appearance of so-called sidebands, which show up, when the radiation field is strong enough ($>10^{10}$ W/cm^2). Due to the high sensitivity to the strength of the applied dressing field, the method is ideally suited to measure precisely the temporal width of the XUV pulses and/or to determine the relative jitter between two femtosecond pulses [12, 18]. It was first observed and studied with high harmonic generation sources [10-12], but the monochromaticity of the Free Electron Laser sources such as FLASH enable us now to perform experiments unperturbed from interference effects between neighboring harmonics, which complicates often the detailed interpretation of the experiments at harmonic sources.

A series of typical spectra recorded on atomic Xe with the photon energy of the FEL at 90.5 eV is presented in Figure 4. The region of the 5p ionization (binding energy of 12.1 and 13.4 eV for the $^2P_{3/2}$ and $^2P_{1/2}$ spin-orbit component, respectively) is shown together with the sidebands belonging to an additional absorption of the optical photon. As a function of the relative orientation of the linear polarization vectors, indicated by the angle θ, the intensity redistribution between the main line and the sideband changes in a characteristic way. For parallel orientation ($\theta = 0°$), the intensity transfer from the main line to the sidebands is the strongest, for perpendicular orientation ($\theta = 90°$) the smallest. The observed variations are well described within the soft-photon approximation [15] and in principle it is possible to extract from the presented data the relative partial cross section for the two-photon ionization of atomic Xe. However, for the case of ionization in the 5p shell of Xenon the detailed analysis is rather difficult, since the one-photon process leads already to the emission of electrons with two different (s and d) symmetries (see Figure 3). The two-photon process is then described by outgoing εp- and εf-electrons, which are produced via different pathways and are therefore subject to interference effects, which are difficult to treat in detail. An additional complication of the analysis arises from the fact it is difficult to separate the contributions from each 5p component. The energy separation of the two 5p spin-orbit components (1.3 eV) is close to the distance between the main photoline and the sidebands (1.5 eV), given by the photon energy of the dressing laser (800 nm).

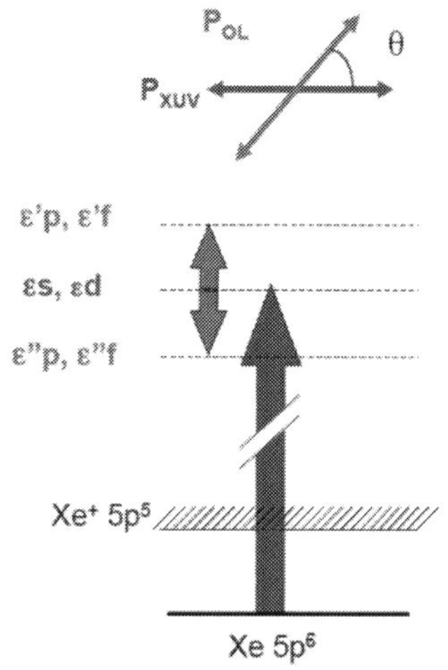

Figure 3. Scheme of the two-color above threshold ionization (ATI) in atomic Xe using an XUV and an intense NIR (800nm) pulse.

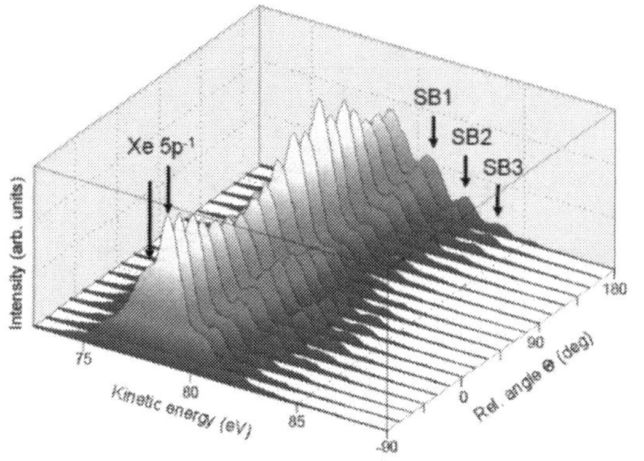

Figure 4. Photoelectron spectra of the 5p ionization in atomic Xe together with the high-energy sidebands for a dressing field of about 6 x10^{11} W/cm^2) upon photoionization at 90.5 eV. Spectra are presented for different relative orientations between the linear polarization vectors of the FEL and the 800nm laser.

For these reasons, the first measurements of polarization effects in the two-color ATI were performed for ionization of the 1s-electron in atomic He [22]. In these spectra, at (relatively) low dressing fields (about 8x10^{10} W/cm^2) only one sideband is seen, which directly reflects the two-photon ionization of the 1s-electron into the εs- and εd-continua. By measuring the polarization dependence of the sideband intensity, a slightly higher (factor of 1.5) emission probability for s than for d electrons was determined from the experimental results. This observation is in contrast to the general propensity rules predicting that, upon absorption of one photon, transitions of the type ($\ell \rightarrow \ell + 1$) are favored with respect to those described by a decrease of the angular momentum ($\ell \rightarrow \ell - 1$). Detailed calculation using the solution of the time-resolved Schrödinger equation are in good agreement with the experiment and explain the behavior by the overlap of the oscillatory asymptotes of the coupled continuum states in the particular case of He 1s two-photon ionization [22].

4. Conclusion

Recent results of two-color photoionization studies combining a XUV and a Visible/NIR photon sources are presented. For the experiments in the low field regime, the action of both photons can be treated separately and spectroscopic investigation with high spectral resolution are possible. The linear dichroism (LADAD and LMDAD) in the 2p photoionization of laser-polarized sodium atoms was chosen as example to illustrate the present state-of-the-art for this type of experiments, with respect to the experimental performances as well as to the theoretical models describing the dichroism. Very detailed information of the photoionization process can be obtained, up to the complete determination of all parameters describing the photoionization process ("complete experiment"). For the future the extension to more complex atomic systems such as the open-shell 3d transition metals is envisaged, necessitating the inclusion of configuration mixing in the theoretical models. For the high-field regime, the combined action of both photons has to be taken into account. A recent example for the study of two-color ATI in rare gases was discussed, where for the first time the dichroic effect in the direct two-photon ionization was investigated. With the advent of the XUV Free Electron Laser sources providing monochromatic, intense, ultra-short pulses for the photoionization process, these studies of non-linear effects will attract more and more interest. In combination with an intense femtosecond laser in the Visible/NIR wavelength regime, multi-photon processes as well as resonant excitation and corresponding electronic relaxation processes in strong dressing fields will be in the center of future investigations.

Acknowledgments
The presented results have been obtained within large collaborations between different experimental and theoretical groups. The author is enjoying these fruitful collaborations and would like to thank especially J. Costello, D. Cubaynes, S. Düsterer, S. Fritzsche, A. N. Grum-Grzhimailo, E. Heinecke, E. Kennedy, W. Li, A. Maquet, R. Taïeb, F. J. Wuilleumier and P. Zimmermann.

References

[1] Wuilleumier F J and Meyer M 2006 *J. Phys.* B **39** R425
[2] Meyer M 2009 *Nucl. Instrum. Meth.* A **601** 88
[3] Cubaynes D, Meyer M, Grum-Grzhimailo A N, Bizau J-M, Kennedy E T, Bozek J, Martins M, Canton S, Rude B, Berrah N and Wuilleumier F J 2004 *Phys. Rev. Lett.* **92** 233002
[4] Schulz J, Määttä M, Heinäsmäki S, Huttula M, Sankari R, Kukk E, Rander T, Svensson S, Aksela S and Aksela H 2006 *Phys. Rev.* A **73** 062721
[5] Meyer M, Müller B, Nunnemann A, Prescher Th, von Raven E, Richter M, Schmidt M, Sonntag B and Zimmermann P 1987 *Phys. Rev. Lett.* **59** 2963
[6] Wedowski W, Godehusen K, Weisbarth F, Zimmermann P, Martins M, Dohrmann T, von dem Borne A, Sonntag B and Grum-Grzhimailo A N 1997 *Phys. Rev.* A **55** 1922
[7] von dem Borne A, Dohrmann T, Verweyen A, Sonntag B, Godehusen K and Zimmermann P 1997 *Phys. Rev. Lett.* **78** 4019
[8] Godehusen K, Zimmermann P, Verweyen A, von dem Borne A, Wernet, Ph and Sonntag B 1998 *Phys. Rev.* A 58 R3371
[9] Grum-Grzhimailo A N and Meyer M 2009 *Eur. Phys. J. Special Topics* **169** 43
[10] Glover T E, Schoenlein R W, Chin A H and Shank C V 1996 *Phys. Rev. Lett.* **76**, 2468
[11] Schins J M, Breger P, Agostini P, Constantinescu R C, Muller H G, Grillon G, Antonetti A and Mysyrowics A 1995 *Phys. Rev.* A **52** 1272
[12] Toma E S, Muller H G, Paul P M, Breger P, Cheret M, Agostini P, Le Blanc C, Mullot G and Cheriaux G 2000 *Phys. Phys.* A **62** 0618015(R)
[13] Véniard V, Taïeb R and Maquet A 1995 *Phys. Rev. Lett.* **74** 4161
[14] Ackermann W *et al* 2007 Nat. Photon. **1** 336
[15] Maquet A and Taïeb R 2007 *J. Mod. Opt.* **54** 1847
[16] Cubaynes D, Guilbaud S, Wuilleumier F J, Meyer M, Heinecke E, Riek K, Zimmermann P, Yalçinkaya M, Fritzsche S, Strakhova S I and Grum-Grzhimailo A N 2009 *Phys. Rev.* A **80** 023410
[17] Radcliffe P, Düsterer S, Azima A, Li W B, Plönjes E, Redlin H, Feldhaus J, Nicolosi P, Poletto L, Dardis J, Gutierrez J P, Hough P, Kavanagh K D, Kennedy E T, Luna H, Yeates P, Costello J T, Delyseries A, Lewis C L S, Glijer D, Cubaynes D and Meyer M 2007 *Nucl. Instrum. Meth.* A **583** 516
[18] Radcliffe P *et al* 2007 *Appl. Phys. Lett.* **90** 131108
[19] Grum-Grzhimailo A N, Gryzlova E V, Cubaynes D, Heinecke E, Yalçinkaya M, Zimmermann P and Meyer M 2009 *J. Phys.* B **42** 171002
[20] Ralchenko Yu, Kramida A E, Reader J, and NIST ASD Team 2008 NIST Atomic Spectra Database, http://physics.nist.gov/asd3
[21] Starace A F 1982 in *Handbuch der Physik* vol 31, ed Mehlhorn W (Berlin: Springer) p 1
[22] Meyer M, Cubaynes D, Glijer D, Dardis J, Hayden P, Hough P, Richardson V, Kennedy E T, Costello J T, Radcliffe P, Düsterer S, Azima A, Li W B, Redlin H, Feldhaus J, Taïeb R, Maquet A, Grum-Grzhimailo A N, Gryzlova E V and Strakhova S I 2008 *Phys. Rev. Lett.* **101** 193002

International Symposium on (e,2e) and the 15th ISPCEAC IOP Publishing
Journal of Physics: Conference Series **212** (2010) 012019 doi:10.1088/1742-6596/212/1/012019

Tracing fingerprints of Young type interferences in angular distributions of ejected electrons from molecular targets

C. R. Stia[1], O. A. Fojón[1], S. Chatterjee[2], D. Misra[2], L. C. Tribedi[2] and R. D. Rivarola[1]

1 Instituto de Física Rosario, CONICET and Universidad Nacional de Rosario, Av. Pellegrini 250, 2000 Rosario, Argentina

2 Tata Institute of Fundamental Research, Homi Bhabha Road, Colaba, Mumbai 400 005, India

Corresponding author's e-mail address: rivarola@fceia.unr.edu.ar

Abstract. Interference patterns in angular distributions of ionized electrons due to coherent emission from the proximities of the nuclei of molecular hydrogen are investigated in (e,2e) collisions. It is shown that double and single differential cross sections show signatures of this effect.

1. Introduction

More than four decades ago, Samson and Cairns [1] obtained experimental photoabsortion total cross sections for N_2 and O_2 molecules, showing that some undulations appeared for photon energies in the 30-60 eV range. Studying the case of H_2^+ targets, Cohen and Fano [2] attributed the possible origin of this behavior, observed for larger molecules, to interference patterns due to coherent electron emission from the proximities of each one of the molecular centers. These studies were followed by other ones on photoionization not only of H_2^+ but also of H_2 molecules [3-4].

Hanssen *et al.* [5] (see also Joulakian *et al.* [6]) investigated theoretically the effect for impact of electron beams, associating the predicted interference patterns in the ionization spectra with the famous two-slit Young experiment [7-10], with the two molecular centers of a H_2^+ molecular ion playing the rôle of the slits. As the considered spectrum corresponds to electrons ejected from the target we have in fact only the second-half of the Young experiment and so the effect is known as the "Young *type* interferences" one.

The effect was measured only thirty five years later, for electron ionization of H_2 by 60 MeV/n Kr^{34+} ions in double differential cross sections (DDCS) as a function of the energy of the emitted electron at fixed values of the solid angle subtended by it [11]. In fact, to enhance the visibility of Young type structures experimental DDCS were divided by theoretical ones corresponding to effective hydrogen atoms. In further work the effect was shown to appear in the ratio of experimental DDCS of H_2 molecules and H atoms [12]. The experimental evidence of coherent electron emission from molecules spurred a lot of theoretical and experimental research with ion beams, but also with photons and electrons beams (for a short review see reference [13]). In particular, an intense activity related to this subject was recently developed for photoionization of H_2 and N_2 molecules [14-18].

We focus our present interest on the case of electron beam impact on H_2 targets. Electrons as projectiles are fundamentally different from equivelocity ions because they are negatively charged and light mass particles with larger de Broglie wave-lengths. Interference patterns were measured for the first time in DDCS for 2.4 keV $e^- + D_2$ collisions [19]. Further evidence of the existence of this effect was also given in triple differential cross sections (TDCS) for H_2 targets in coplanar asymmetric kinematics [20]. TDCS were presented as a function of the polar angle of the emitted electron at fixed emission energy and for a given projectile scattering angle in the forward direction. In fact, the ratio between experimental TDCS for hydrogen molecules [21] and theoretical TDCS for an *effective* atomic hydrogen showed an oscillatory behaviour associated with two-center coherent electron emission. Further works were also developed but investigating the comportment of the ratio of experimental TDCS for H_2 and He. In the first of them a supression of the recoil peak with respect to the binary encounter one was shown to appear as a consequence of partial destructive interference [22]. In a second one, for the same collision system and also within a coplanar asymetric geometry, it was shown that depending of the kinematical conditions describing the emitted and scattered electrons in the exit channel, constructive patterns can be also observed, as an increase of the recoil peak with respect to the binary one instead of the supression measured for destructive interference [23]. In all these investigations the theoretical analysis of the data to support the constructive and destructive interference hypothesis was based on an interference factor previously predicted by Stia *et al.* [24]. It should be also mentioned that in general the experimental TDCS were well described using distorted wave [24-25] and convergent close-coupling [26] models.

Very recently, the Young-type interference effect was measured in DDCS for impact of 8 keV-e^- on hydrogen molecules as a function of the momentum coordinates of the ejected electron [27-28]. In particular, ionized electron angular distributions were determined. Moreover, fingerprints of the effect were found also in angular- and energy-dependent single differential cross sections (SDCS). The scope of the present work is to give a theoretical analysis of the angular spectrum giving some physical explanations of the experimental data behaviour.

Atomic untis will be used in the following except where otherwise stated.

2. Theory

Let us study the ionization process produced by the impact of a fast electron beam on a hydrogen molecule,

$$e^- + H_2\left({}^1\Sigma_g^+\right) \rightarrow 2e^- + H_2^+\left({}^2\Sigma_g^+\right), \quad (1)$$

Highly asymmetric kinematics is considered, so that exchange effects can be neglected. Furthermore, the internuclear molecular axis is considered to remain as frozen during the collision, so that vibration and rotation of the target are not taken into account. This approximation, supported by the fact that the collision time is much shorter than the vibrational and rotational ones, has been previously studied in detail [29] and applied with success [30-32]. Moreover, the electronic reaction is assumed to be produced at the fixed equilibrium internuclear distance ρ_0 of the initial H_2 state.

The process is described from the center of mass of the target and the initial wavefunction is represented as

$$\psi_i\left(\vec{r}_1,\vec{r}_2,\vec{R}\right) = \frac{\exp\left(i\vec{k}_i\cdot\vec{R}\right)}{\left(2\pi\right)^{3/2}}\varphi_i\left(\rho_o,\vec{r}_1,\vec{r}_2\right) \quad (2)$$

where $\vec{k}_i$ is the reduced mass of the projectile electron and $\vec{R}$ its position, $\vec{r}_j$ (j=1,2) being the position of the jth-electron and φ_i the Heitler-London type wavefunction given by

$$\varphi_i\left(\rho_o,\vec{r}_1,\vec{r}_2\right) = N_i\left(\rho_0\right)\left(e^{-\alpha r_{1a}}e^{-\alpha r_{2b}} + e^{-\alpha r_{1b}}e^{-\alpha r_{2a}}\right), \quad (3)$$

with $N_i(\rho_0)$ a normalization factor, $\rho_0 = 1.406$, $\alpha = 1.166$ and, $\vec{r}_{ja}(\vec{r}_{jb})$ the position of the jth-electron with respect to the molecular nucleus a (b).

A two-effective center approximation is adopted to describe the final wavefunction [29]. In such approximation the single ionization process is assumed to be produced in the proximities of each one of the molecular centers, while the non-ionized electron completely screens the charge of the nucleus from which ionization is not produced. Without loss of generality, let us take the electron 1 as the one ionized. Thus, if it is ionized from the proximities of the center a (b), the electron 2 is considered to completely screen the center b (a). Thus, the final wavefunction in the exit channel is chosen as

$$\psi_f\left(\vec{r}_1,\vec{r}_2,\vec{R}\right) = \frac{\exp\left(i\vec{k}_s\cdot\vec{R}\right)}{(2\pi)^{3/2}}\phi_f\left(\rho_0,\vec{r}_2\right)C\left(\vec{k}_e,\vec{r}_{1i},\gamma\right);\ i = a,b \quad (4)$$

where $\vec{k}_s$ and $\vec{k}_e$ are the linear momenta of the scattered and ejected electrons respectively, $\gamma = -Z_T/k_e$ with Z_T the nuclear charge associated with each one of the molecular centers and φ_f the wavefunction corresponding to the fundamental bound state of the residual target given by

$$\phi_f\left(\rho_0,\vec{r}_2\right) = N_f\left(\rho_0\right)\left(e^{-\beta r_{2a}} + e^{-\beta r_{2b}}\right) \quad (5)$$

In Equation (5), $N_f(\rho_0)$ is a normalization factor, the parameter $\beta = 1.3918$ and $C\left(\vec{k}_e,\vec{r}_{1i},\gamma\right)$ is the Coulomb wavefunction describing the electron in a continuum state of the i-molecular center,

$$C\left(\vec{k}_e,\vec{r}_{1i},\gamma\right) = \frac{\exp\left(i\vec{k}_e\cdot\vec{r}_1\right)}{(2\pi)^{3/2}}\Gamma(1-i\gamma)\exp\left(\frac{\pi\gamma}{2}\right){}_1F_1\left[i\gamma,1,-i\left(k_e r_{1i} + \vec{k}_e\cdot\vec{r}_{1i}\right)\right];\ i = a,b. \quad (6)$$

Neglecting indirect contributions which consider that the electron is ionized from one of the molecular center due to the interaction of the projectile with the other center or with the passive electron (see reference [27]) and after some algebra, the four-fold differential cross section (FDCS) can be obtained,

$$\sigma^{(4)} = \frac{d^4\sigma}{d\Omega_\rho d\Omega_e d\Omega_s d\left(k_e^2/2\right)} \cong 2\left[1+\cos\left(\vec{\chi}\cdot\vec{\rho}_0\right)\right]\sigma_{eff}^{(3)}, \quad (7)$$

with Ω_ρ, Ω_e and Ω_s the solid angles subtended by the molecular orientation and by the ejected and scattered electrons, respectively and, $\vec{\chi} = \vec{k}_e - \vec{K}$, where $\vec{K} = \vec{k}_i - \vec{k}_s$ is the momentum transferred to the target and $\sigma_{eff}^{(3)}$ represents an effective one-center triple differential cross section. Moreover, a TDCS can be calculated after integration on Ω_ρ, so that

$$\sigma^{(3)} = \frac{d^3\sigma}{d\Omega_e d\Omega_s d\left(k_e^2/2\right)} \cong 2I\sigma_{at}^{eff} = 2\left[1+\frac{\sin\left(\chi\rho_0\right)}{\chi\rho_0}\right]\sigma_{eff}^{(3)} \quad (8)$$

In Equation (8), I is known as the *interference factor*. Further integration on Ω_s allows us to obtain DDCS under the expression

$$\sigma^{(2)} = \frac{d^2\sigma}{d\Omega_e d\left(k_e^2/2\right)} \cong 2\int d\Omega_s\left\{\left[1+\frac{\sin\left(\chi\rho_0\right)}{\chi\rho_0}\right]\sigma_{eff}^{(3)}\right\} \quad (9)$$

Moreover, SDCS, $\sigma^{(1)} = d\sigma/d\Omega_e$, can be calculated integrating $\sigma^{(2)}$ on the energy of the ejected electron.

3. Results and discussions

According to expression (8), constructive interferences will be found in TDCS for $\chi \cong 0$ values. It corresponds to soft collisions (for which $K \cong 0$ and $k_e \cong 0$) or to electrons ejected in the Bethe region ($\vec{k}_e \cong \vec{K}$).

In the left side of Fig. 1, absolute experimental [26]and theoretical DDCS in a 8 keV $e^- + H_2$ collision are presented as a function of the polar emission angle at fixed emission energies of 9, 70 and 280 eV. Effective one-center double differential $\sigma_{eff}^{(2)}$ calculations, obtained after integration of $\sigma_{eff}^{(3)}$ on Ω_s are included. Theoretical molecular results are in good agreement with measured data, particularly for low-energy ejected electrons (soft collisions) as well as for binary electrons (which correspond to the peaks in DDCS). In the right side of Fig. 1, the corresponding DDCS ratios obtained dividing experimental and theoretical DDCS by the effective one-center calculations are presented. Partial constructive interferences due to coherent emission from the molecular centers are evidenced by experiments and supported by theoretical results so well for soft collisions as for binary electrons, for which the ratios take values larger than 1.5 and smaller than 2.

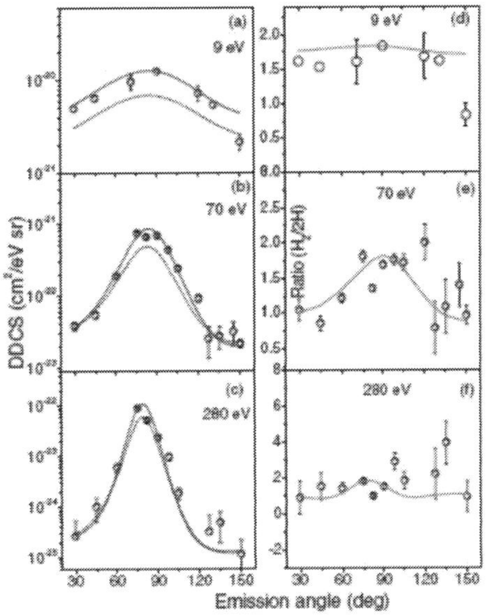

FIGURE 1. (a), (b) and (c) show angular distributions of secondary electrons. Open circles, experimental data from [27]; solid lines, theoretical molecular cross sections; dashed lines, twice the calculations for effective one-center DDCS. (d), (e) and (f) show the corresponding ratios (for details see the text).

As the result of integration of DDCS on emitted electron energies ranging from 9 eV to 400 eV, experimental [28] and theoretical single differential cross sections (SDCS) are determined and presented in Fig. 2 together with the corresponding SDCS ratios. Again experimental and theoretical ratios are obtained by dividing the cross sections with effective one-center SDCS predictions. A reasonable agreement between theory and experiment is found for SDCS as well as for SDCR ratios.

It is very interesting to observe that fingerprints of Young-type interferences are preserved in SDCS. This behaviour can be explained by the fact that the contributions of low energy and binary encounter electrons dominate SDCS.

Finally, we must mention that for (e,2e) collisions an oscillatory behaviour in the angular asymmetry parameter

$$\alpha(k_e,\theta_e)=\frac{\sigma^{(2)}(k_e,\theta_e)-\sigma^{(2)}(k_e,\pi-\theta_e)}{\sigma^{(2)}(k_e,\theta_e)+\sigma^{(2)}(k_e,\pi-\theta_e)} \quad (10)$$

associated with coherent emission from the centers of the target has been obtained, as it was previously observed for ion impact [33]. In Equation (10), θ_e is the polar angle of the ejected electron. A comparison of the asymmetry parameter $\alpha(k_e,\theta_e)$ for electron and ion impact is given in reference [27].

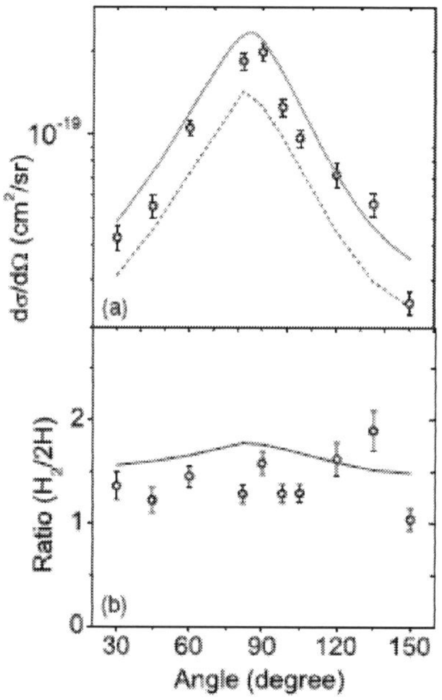

FIGURE 2. (a) Angular distribution of SCDS. Open circles, experimental data from [28]; solid line, theoretical molecular SDCS; dashed line, twice the calculations for effective one-center SDCS. In (b) the corresponding ratios are shown (for details see the text).

4. Conclusions

Young type interference effects are studied in (e,2e) collisions for impact of fast electron beams on H_2 targets. Angular distributions of ejected electrons both in double and single differential cross sections present evidence of partial constructive interferences for low-energy emitted electrons and for binary encounter ones, due to coherent emission from the proximities of the centers of the molecular target. The present theoretical description takes account of recent experimental data.

5. Acknowledgments

CRS, OAF and RDR acknowledge support from the Agencia Nacional de Promoción Científica y Tecnológica and from the Consejo Nacional de Investigaciones Científicas y Técnicas, both from the República Argentina.

References

[1] Samson J A R and Cairns R B 1965 *J. Opt. Soc. Am.* **55** 1035.
[2] Cohen H D and Fano U 1966 *Phys. Rev.* **150** 30.
[3] Kaplan I G and Markin A P 1969 *Sov. Phys. Dokl.* **14** 36.
[4] Walter M and Briggs J 1999 *J. Phys. B: Atom. Opt. Mol. Phys.* **32** 2487.
[5] Hanssen J, Joulakian B, Rivarola R D and Allan R J 1996 *Phys. Scr.* **53** 41.
[6] Joulakian B, Hanssen J, Rivarola R D and Motasim A 1996 *Phys. Rev. A* **54** 1473.
[7] Young T 1803 *Lecture at the Royal Society of London.*
[8] Young T 1804 *Phil. Trans. R. Soc.* **94** 1.
[9] Young T 1807 *A Course of Lectures on Natural Philosophy and the Mechanical Arts* (London: J. Johnson).
[10] Jönsson C 1961 *Z. Phys.* **161** 454.
[11] Stolterfoht N *et al.* 2001 *Phys. Rev. Lett.* **87** 023201.
[12] Misra D, Kadhane U, Singh Y P, Tribedi L C, Fainstein P D and Richard P 2004 *Phys. Rev. Lett.* **92** 153201.
[13] Rivarola R D 2007 *Nucl. Instrum. Methods Phys. Res.* **B 261** 161.
[14] Rolles D *et al.* 2005 *Science* **437** 711.
[15] Liu X-J *et al.* 2006 *J. Phys. B: Atom. Mol. Opt. Phys.* **39** 4801.
[16] Akouri D *et al.* 2007 *Science* **318** 949.
[17] Zimmermann B *et al.* 2008 *Nature* **4** 649.
[18] Schöffler M S *et al.* 2008 *Science* **320** 920.
[19] Kamalou O, Chesnel J-Y, Martina D, Frémont F, Hanssen J, Stia C R, Fojón O A and Rivarola R D 2005 *Phys. Rev. A* **71** 010702.
[20] Fojón O A, Stia C R and Rivarola R D 2006 *AIP Conference Proceedings* **811.**
[21] Chérid M *et al.* 1989 *J. Phys. B: Atom.. Mol. Opt. Phys.* **36** L257.
[22] Milne-Brownlie D S, Foster M, Gao J, Lohmann B and Madison D H 2006 *Phys. Rev. Lett.* **96** 233201.
[23] Staicu-Casagrande E M *et al.* 2008 *J. Phys. B: Atom. Opt. Mol. Phys.* **41** 025204.
[24] Stia C R, Fojón O A, Weck P F, Hanssen J and Rivarola R D 2003 *J. Phys. B: Atom. Mol. Opt. Phys.* **36** L257.
[25] Prideaux A and Madison D H 2003 *Phys. Rev. A* **67** 052710.
[26] Bray I, Fursa D V, Kheifets A and Stelbovics A T 2002 *J. Phys. B: Atom. Opt. Mol. Phys.* **35** R117.
[27] Chatterjee S, Misra D, Kelkar A H, Tribedi L C, Stia C R, Fojón O A and Rivarola R D 2008 *Phys. Rev. A* **78** 052701.
[28] Chatterjee S, Kasthurirangan S, Kelkar A H, Stia C R, Fojón O A, Rivarola R D and Tribedi L C 2009 *J. Phys. B: Atom. Opt. Mol. Phys.* **42** 065201.
[29] Weck P, Joulakian B, Hanssen J, Joulakian B, Fojón O A and Rivarola R D 2000 *Phys. Rev. A* **62** 014701.
[30] Weck P, Fojón O A, Hanssen J, Joulakian B and Rivarola R D 2001 *Phys. Rev. A* **63** 042709.
[31] Weck P F, Fojón O A, Joulakian B, Stia C R, Hanssen J and Rivarola R D 2002 *Phys. Rev. A* **66** 012711.
[32] Stia C R, Fojón O A, Weck P F, Hanssen J, Joulakian B and Rivarola R D 2003 *Phys. Rev. A* **66** 052709.
[33] Misra D, Kelkar A, Kadhane U, Kumar A, Tribedi L C and Fainstein P D 2006 *Phys. Rev. A* **74** 060701 (R).

Journal of Physics: Conference Series **212** (2010) 012020 doi:10.1088/1742-6596/212/1/012020

Probing Electron Correlation and Nuclear Dynamics in Momentum Space

M. S. Deleuze[1], **B. Hajgató, F. Morini, S. Knippenberg**[2]
Research Group of Theoretical Chemistry, Department SBG, Hasselt University, Agoralaan, Gebouw D, B3590 Diepenbeek (Belgium)

E-mail: michael.deleuze@uhasselt.be

Abstract. Orbital imaging experiments employing Electron Momentum Spectroscopy are subject to many complications, such as distorted wave effects, conformational mobility in the electronic ground state, ultra-fast nuclear dynamics in the final state, or a dispersion of the ionization intensity over electronically excited (shake-up) configurations of the cation. The purpose of the present contribution is to illustrate how a proper treatment of these complications enables us to probe in momentum space the consequences of electron correlation and nuclear dynamics in neutral and cationic states.

1. Introduction

An everlasting dilemma in quantum mechanics pertains to the physical relevance of orbitals, or orbital densities. Assuming that the Born (sudden), binary encounter, and plane wave impulse approximations are valid, spherically averaged Dyson orbital densities relating to transition amplitudes between neutral and ionized states (eq. 1) can be experimentally reconstructed in momentum space by means of Electron Momentum Spectroscopy (EMS) [1], from the angular dependence (eq. 2) of cross sections in (e, 2e) electron-impact ionization experiments at high kinetic energies (Figure 1). In most applications, (normalized) Dyson orbitals are replaced by standard Kohn-Sham orbitals. However, as with any ionization experiment, analyses of EMS measurements are subject to many complications, such as distorted wave effects, conformational mobility in the neutral ground state along with the usually strong influence of the molecular conformation on electron binding energies, electron correlation and relaxation, dispersion of the ionization intensity over electronic (shake-up) excited states of the cation, and ultra-fast nuclear dynamics in the final ionized state. The main conclusions of two "school case" studies are summarized in the present contribution, in order to illustrate the assertion that, if it has to have any value at all, the interpretation of EMS experiments requires extensive theoretical work.

$$\sigma_n(p) \propto \int d\Omega \left|g_n(\vec{p})\right|^2 = \int d\Omega \left|\left\langle e^{i\vec{p}\bullet\vec{r}}\,\Psi_n^{N-1} \middle| \Psi_0^N \right\rangle\right|^2 \quad (1)$$

$$p = \sqrt{(2p_1\cos\theta - p_0)^2 + (2p_1\sin\theta\sin(\phi/2))^2} \quad (2)$$

[1] To whom any correspondence should be addressed.

[2] Present address: Institut für Physikalische und Theoretische Chemie, Johann Wolfgang Goethe Universität Frankfurt, Max-von-Laue-Str. 7, 60438 Frankfurt am Main, Germany.

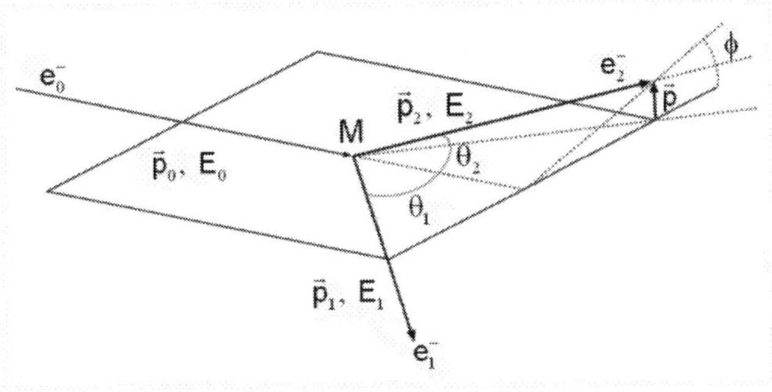

Figure 1. Kinematic set-up for an EMS experiment

We first consider a most puzzling discrepancy [2, 3] between the results of a model assuming thermal equilibrium between the *gauche* and *trans* conformers (relating to C_1 and C_s energy minima) and the experimentally inferred momentum distribution (MD) for the highest occupied molecular orbital (HOMO) of ethanol (Figure 2). This discrepancy cannot be due [4] to (1) a breakdown of the Plane Wave Impulse Approximation (the experimentally inferred MD does not depend much upon E_0, the kinetic energy of the impinging electron); (2) a defect in the treatment of electronic correlation and relaxation (comparison with benchmark ADC(3) Dyson orbitals shows that the target Kohn-Sham approximation is valid); (3) a breakdown of the one-electron picture of ionization (the corresponding ADC(3) pole strength exceeds 0.90); (4) some energy overlap with higher lying bands (the energy interval with the next band exceeds 1.6 eV); (5) tunneling effects across a shallow but wide hydroxyl torsion barrier of ~1.15 kcal/mol (the rate of which is estimated to be of one transition per 333 *ps*, compared with a thermal conversion rate around 5.6 ps^{-1} at 298K); (6) a dimerization of the molecule under high vacuum conditions (a process which should affect all momentum distributions, at odds with the experimental evidence). In view of the flatness of the conformational energy map characterizing ethanol, one of the most tempting explanations resides in thermally induced molecular motions and conformational disorder in the ground state [2, 3]. As shall be shown, molecular relaxation processes induced by ionization have also to be accounted for [4], in order to reconcile theory with experiment.

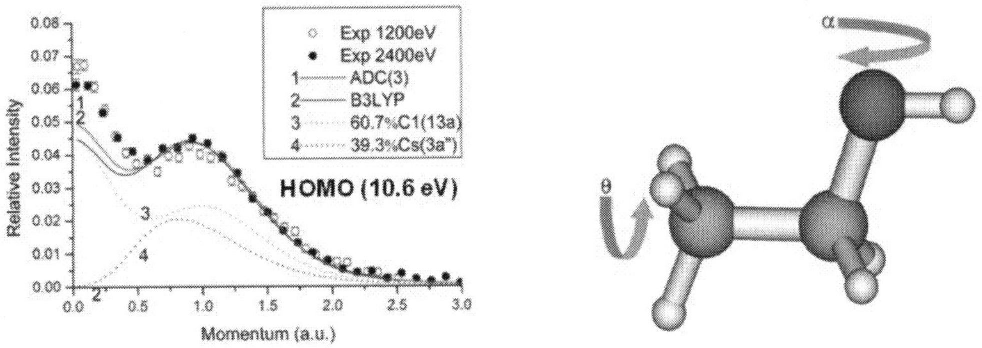

Figure 2. Theoretical versus experimental (e, 2e) outermost momentum distributions of ethanol (left) and conformational degrees of freedom of the molecular target (right).

Our second scope is to refute conclusions drawn by Saha, Wang, *et al* in [5] regarding fingerprints of the *gauche* (C_2) conformational isomer of 1,3-butadiene (CH_2=CH-CH=CH_2) in electron momentum distributions that were experimentally inferred from gas phase (e, 2e) measurements on this compound [6]. Because of enthalpy and entropy difference of 2.93 kcal mol^{-1} and 3.98 cal mol^{-1} K^{-1} in favour of the *trans*-conformer, of C_{2h} symmetry, this conclusion implies that, if we assume that EMS experiments remain subject to the laws of statistical thermodynamics, and that, *prior* to ionization, the molecule resides in its electronic ground state, some of the EMS experiments which have been analyzed by Saha, Wang *et al* should have been conducted at temperatures far above 1000K. This most astonishing conclusion was the result [7] of an erroneous assignment of spectral bands due to the neglect of shake-up satellites as well as of the strong influence of the molecular conformation on the ionization spectrum.

2. Ethanol

2.1. Methodology

Spherically averaged momentum distributions (SAMD's) have been computed for the Highest Occupied Molecular Orbital (HOMO) of ethanol, according to the target Kohn-Sham (B3LYP/aug-cc-pVTZ) approximation. Resolution folding was accounted for, according to the characteristics of the employed (e, 2e) spectrometer (Δp~0.16 a.u. FWHM) at Tsinghua University, Beijing [2]. Three models [3, 4] have been considered for evaluating the outcome of nuclear dynamics in the ground state, on the basis of a thermal averaging of internal molecular motions and of their influence on the SAMD's: (1) statistical thermodynamics (beyond the RRHO level), assuming thermal equilibrium between the C_1 and C_s conformers; (2) thermal averaging over a "one-dimensional" (1D) subset of 72 nuclear configurations characterized by α=-180°, -175°, ... +170°, +175°}, and (3) thermal averaging over a "two-dimensional" (2D) subset of 1728 nuclear configurations characterized by α=-180°, -175°, ... +170°, +175°, along with θ=-60°, -55°, ... +50°, +55°}. Assuming that these configurations are independent enables us to correspondingly calculate their weight at T=298K according to:

$$w_i(\alpha)=\frac{e^{-E(\alpha_i)/k_B T}}{\sum_{i=1}^{72} e^{-E(\alpha_i)/k_B T}} \quad (3) \qquad \text{or } w_i(\alpha,\theta)=\frac{e^{-E(\alpha_i,\theta_i)/k_B T}}{\sum_{i=1}^{1728} e^{-E(\alpha_i,\theta_i)/k_B T}} \quad (4)$$

The influence of nuclear dynamics (i.e. structural relaxation effects) in the final ionized state has been modelled using the HOMO of the neutral upon the B3LYP/aug-cc-pVTZ geometry of the radical cation. Furthermore, the evolution of the system immediately after ionization has been studied according to Born-Oppenheimer Molecular Dynamics (BOMD) simulations, at the B3LYP/aug-cc-pVDZ level, using trajectory and integration step sizes of 0.250 a.u. and 0.2 *fs*, the Bulirsch-Stoer integration scheme along with an integration correction scheme using fifth-order polynomial fit, and assuming a temperature of 0K, that is, the only source of kinetic energy immediately after ionization is assumed to exclusively arise from potential energy gradients. Changes in chemical connectedness have been correspondingly studied using Natural Bond Orbitals.

3.2. Results and Discussion

The flatness of the conformational energy map of ethanol is found to result into significant broadening of conformer populations (Figure 3a). We note that the HOMO of ethanol pertains to a localized oxygen lone pair. Therefore (Figure 3b), significant changes in the outermost momentum distributions reflecting the development of anomeric interactions occur upon rotations of the hydroxyl group (α torsion angle), whereas only marginal changes are induced by methyl rotations (θ). Thermal fluctuations in the electronic ground state alone do not improve agreement between theory and experiment (Figure 4), but are found to significantly populate the nuclear configurations around α=±90°. It is worth noticing therefore that structural relaxation in the adiabatic cationic state increases by a factor 4 (!) the outermost electron momentum density at $p \rightarrow 0$, due to a rotation of the OH bond to α~90°, along with an increase of the C-C bond length from ~1.50 to 1.76 Å, and an enhancement thereby of anomeric $[n_O - \sigma^*_{C\text{-}C}]$ interactions (Figure 4). Lacking any other satisfactory explanation, one has to wonder therefore whether the experimental turn-up of (e, 2e) intensities at $p \rightarrow 0$ reflects nuclear dynamics induced by the ionization process (i.e. nuclear dynamics in the final state), and whether these structural changes occur fast enough to be detectable in electron impact ionization experiments at high kinetic energies [E_0=1.2 keV].

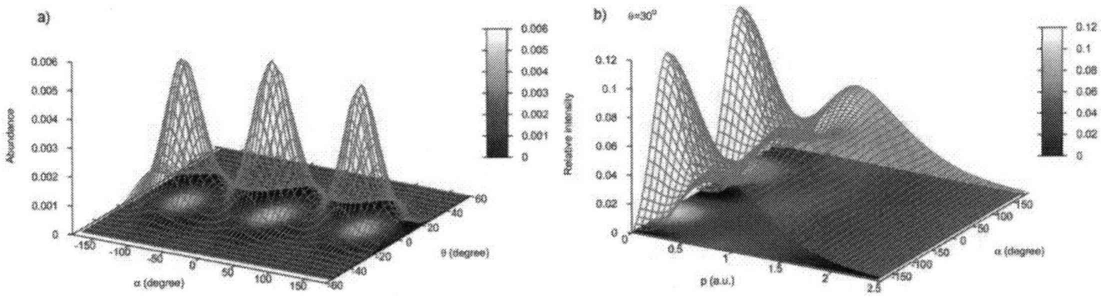

Figure 3. Conformer distributions (a) and influence of hydroxyl (α) rotations upon the outermost electron momentum distribution (b) of ethanol.

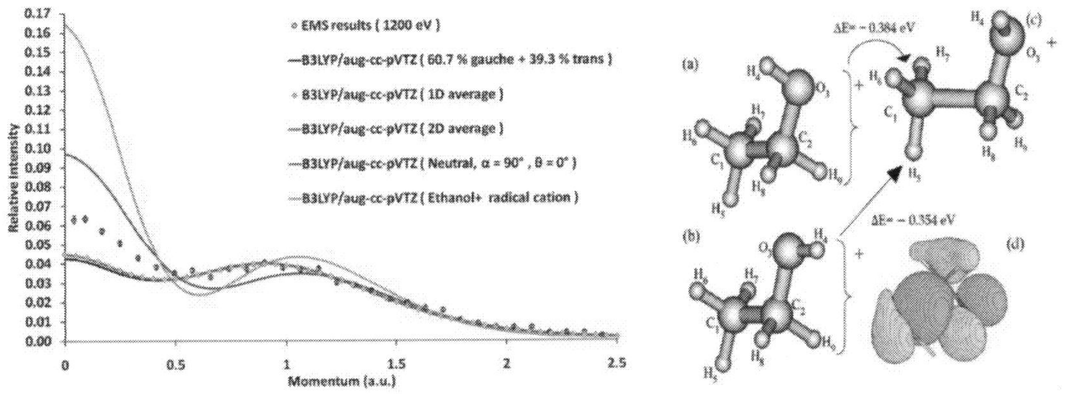

Figure 4. Theoretical versus experimental (e, 2e) outermost momentum distributions of ethanol (left) and structural changes induced by an adiabatic ionization process (right).

Upon analyzing the results of BOMD calculations for the radical cation (Figure 5), it appears that significant nuclear dynamics is impeded by C_s symmetry when α=0° (*trans* conformation), whereas hydroxyl rotations are a prerequisite for ultra-fast nuclear dynamics in the final ionized state, in the form of a stretching of the C-C bond by ~0.25 Å and ~0.60 Å upon time intervals of 13 *fs* and 40 *fs*, respectively, when $\alpha(t=0)$ is around ±90°. Therefore, *incomplete* relaxation processes ought to be sufficient for unravelling the discrepancy between the theoretical and experimental momentum distributions characterizing the outermost orbital of ethanol. In particular, according to the computed BOMD trajectories and detailed B3LYP/aug-cc-pVTZ calculations of the outermost momentum distributions of the ethanol radical cation under the constraints of fixed torsion angles (α=90° and θ=0°), the missing fraction (40%, see Figure 4) of relative ionization intensity at $p \rightarrow 0$ could be recovered upon an inferred time scale of the order of ~8 *fs* ***only*** for the (e, 2e) ionization process, considering the dependence (eq. 5) of the momentum density for the HOMO at $p \rightarrow 0$ as a function of the length of the C-C bond (x):

$$\sigma_{HOMO}(p \rightarrow 0) = 4.9143x^4 - 33.013x^3 + 82.279x^2 - 89.93x + 36.399 (R^2 = 0.994) \qquad (5)$$

Immediately after ionization, strong oscillations by ~120° of the hydroxyl torsion angle are also noted when ~60° < $\alpha(t{=}0)$ < ~120°, along with rapid oscillations in the length of the C-O bond, by 0.12 Å, with periods around 40 *fs*. These observations are consistent with a breaking of the C-C bond and with the formation of a double bond between the C and O atoms.

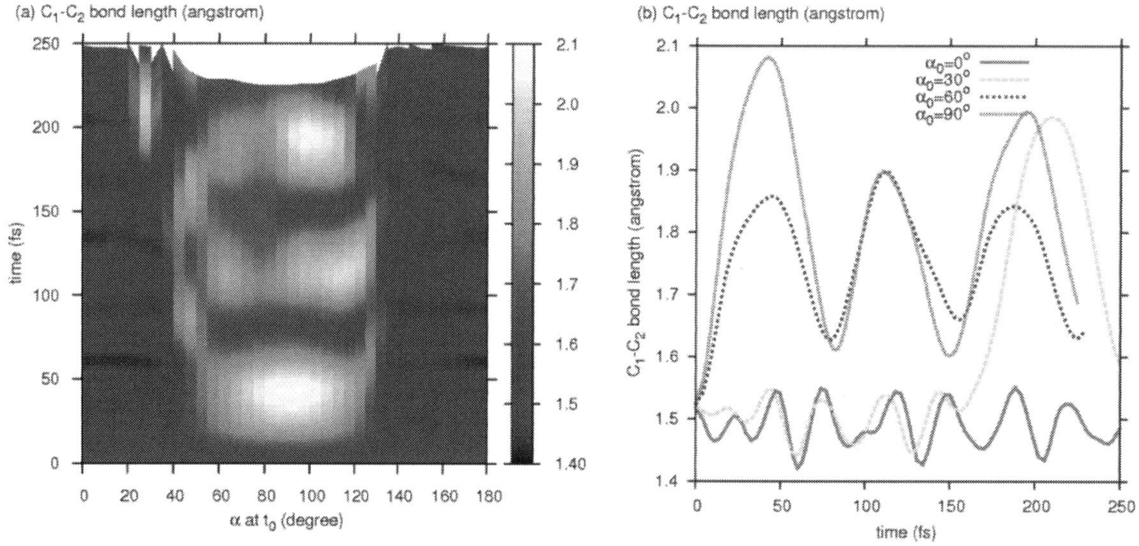

Figure 5. Evolution of the length of the C-C bond immediately after ionization, according to an analysis of BOMD trajectories for various initial configurations.

Table 1. Evolution of atomic charges and Wiberg bond indices as a function of the length of the C-C bond in CH_3-CH_2-OH^+

C-C bond	**Atomic charge**				**Electric charge**		**Wiberg indices**		
length (Å)	**C_1**	**C_2**	**O_3**	**H_4**	**$CH_3^{\cdot}$**	**CH_2OH^+**	**C_1-C_2**	**C_2-O_3**	**O_3-H_4**
1.52[a]	-0.676	-0.080	-0.768	0.473		-	1.033	0.929	0.760
1.51[b]	-0.666	-0.079	-0.776	0.479		-	1.037	0.922	0.759
1.76	-0.470	-0.010	-0.428	0.543	0.340	0.660	0.617	1.148	0.686
2.26	-0.364	0.160	-0.542	0.544	0.344	0.656	0.306	1.312	0.682
2.76	-0.377	0.252	-0.542	0.550	0.267	0.733	0.187	1.396	0.675
3.26	-0.390	0.296	-0.535	0.553	0.223	0.777	0.145	1.430	0.671
3.76	-0.388	0.313	-0.533	0.554	0.207	0.793	0.131	1.444	0.669
4.26	-0.377	0.317	-0.534	0.555	0.207	0.793	0.129	1.444	0.669
4.76	-0.363	0.315	-0.536	0.554	0.216	0.784	0.132	1.437	0.670
5.26	-0.350	0.310	-0.538	0.553	0.226	0.774	0.135	1.430	0.671
∞[c]	-0.533	0.438	-0.495	0.571	0	1	-	-	-

[a] *Gauche* conformer (B3LYP/aug-cc-pVTZ geometry); initial neutral ground state.
[b] *Trans* conformer (B3LYP/aug-cc-pVTZ geometry); initial neutral ground state.
[c] Results from separate B3LYP/aug-cc-pVDZ calculations upon $CH_3^{\cdot}$ and CH_2OH^+

B3LYP/aug-cc-pVDZ Natural Bond Orbital calculations of Wiberg bond indices and atomic charges (Table 1) correspondingly indicate that the above reorganization of the molecular structure immediately after ionization corresponds to the very first stage of a dissociation of the ethanol radical cation ($[CH_3$-CH_2-$OH]^{\cdot+}$) into a methyl radical ($CH_3^{\cdot}$) and a protonated form of formaldehyde (H_2C=O-H^+). Indeed, due to the lack of resolution in energy in EMS experiments, as well as natural and vibrational broadenings, there will always be huge excesses of energy available, in the form of vibrations, which can easily convert into dissociation over an energy barrier of ~18.2 kcal/mol only. A residual Wiberg bond order of ~0.1 remains at *R*[C-C]=5.26 Å (Table 1), reflecting a charge transfer interaction between the HOMO (C_{2p} orbital) of the methyl radical and the LUMO (π^* orbital) of $H_2C{=}O{-}H^+$ (Figure 6). The displacement of the outermost electron densities towards the C-C bond and methyl group, which the rise of experimental (e, 2e) ionization intensities at $p \rightarrow 0$ most presumably reflects as a result of the enhancement of $n_O - \sigma^*_{C\text{-}C}$ anomeric interactions, therefore images a charge (electron) transfer, from the oxygen lone pair to the methyl group, during an ultra-fast stretching, and possibly cleavage, of the C-C bond.

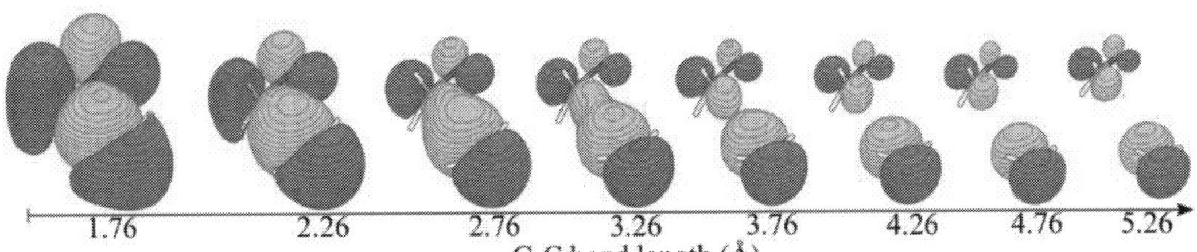

Figure 6. Evolution of the HOMO of CH_3-CH_2-OH^+ upon a stretching of the C-C bond

3. 1,3-butadiene

3.1.Methodology.

Spherically averaged Dyson orbitals momentum distributions have been computed for all resolvable ionization bands, using Feynman-Dyson transition amplitudes obtained as by-products of one-particle Green's Function calculations of the valence ionization spectrum of *trans*-1,3-butadiene, using the benchmark third-order Algebraic Diagrammatic Construction scheme [ADC(3)], in conjunction with Roos's augmented double-ζ basis set of atomic natural orbitals, on the basis of B3LYP/TZVP geometrical parameters. The molar fraction of the *gauche* form of 1,3-butadiene is known to be around 5% at 298K, and can be neglected therefore.

3.2 Results and Discussion.

The ADC(3) valence ionization spectrum of the gauche conformer is totally incompatible [7] with high-resolution photoelectron measurements employing a synchrotron radiation beam [8]. On the other hand, the ADC(3) calculations provide quantitative insights, within ~0.3 eV accuracy, into the available experimental ionization energies and related electron momentum distributions, and this throughout the (outer- and inner-) valence region. In particular, we note (Figure 7) a dramatic improvement in the quality of the description of the (e, 2e) momentum profile inferred from an ionization band at 13.9 eV, when taking into account the contribution of an intense shake-up satellite (S1), with a spectroscopic strength of 0.29, which corresponds to the $1b_g^{-2}\ 2a_u^{+1}$ excited configuration of the cation. The conclusion by Saha, Wang *et al* that several experimentally inferred (e, 2e) momentum distributions "are, in fact, engaged with" the gauche conformer of 1,3-butadiene [5] was the result of many redhibitory mistakes in their assignment of ionization bands (see Figure 8 and ref. 7 for details), due to the neglect of the influence of the conformation on the energy order of ionization lines. We note in particular that the outermost and second ionization bands at 9.2 and 11.5 eV were related to the 7b and 8a ionization lines of the gauche conformer, at 11.2 and 9.1 eV, respectively! Further similar mismatches were noted [7] for two bands (4 and 5) at higher electron binding energies. Also, the presence of the low-lying $1b_g^{-2}\ 2a_u^{+1}$ shake-up line at ~13.1 eV was neglected. At last, Saha, Wang *et al* did not consider that the gaussian bands 2 and 3 in the deconvoluted spectrum (Figures 8a and b) too strongly overlap to permit a reliable enough separate analysis of the underlying intensities.

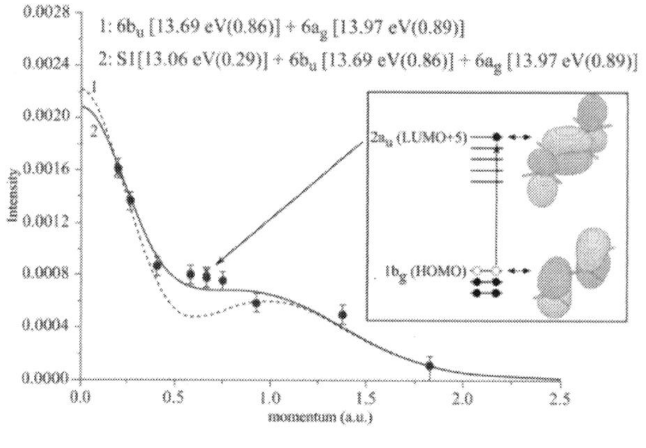

Figure 7. ADC(3)/aug-ANO Dyson orbital analysis of the momentum density distribution characterizing the band at 13.9 eV in the EMS ionization spectrum [6] of 1,3-butadiene.

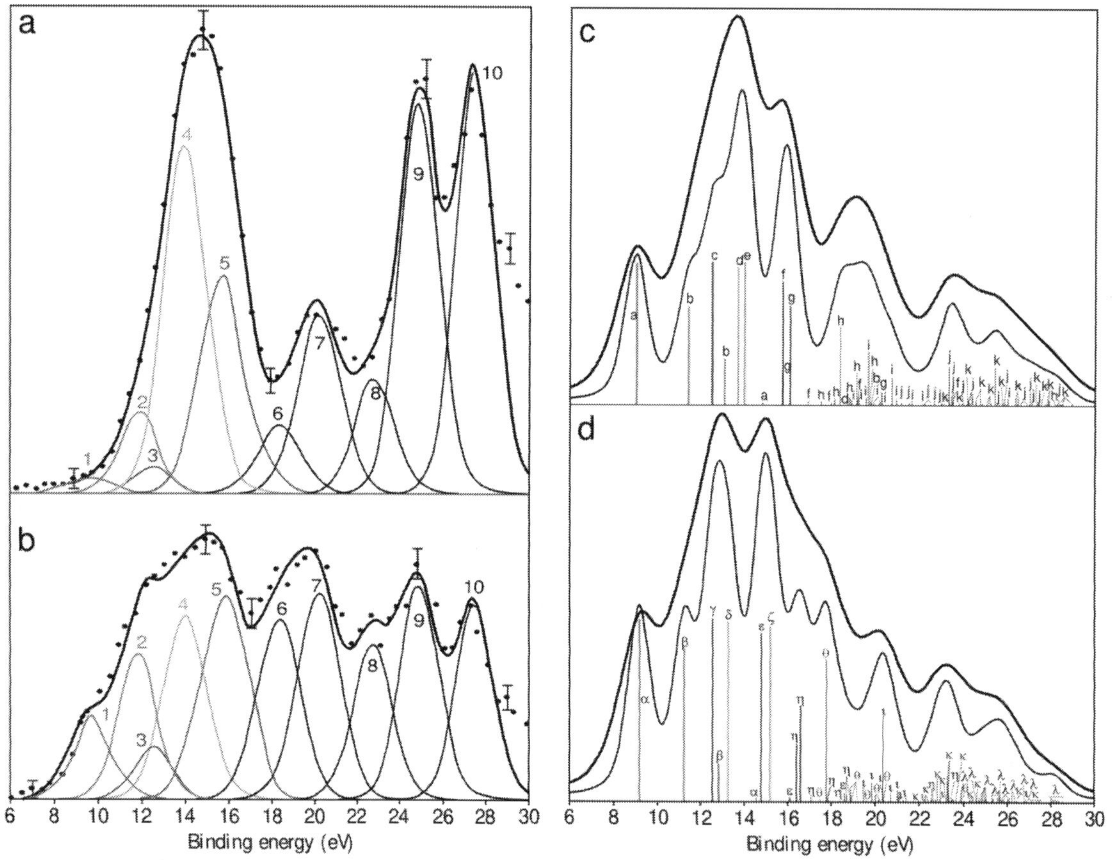

Figure 8. Comparison of the EMS ionization spectra at ϕ=0° (a) and ϕ=10° (b) with ADC(3)/aug-ANO simulations for the *trans* (c) and *gauche* (d) forms of 1,3-butadiene.
Colors refer to the assignment proposed by Saha, Wang *et al* in ref. 5.

Conclusion and outlook for the future

A review of EMS measurements on 1,3-butadiene and ethanol demonstrates that thorough theoretical studies are mandatory for reliably interpreting "orbital imaging" experiments that may be subject to many complications, such as electron correlation and relaxation, and a dispersion thereby of the ionization intensity over electronically excited (shake-up) configurations of the cation, distorted wave effects, conformational mobility in the electronic ground state, and the influence thereby of the conformation on ionization energies, as well as nuclear dynamics in the final state. Further improved descriptions of nuclear dynamics in the initial and final ground states would imply a quantization of vibrational energies, at least for the most relevant torsional degrees of freedom, and a computation thereby of Franck-Condon factors if we assume that the Born-Oppenheimer approximation remains valid. On longer terms, coupled electron-nuclear dynamical simulations of (e, 2e) experiments should also be considered for quantitatively describing ultra-fast dissociation processes that may be subject to a breakdown of the Born-Oppenheimer approximation. Considering the number of trajectories that one would need to compute for enabling reliable enough statistics, this goal will probably remain beyond reach for a number of years. Meanwhile, effort is required on the experimental side for a better monitoring of important parameters such as the temperature, and the time scale characterizing (e, 2e) electron impact ionization processes.

References

[1] Weigold E and McCarthy I.E., Electron Momentum Spectroscopy (Kluwer, New York, 1999).
[2] Ning CG, Luo ZH, Huang YR, Hajgató B, Morini F, Liu K, Zhang SF, Deng JK, Deleuze MS, J. Phys. B, 41 (2008) 175103.
[3] Morini F, Hajgató B, Deleuze MS, Ning CG, Deng JK, J. Phys. Chem. A 112 (2008), 9083.
[4] Hajgató B, Deleuze MS, Morini F, J. Phys. Chem. A, 113 (2009), 7138.
[5] Saha S, Wang F, Falzon CT, Brunger MJ, J. Chem. Phys. 123 (2005) 124315.
[6] Brunger MJ, Winkler DA, Michalewicz MT, Weigold E, J. Chem. Phys, 108 (1998) 1859.
[7] Deleuze MS, Knippenberg S. J. Chem. Phys, 125 (2006), 104309.
[8] Holland DMP, MacDonald MA, Hayes MA, Baltzer P, Wannberg B, Lundqvist M, Karlsson L, von Niessen W, J. Phys. B. 29 (1996), 3091.

Acknowledgments

This work has been supported by the FWO-Vlaanderen, the Flemish branch of the Belgian National Science Foundation, and by the "Bijzonder Onderzoeksfonds" (special research fund) of Hasselt University.

Journal of Physics: Conference Series **212** (2010) 012021 doi:10.1088/1742-6596/212/1/012021

Alignment and orientation of diatomic molecules induced by polarized electron impact

J W Maseberg and T J Gay

Behlen Laboratory of Physics, University of Nebraska, Lincoln, Nebraska 68588-0111, USA

E-mail: `maseberg@huskers.unl.edu`

Abstract. We report measurements of linear and circular fluorescence polarizations for molecular transitions in H_2, D_2, and N_2 induced by spin-polarized electron impact. Circular polarizations resulting from some Fulcher-α transitions in H_2 and D_2 are found to be significant, while the null results from nitrogen's second positive system are consistent with an earlier measurement by the Münster group. We compare this nitrogen data to that from our previous study of nitrogen's first negative system. Emphasis is placed on understanding the mechanisms that cause the values of circular polarization from N_2 to be relatively small compared to those observed from H_2 and D_2 molecules.

1. Introduction

Measurements of molecular fluorescence polarization resulting from spin-polarized electron impact excitation are rare compared to the number of existing studies for atomic targets. Investigation of such polarizations (Stokes parameters) has proven to be a useful means with which to probe the angular momentum coupling dynamics in excited states [1, 2]. The earliest experiment to measure the Stokes parameters for molecular fluorescence due to spin-polarized electron excitation was performed by the Münster group in Germany [3]. The integrated Stokes parameters (scattered electrons not detected) were reported for the second positive system in molecular nitrogen,

$$\mathrm{e} + \mathrm{N}_2\, X\,{}^1\Sigma_g^+(v=0) \rightarrow C\,{}^3\Pi_u(v'=0) + \mathrm{e} \rightarrow B\,{}^3\Pi_g(v''=0) + \mathrm{e} + \gamma(\sim 337.1\ \mathrm{nm}). \quad (1)$$

Interestingly, the observed circular polarization (spin-normalized Stokes parameter P_3/P_e) was found to be zero at ~14.5 eV within an uncertainty of 2×10^{-3}. This result stood in contrast with the atomic situation, where sizeable P_3/P_e values result from triplet excitation [4]. In light atoms, LS-coupling is responsible for the conversion of spin-polarization into fluorescence circular polarization. This situation is depicted schematically in Fig. 1.1. Exchange excitation of a multiplet state having $S > 0$ allows transfer of spin polarization (or orientation) from the polarized incident beam to the excited system. Then, LS-coupling of the oriented $\boldsymbol{S}$ with the initially unoriented $\boldsymbol{L}$ results in orientation of the total angular momentum $\boldsymbol{J}$ (neglecting nuclear spin). The Stokes parameter P_3 for the subsequent spontaneous emission of the excited state is then typically non-zero when observed in a direction parallel to the incident spin polarization. Angular momentum coupling schemes for molecules are generally more complex than the LS-coupling in atoms, and vector diagrams for some selected states are illustrated in Fig. 1. Hund's

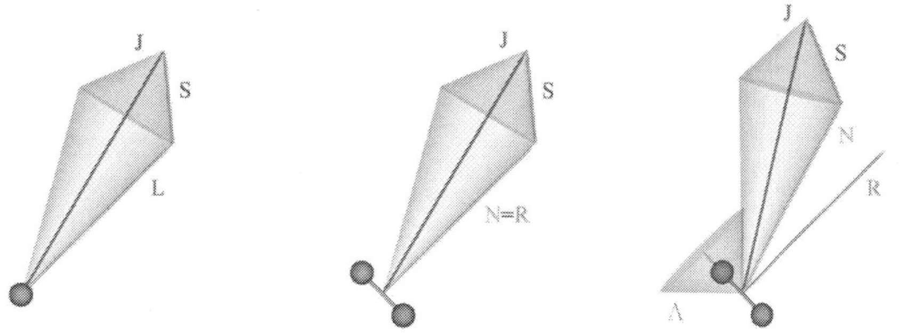

1.1: Atomic LS coupling. 1.2: Hund's case (b), $\Lambda = 0$. 1.3: Hund's case (b), $\Lambda = 1$.

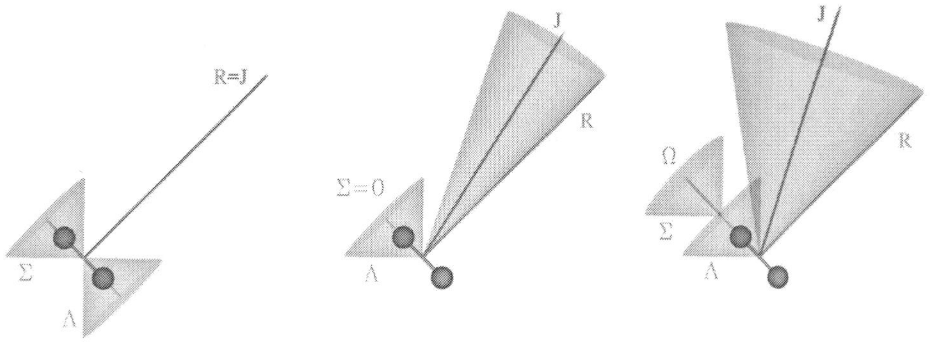

1.4: Hund's case (a), $\Omega = 0$. 1.5: Hund's case (a), $\Omega = 1$. 1.6: Hund's case (a), $\Omega = 2$.

Figure 1. Angular momentum coupling in atoms and molecules. The atomic LS coupling scheme is shown in (1.1). Molecular Hund's case (b) situations for Σ ($\Lambda = 0$) and Π ($\Lambda = 1$) states are shown in (1.2) and (1.3), respectively. Hund's case (a) situations for Π states with $\Omega = 0$, 1, and 2 are shown in (1.4), (1.5), and (1.6), respectively.

cases (b) and (a) are depicted in Figs. 1.2–1.6 as they are relevant for H_2 (or D_2) and N_2 molecules, respectively. Figures 1.2, 1.3 and 1.6 indicate that orientation of $\boldsymbol{S}$ results in orientation of $\boldsymbol{J}$. In simple analogy with the atomic case, one might expect the N_2 second positive band to exhibit non-zero values of P_3/P_e (Fig. 1.6). Thus the null values observed were a surprise.

Our group has done a number of experiments with both H_2 and N_2 targets to try understand this atomic/molecular difference. In an initial report, we presented the first measurements of non-zero circular polarization for molecular fluorescence from H_2 due to spin exchange [5]. In this paper we report recent extensions of these results to include the linear polarization (Stokes parameter P_1) and data for D_2. We also present our verification of the Münster result for the second positive system of N_2, and discuss mechanisms which could account for the null measurements of P_3/P_e in this case.

2. Experiment

Our experimental apparatus has been discussed in detail elsewhere [6, 7]. An unstrained Cs-O activated GaAs crystal is irradiated with 785 nm light to produce an electron beam with $\Delta E \sim 0.3$ eV FWHM and a typical transverse spin polarization of $P_e \sim 0.25$ as measured by optical polarimetry [4, 8, 9]. The beam is directed through a target gas cell; emission is collected perpendicular to the beam. A rotatable retarder followed by a polarizer and optical bandpass filter allow for determination of Stokes parameters at desired wavelengths [10].

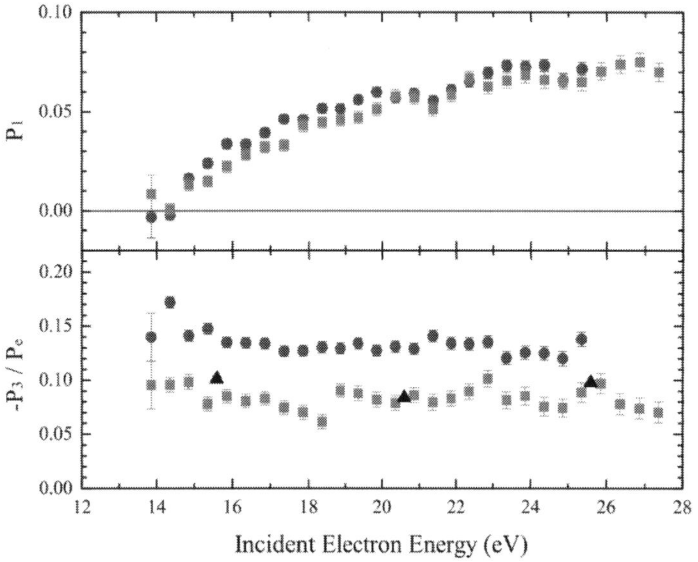

Figure 2. Linear and circular polarization fractions (P_1 and $-P_3/P_e$) for molecular transitions in H_2 (circles) and D_2 (squares). Our earlier data for H_2 is shown in triangles [5]. Data was acquired using a 600 nm (10 nm FWHM) bandpass filter; the gas pressure was 0.5 mTorr for the current results and 12 mTorr for the earlier measurements. Polarizations corresponding to the low intensities observed below 13.9 eV are omitted.

3. Linear and circular polarizations for H_2 and D_2

Our results for the linear (Stokes parameter P_1) and circular polarizations are shown in Fig. 2. The polarization P_1 is defined with respect to the electron beam and is independent of the value of P_e. The spin-normalized circular polarization is reported with a minus sign; this is done to eliminate possible confusion arising from the definition of P_3, which has a sign opposite that of the photon helicity [2]. Positive values of $-P_3/P_e$ occur when the emitted photon angular momentum points in the same direction as the incident electron spin. The spin-normalized linear polarization P_2/P_e is not shown in Fig. 2 but is consistent with zero for both the H_2 and D_2 data sets. (The Stokes parameter P_2 is the linear polarization with respect to an axis offset by 45° from the electron beam direction.)

Due to the relatively wide bandpass of the optical interference filter used for this experiment (600 ± 5 nm FWHM), several different ro-vibrational transitions contribute to the measured Stokes parameters [11, 12, 13, 14]. However, in this wavelength region the transitions in H_2 and D_2 are largely dominated by the Fulcher-α $d^3\Pi_u(v' = 0) \rightarrow a^3\Sigma_g^+(v'' = 0)$ band. The excitation threshold for the $d\,^3\Pi_u$ state is 13.9 eV for both H_2 and D_2 [12, 15].

In a recent paper, Aguilar *et al* [14] studied the spectra of H_2 with a spectrometer and reported the cross sections of ro-vibrationally isolated lines for an incident electron energy of 20 eV. Their data indicates that the brightest hydrogen line in our filter bandpass is the $Q(1)$ transition with a cross section of 0.77×10^{-19} cm^2. Labeling of the branches is done using the total orbital angular momentum quantum number N of the final state for Hund's case (b), e.g. $P(N'')$, $Q(N'')$ and $R(N'')$. (For the P-branch, $\Delta N = N' - N'' = -1$; for the Q-branch, $\Delta N = 0$; for the R-branch, $\Delta N = 1$.) The $Q(1)$ transition provides ~50% of the total measured emission. Other significant contributions come from the $Q(2,3)$ and $R(0,1,2)$ lines. For D_2, significant contributions come from the $P(2,3)$, $Q(1,2,3,4)$ and $R(0,1,2,3)$ lines [11, 13].

The appreciable values of circular polarization for both H_2 and D_2 shown in the lower panel of Fig. 2 indicate that significant transfer of spin-to-orbital angular momentum is achieved for these molecules. The earlier $-P_3/P_e$ data for H_2 appears low when compared with the current results; we attribute this to the fact that the earlier measurements were taken at a higher pressure (12 mTorr *vs* 0.5 mTorr) and in the presence of a 10^{-2} T magnetic field [5]. Direct comparison of the H_2 and D_2 values is problematic due to the fact that several rotational states are contributing to the measured fluorescence, and the different states will generally exhibit varying degrees of

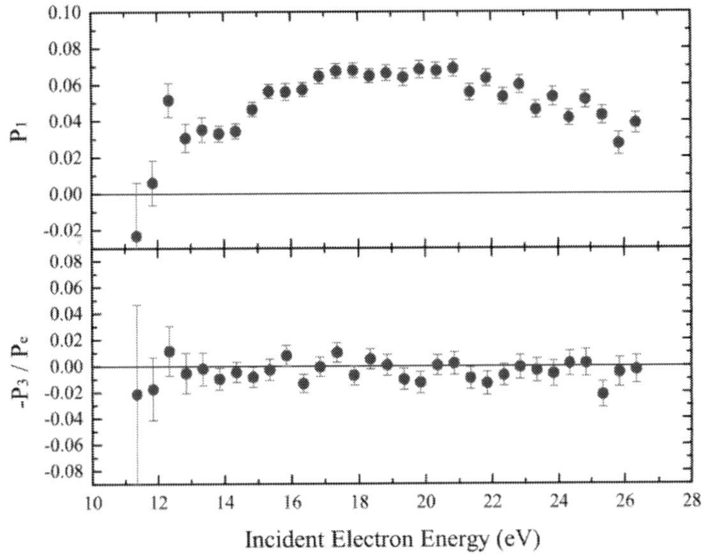

Figure 3. Linear and circular polarization fractions (P_1 and $-P_3/P_e$) for the second positive band in N_2. Data was acquired at a pressure of 0.3 mTorr with a 380.1 nm (3.5 nm FWHM) bandpass filter.

polarization. However, this data does represent a lower bound for $-P_3/P_e$ values from the individual lines with the largest circular polarizations. We are presently preparing a study of the Stokes parameters from several rotationally-resolved states in H_2 and D_2.

4. Linear and circular polarizations from N_2

Given that fluorescence from H_2 and D_2 was found to possess a non-zero circular polarization component, we set out to confirm the previously mentioned null values obtained by the Münster group for the second positive system in N_2. We have measured the ($v' = 0$, $v'' = 2$) vibrational band whereas the Münster group investigated the ($v' = 0$, $v'' = 0$) band, but we do not expect this difference in vibrational quantum numbers to significantly affect the Stokes parameters. The process is

$$e + N_2\, X\, {}^1\Sigma_g^+(v = 0) \rightarrow C\, {}^3\Pi_u(v' = 0) + e \rightarrow B\, {}^3\Pi_g(v'' = 2) + e + \gamma(\sim 380.5\ \text{nm}). \quad (2)$$

Figure 3 shows our results for P_1 and $-P_3/P_e$. Our values for the circular polarization confirm those reported by the Münster group [3]. The energy-averaged value of $-P_3/P_e$ is -0.004 ± 0.002.

To understand why the circular polarization is zero, a consideration of the measured rotational transitions is in order. For this experiment, the interference filter bandpass (380.1 ± 3.5 nm FWHM) transmitted all the rotational lines (populated at room temperature, $J_{max} \sim 25$), with only a slight discrimination of R-branch intensities corresponding to large J. Figure 4 depicts the spectrum as computed following the method described in Ref. [16], using the updated C and B state molecular constants reported in Ref. [17]. The upper panel shows the wavelength dependence of the $P(J'')$-, $Q(J'')$- and $R(J'')$-branches. The branches have three sub-branches, corresponding to ${}^3\Pi_0 \rightarrow {}^3\Pi_0$, ${}^3\Pi_1 \rightarrow {}^3\Pi_1$ and ${}^3\Pi_2 \rightarrow {}^3\Pi_2$ transitions (the ${}^3\Pi_0 \rightarrow {}^3\Pi_0$ Q-branch is forbidden [20]). The lower panel shows the intensities as a function of wavelength, and for this plot the transmission profile of the optical filter we used has been taken into account.

Hanne has pointed out that orientation of $\boldsymbol{S}$ cannot lead to orientation of $\boldsymbol{J}$ for the Hund's case (a) situations where $\Omega = 0, 1$ (see Figs. 1.4 and 1.5) [18]. Thus, no circular polarization can be produced from the Hund's case (a) ${}^3\Pi_0 \rightarrow {}^3\Pi_0$ and ${}^3\Pi_1 \rightarrow {}^3\Pi_1$ sub-bands. However, orientation of J can be achieved for the case where $\Omega = 2$, thus non-zero P_3 was expected [3]. (Due to spin uncoupling, or the transition of the C and B states from Hund's case (a) to Hund's

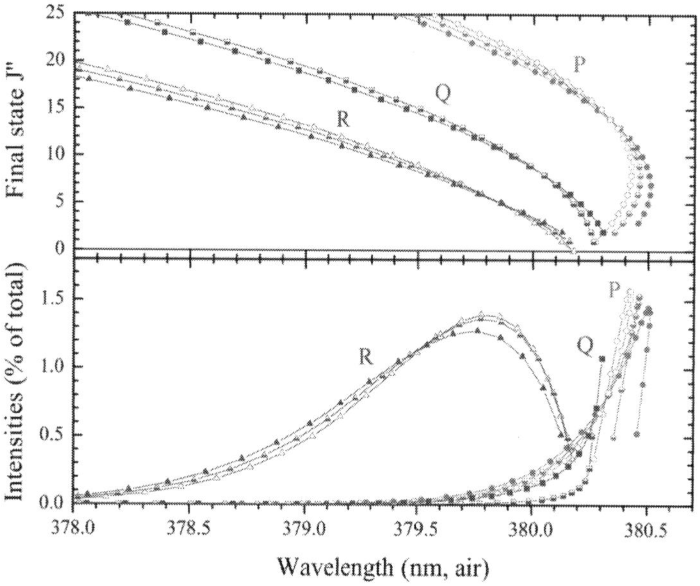

Figure 4. Upper panel: $P(J'')$, $Q(J'')$ and $R(J'')$ branches for the N_2 second positive $v' = 0 \rightarrow v'' = 2$ system. The empty, half-filled and filled points correspond to the $\Omega = 0, 1, 2$ sub-bands, respectively (see text). Lower panel: Intensities for branch J'' values as computed using the method described in Ref. [16] for room-tempurature N_2. This simplified spectra neglects Λ-doubling and weak $\Delta\Omega \neq 0$ transitions [19].

case (b) with increasing J [20], it is likely that the ${}^3\Pi_0 \rightarrow {}^3\Pi_0$ and ${}^3\Pi_1 \rightarrow {}^3\Pi_1$ sub-bands can contribute non-zero circular polarization as well for large J.) The Q-branches are weak: their relative intensity contributions are only 5% of the total emission. Thus, the measured Stokes parameters are mainly due to the R- and P-branches. The P- to R-branch intensity ratios are $\sim$1.3, meaning they have similar contributions.

We now consider the sign of $-P_3/P_e$ values for P- *vs* R-branch fluorescence. The definition of P- and R-branches in conjunction with conservation of angular momentum suggests that $-P_3/P_e$ for the P- and R-branches should have opposite signs. For a P-branch transition, ΔJ(or ΔN) $= -1$, while the R-branch case has ΔJ(or ΔN) $= 1$. The molecular angular momentum $\boldsymbol{J}$ is oriented due to $\boldsymbol{S}$ in a direction parallel to that of the spin polarization of the incident electron beam. Figure 5 illustrates an initially oriented $\boldsymbol{J}$ which yields outgoing photon helicities of opposite sign for the P- and R-branches upon fluorescence. This indicates that the sign of $-P_3/P_e$ should be positive for R-branch transitions and negative for P-branch fluorescence. Our recent study of the Stokes parameters for the first negative system in N_2^+ confirms this concept [7]. For this N_2^+ Hund's case (b) $B\,{}^2\Sigma_u^+(v' = 0) \rightarrow X\,{}^2\Sigma_g^+(v'' = 0)$ band, our interference filter isolated the P-branch fluorescence from the R-branch contribution and we measured a non-zero energy-averaged value of $-P_3/P_e = -0.0133(8)$. Furthermore, the angular momentum coupling scheme for the excited and final states closely resembles atomic LS-coupling (compare Figs. 1.1 and 1.2). With the substitution $N \rightarrow L$, we use the formalism for atoms provided in Ref. [21] to compute the threshold circular polarization (neglecting hyperfine depolarization) for the $R(N'' = 0)$ and $P(N'' = 2)$ transitions excited from the $N = 0$ molecular ground state. Doing this results in $-P_3/P_e = 0.29$ for the $R(0)$ case and $-P_3/P_e = -0.16$ for the $P(2)$ line, which again confirms that the circular polarizations from P- and R-branches have opposite signs.

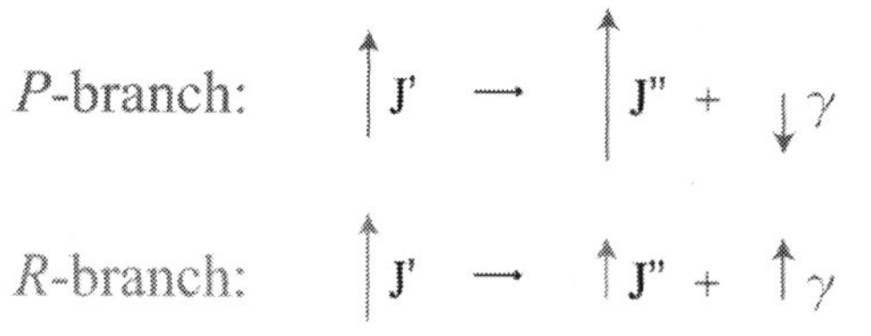

Figure 5. Initally oriented angular momentum $\boldsymbol{J}$ yielding photons with opposite helicites from the P- and R-branches upon fluorescence.

Returning to the discussion of the circular polarization from the second positive band, the total degree of circular polarization can now be expected to tend towards zero if the magnitude of the P- and R-branch polarizations are similar but have opposite signs, since the total measured fluorescence is comprised of similar amounts of P- and R-branch emission. The relatively large J values in N_2 also play a role in the reduction of observed circular polarization compared to that from H_2 and D_2. For N_2, the room-temperature intensity mean occurs at $J \sim 9$, whereas for H_2 $J \sim 1$ and for D_2 $J \sim 2$. Larger J values are less oriented by the electron spin polarization, which leads to smaller values of $-P_3/P_e$. This idea also provides a qualitative explanation as to why the D_2 circular polarization in Fig. 2 is found to be less than that for H_2. Also, for the H_2/D_2 comparison, D_2 P-branch fluorescence (not seen in the H_2 spectra), which tends to yield negative circular polarization (Fig. 5), should contribute to the reduction of $-P_3/P_e$ values.

5. Conclusion

We have observed significant values of fluorescence circular polarization from Fulcher-α transitions in both H_2 and D_2 excited by spin polarized electron impact. The null circular polarization observed by the Münster group for the N_2 second positive band has been confirmed, and we attribute this effect to rotational-branch averaging. We note that further measurements of the Stokes parameters for the second positive system of N_2 using optical filters to isolate the P- or R-branches, or both, would be useful to confirm the suggested rotational averaging mechanism.

Acknowledgments

This work was supported by NSF Grant No. PHY-0653379. The authors wish to thank Gordon Gallup and especially Friedrich Hanne for useful discussions. The authors are also indebted to Alejandro Aguilar and Joseph Ajello for providing H_2 spectra data from Aguilar *et al* [14].

References

[1] Blum K 1996 *Density Matrix Theory and Applications* 2nd ed. (New York: Plenum)
[2] Andersen N and Bartschat K 2001 *Polarization, Alignment, and Orientation in Atomic Collisions* (New York: Springer-Verlag)
[3] Hanne G F 1998 *Novel Aspects of Electron-Molecule Collisions*, ed K Becker (Singapore: World Scientific); Mette C *et al* 1994 *Verh. Dtsch. Phys. Ges.* (IV) **29** 462
[4] Furst J E, Wijayaratna W M K P, Madison D H and Gay T J 1993, *Phys. Rev.* A **47** 3775–87
[5] Green A S, Gallup G A, Rosenberry M A and Gay T J 2004 *Phys. Rev. Lett.* **92** 093201
[6] Maseberg J W and Gay T J 2006 *J. Phys.* B **39** 4861–4870
[7] Maseberg J W and Gay T J 2009 *Phys. Rev.* A **79** 022705
[8] Gay T J 1983 *J. Phys.* B **16** L553–6
[9] Gay T J, Furst J E, Trantham K W and Wijayaratna W M K P 1996 *Phys. Rev.* A **53** 1623–9
[10] Berry H G, Gabrielse G and Livingston A E 1977 *Appl. Opt.* **16** 3200–5
[11] Dieke G H 1958 *J. Mol. Spectrosc.* **2** 494–517
[12] Crosswhite H M 1972 *The Hydrogen Molecule Wavelength Tables of G. H. Dieke* (New York: Wiley Interscience)
[13] Freund R S, Schiavone J A and Crosswhite H M 1985 *J. Phys. Chem. Ref. Data* **14** 235–383
[14] Aguilar A, Ajello J M, Mangina R S, James G K, Abgrall H, and Roueff E 2008 *ApJS* **177** 388–407
[15] Lavrov B P and Umrikhin I S 2008 *J. Phys.* B **41** 105103
[16] Hartmann G and Johnson P C 1978 *J. Phys.* B **11** 1597–1612
[17] Roux F, Michaud F and Vervloet M 1993 *J. Mol. Spectrosc.* **158** 270–7
[18] Hanne G F 2005, private communication
[19] Roux F, Michaud F and Vervloet M 1989 *Can. J. Phys.* **67** 143–7
[20] Herzberg G 1967 *Molecular Spectra and Molecular Structure* vol 1, 2nd ed. (Princeton, New Jersey: D. Van Nostrand Company)
[21] Fischer T and Kessler J 1995 *Rev. Sci. Instrum.* **66** 4885–93

Journal of Physics: Conference Series **212** (2010) 012022 doi:10.1088/1742-6596/212/1/012022

Atomic systems with one and two active electrons in electromagnetic fields: ionization and high harmonics generation

I A Ivanov and A S Kheifets

Research School of Physics and Engineering, The Australian National University, Canberra ACT 0200, Australia

E-mail: igor.ivanov@anu.edu.au, A.Kheifets@anu.edu.au

Abstract. We describe a theoretical procedure for solving the time-dependent Schrödinger equation (TDSE) for atomic systems with one or two valence electrons. Motion of the valence electrons is described by means of the Hartree-Fock potential including the exchange interaction. We apply the procedure to various physical phenomena occurring in atoms exposed to strong electromagnetic fields. As an illustration, we consider below the processes of high harmonics generation and attosecond pulses production.

1. Introduction

Nonlinear optical phenomena, such as above-threshold ionization (ATI) [1] or high harmonics generation (HHG) [2], occur when atomic systems are exposed to an intense electromagnetic (EM) field. Qualitatively, many features of these phenomena can be understood using simple pictures of classical physics, which makes their study particularly interesting. For the HHG process such a picture is provided by the well-known 3-step model [3]. In the framework of this model, atomic electron emerges into the continuum with zero velocity at a particular moment of time t_0 within the optical cycle. For some selected values of t_0 classical electron trajectory returns to the nucleus, where the electron can recombine and emit a photon. The frequency of the emitted photon is determined by the amount of energy acquired by the electron, and the atomic ionization potential $I_{\rm p}$, assuming that the electron recombines to the ground state. A simple analysis of the classical trajectories shows, that the electron kinetic energy cannot exceed the value of $3.17U_{\rm p}$, where $U_{\rm p}$ is the ponderomotive potential. This leads to the well-known $I_{\rm p} + 3.17U_{\rm p}$ cut-off rule for the maximum energy of a photon emitted in the recombination process. A classical picture of the ATI process differs in the respect that the released electron rescatters instead of recombining, thereby acquiring additional energy of up to $10U_{\rm p}$ [4].

A quantum mechanical description [5] explains the reason of the success of the classical approach. Instead of the classical electron trajectories in the EM field, this approach uses the quantum-mechanical description of the motion of the released electron. Such a description is greatly simplified if the so-called strong-field approximation (SFA) is used. SFA neglects the influence of the atomic potential on the motion of the released electron during the second stage of the HHG process. In the framework of this approximation the well-known Volkov states can be employed for the quantum mechanical description of the electronic motion in presence of the EM field, and the problem becomes analytically tractable. Calculation of the transition amplitudes in this approach can be performed with the help of the saddle-point method, and the classical trajectories naturally emerge as extrema in the saddle-point analysis of the quantum-mechanical amplitudes computed within the SFA.

The quantum mechanical description based on the SFA takes into account only the influence of the EM field on the electron and neglects the atomic potential. Yet, it is known that the atomic structure may considerably modify the observed spectra. In the HHG process, for instance, resonant enhancement of harmonics in the spectrum may take place if a multiphoton resonance occurs between the ground and an excited level [6]. Such phenomena can, in principle, be described by employing the exact propagator [7] instead of the Volkov propagator used in the SFA theory. However, this approach is very difficult to implement in practical calculations. Below, we present an alternative procedure describing the atom – EM field interaction, which is based on solving the TDSE for an atomic system described by means of the Hartree-Fock and polarization potentials.

For illustration purposes, we shall consider below only the systems with one valence electron. All formulas presented below can easily be modified for the case of two valence electrons. We illustrate the proposed computational procedure by applying it to the HHG process in the lithium atom.

2. Theory

Our goal is to find a weak solution of the TDSE for the atom in the external EM field:

$$\langle\Phi|\left(i\frac{\partial}{\partial t}-\hat{H}_{\rm atom}-\hat{H}_{\rm int}(t)\right)\Psi\rangle=0, \tag{1}$$

where $|\Phi\rangle$ is any vector from a subspace $\mathcal{V}$ of the Hilbert space $\mathcal{H}$. The subspace $\mathcal{V}$ is constructed using the set of pseudostates, obtained by diagonalizing the field-free atomic Hamiltonian in a square integrable basis [8]:

$$\langle f_j^N|\hat{H}_{\rm atom}|f_{j'}^N\rangle=E_j\delta_{jj'}\ . \tag{2}$$

In Eq.(2), $\hat{H}_{\rm atom}$ is the Hamiltonian of the field-free atom. We describe the field-free atom in the ground state by solving the set of self-consistent Hartree-Fock equations [9], which enables us to define direct and exchange potentials for the valence electrons. The core polarization effects are included by means of the polarization potential.

For systems with a non-empty core, field-free atomic Hamiltonian $\hat{H}_{\rm atom}$ has low-lying eigenstates corresponding to the core states, which must be excluded from consideration. The subspace $\mathcal{V}$ in which the solution of Eq.(1) is sought, is thus a span of all pseudostates defined in Eq.(2), which do not belong to the atomic core:

$$\Psi(\boldsymbol{r},t)=\sum_{j\notin{\rm core}}b_j(t)f_j(\boldsymbol{r}). \tag{3}$$

To describe the interaction of the atom and the external EM field we employ either velocity or length gauges.

To construct the set of pseudostates satisfying Eq. (2), we use the Laguerre basis, or (for large angular momenta) a set of B-splines, confined to a box of the size $R_{\rm max}=200$ a.u. The B-splines of the order $k=7$ with the knots located at the (approximately equidistant) sequence of points lying in $[0,R_{\rm max}]$ are used in the calculation presented below.

The fact that the system is confined to a box of a finite size, may lead to spurious effects, such as appearance of extra harmonics in the HHG spectrum due to reflection of the wavepackets from the boundaries of the box [2]. One can minimize this effect by using a mask function or an absorbing potential. We use the absorbing potential $-iW(\boldsymbol{r})$, which is a smooth function, zero for $r \leq 180$ a.u. and continuously growing to a constant $-iW_0$ with $W_0 = 2$ a.u. outside this region.

In the calculations relying on the length gauge to describe atom – EM field interaction, the pseudostates with angular momenta $l < 120$ were included in Eq. (3). This implies, that for EM field strength of the order of 0.01 a.u., corresponding to 3.5×10^{12} W/cm^2 field intensity, we may describe accurately formation of harmonics of the order of a hundred. Considerably smaller number of the angular momentum states is needed if the velocity gauge is used [10].

With the total Hamiltonian thus defined, equations (1) and (3) lead to a system of differential equations for the coefficients $b_j(t)$. We find solution of this system for the time interval $(0, 30T)$, where T is the period of the field oscillation, by using the Crank-Nicholson method [11]. With the solution of the TDSE thus obtained, the harmonics spectrum can be calculated as Fourier image of the time-dependent dipole operator [2].

3. Results

3.1. Optimization of the high harmonics generation.

In this section, we study the effect of the form of the driving laser pulse on the cut-off of the HHG process. We consider a special class of the waveforms, which can be represented as a superposition of a number of the harmonics of a base frequency $\Omega = 2\pi/T$, corresponding to a given period T, and which do not contain the DC component:

$$F(t) = 2\mathrm{Re}\sum_{k=1}^{K} a_k e^{ik\Omega t} \tag{4}$$

Remarkable success of the classical model of the HHG process in predicting the cut-off values of the harmonics generation suggests the following strategy. We first consider a problem of classical mechanics of finding a set of coefficients a_k in Eq. (4) such, that the maximum classical kinetic energy of the recombining electron is achieved. To show, that the increase of the HHG cut-off value obtained classically is not an artifact of a simplified treatment, we supplement the classical trajectory analysis by a quantum mechanical TDSE calculation.

Thus, we begin with a purely classical study of the returning trajectories of an electron moving in a periodic EM field given by Eq. (4). Our task is to find the set of coefficients a_k in Eq. (4), for which electron returning to the nucleus possesses the highest possible kinetic energy. To make the problem well-defined, we impose some restrictions on the possible choice of this set. A natural requirement is that only the fields $F(t)$ with the fixed fluency are to be considered. This implies that $4\sum\limits_{k=1}^{K}|a_k|^2 = F_0^2$, where the field strength F_0 is related to the intensity via $W = 3.5 \times 10^{16} F_0^2$. Here the field intensity W is measured in W/cm^2 and the field strength is expressed in the atomic units. Another restriction is that we consider pulses with only low order harmonic terms $k \leq 5$ in Eq. (4), which is motivated primarily by practical feasibility of the pulse creation.

Following the procedure adopted in the classical 3-step model theory of HHG, we neglect the influence of the atomic core on the electron motion. Classical equations of motion of an electron in the EM field given by Eq. (4) are solved with the initial conditions $x(t_0)=0$, $\dot{x}(t_0) = 0$, where t_0 is a moment of time at which atomic ionization event occurs. Among these trajectories we are interested only in the returning trajectories, for which we compute electron kinetic energy E at the moment of return to the nucleus.

We use the following field parameters: $F_0 = 0.0053$ a.u. (corresponding to the intensity of 10^{12} W/cm^2), $\Omega = 0.185$ eV. For this set of the field parameters, the value of the Keldysh parameter $\gamma = \sqrt{I/2U_p}$=0.8, where $I = 0.196$ a.u. is the ionization potential of the Li atom.

Table 1. Coefficients (multiplied by 10^3) for cosine wave and the optimized pulse.

k	Cosine wave	Optimized pulse
1	2.665	$2.123 - 1.033i$
2	0	$0.403 + 0.754i$
3	0	$-0.558 + 0.271i$
4	0	$-0.302 - 0.358i$
5	0	$0.224 - 0.248i$

For the pure cosine form of the EM field, the classical procedure described above gives the well-known curve of the dependence of the kinetic energy E at the moment of return on the time of release, shown in Figure 1 by the solid (red) line. To find the set of parameters in Eq. (4) for which recollision energy is maximized (with the constraints of a fixed fluency and $K = 5$ in Eq. (4)), we look for a maximum of the quantity $(E + I)/\Omega$, considered as a function of the coefficients a_k in the Eq. (4), imposing the two constraints which we mentioned above. This strategy gives us the pulse composition summarized in Table 1, and dependence of the kinetic energy on the time of release, shown in Figure 1 by the dashed (green) line.

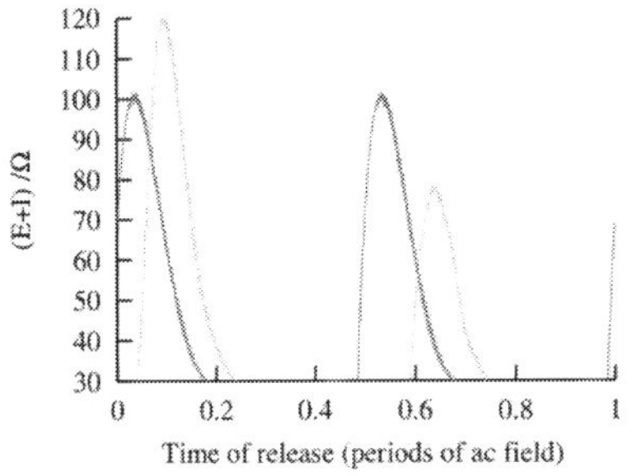

Figure 1. Classical dependence of the quantity $(E + I)/\Omega$ on the time of electron release within an optical cycle. Pure cosine wave – solid (red) line; optimized pulse dashed (green) line.

In Figure 2 we show results of the TDSE calculations of the HHG yield for both pulses formed with the coefficients from Table 1. Harmonics spectrum is computed as Fourier image $d(\omega)$ of the time-dependent dipole operator in the length and velocity gauges. Results of the quantum calculation show, in agreement with the classical picture of Figure 1, that optimization of the pulse given by Eq. (4) with $K = 5$ allows to increase the cut-off position by about 20%.

On the right panel of Figure 2 we show the results obtained for the pure cosine waveform of a driving pulse for various forms of the Hamiltonian representing atom-field interaction, and various forms of the dipole operator. As we noted above, we can employ either length or velocity gauges to represent atom-field interaction when solving the TDSE. We may also use either length or velocity forms of the dipole operator when computing the harmonics spectrum. In Figure 2 we refer to these calculations, using combinations of letters L and V. For example, the LV calculation uses the length gauge to solve the TDSE, and the velocity form of the dipole operator, the VL calculation uses the velocity gauge for the solution of the TDSE, and the length form of the dipole operator to compute the harmonics spectrum. Comparing the results of the calculations using different gauges to solve the TDSE, and different dipole operator forms, gives us a reliable indicator of the accuracy with which the TDSE has been solved.

There are waveforms which allow to achieve more substantial increase in the HHG cutoff position. Such are the ideal waveform proposed in [12], for which we should allow the term

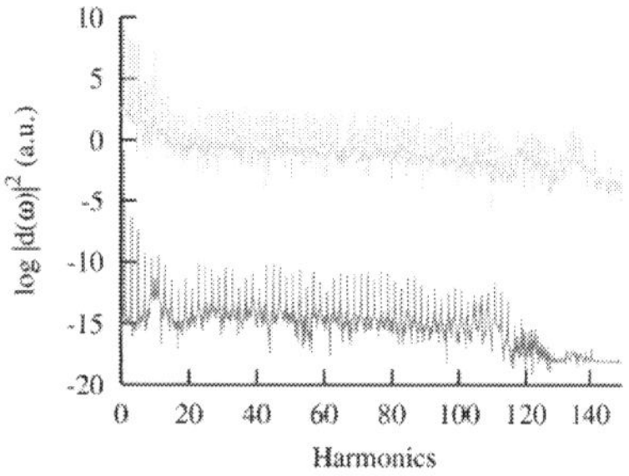

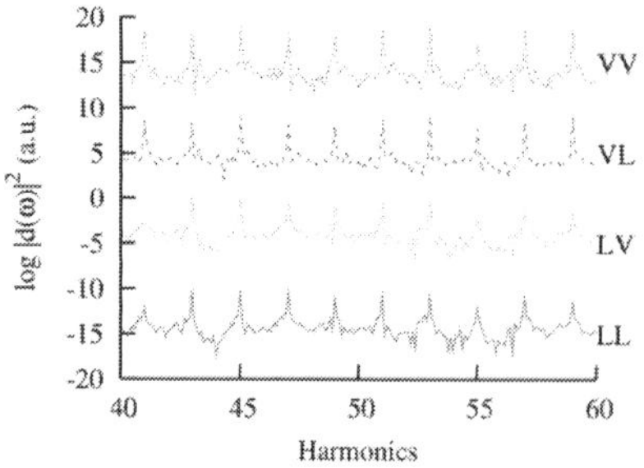

Figure 2. Left panel: Harmonics spectra of Li for the EM fields constructed with coefficients from Table 1. Pure cosine wave ((red) solid line), optimized pulse (green) dashed line, for convenience of comparison the quantity $\log|d(\omega)|^2 + 10$ is shown. Right panel: LL: ($\log|d(\omega)|^2$), LV: ($\log|d(\omega)|^2 + 10$), VL: ($\log|d(\omega)|^2 + 20$), and VV: ($\log|d(\omega)|^2 + 30$) calculations for pure cosine wave in the interval $40\Omega < \omega < 60\Omega$.

with $k = 0$ in the sum in Eq. (4)), or the field configurations containing subharmonic fields with frequencies $\Omega/2$ [13, 14]. Such waveforms cannot be described by Eq. (4). Our result presents an upper limit for the increase of the HHG cutoff achieved for the class of the waveforms given by Eq. (4), i.e. for the waveforms which are periodic with a given period T and do not contain the DC components.

3.2. Generation of short pulses of EM radiation.

In this section, we consider application of the so-called polarization gate technique [15, 16] for generation of very short bursts of EM radiation. This technique employs a field configuration consisting of two pulses of left and right circular polarizations, separated by a time delay. For such a configuration, the polarization of the total EM field is close to linear only for the time interval on which both pulses significantly overlap. The reason why this field configuration turns out to be so efficient for the generation of the short pulses is clear, if one takes into account the role played by the classical returning trajectories in the process of HHG. Such trajectories exist only for linearly polarized EM field. Thus, the field configuration used in the polarization gating method, reduces the time interval during which the return to the nucleus is possible to the interval of overlap of both pulses. Quantum-mechanical simulations relying on the SFA approach, performed in works [15, 16], confirmed that this field configuration could be used to produce single attosecond pulses.

Below we study the harmonics formation for this field configuration by means of the TDSE approach, which takes into account the atomic structure. As a target, we again choose the lithium atom. For the present setup, when the atom interacts with two pulses of right and left circular polarizations, arriving with the time-delay Δ, the interaction Hamiltonian in the length gauge can be written as:

$$\hat{H}_{\rm int}(t) = E_0\Big[xf_1(t)\cos\omega t + yf_2(t)\sin\omega t\Big] \ , \qquad (5)$$

where $f_1(t) = f(t)+f(t-\Delta)$, $f_2(t) = f(t)-f(t-\Delta)$, and $f(t)$ is the function describing envelope of the pulse. The envelope function $f(\tau)$ is constant for $\tau \in (T, 4T)$, where $T = 2\pi/\omega$ is a cycle duration corresponding to the carrier frequency in Eq. (5). The fields are ramped on and off smoothly, so that the field strength, and its first time-derivative remain continuous. The total duration of the left- and right-polarized pulses of EM radiation is therefore $T_1 = 6T$. The atom thus interacts with the combined pulse during the interval of time $(0, T_2)$, where $T_2 = 6T + \Delta$. In this interval the left and right polarized pulses overlap within the time interval $(\Delta, 6T)$. On the interval $(\Delta + T, 5T)$ of duration $4T - \Delta$ both overlapping pulses are completely ramped on, the resulting EM field is therefore polarized linearly along the x- axis. Field parameters in Eq. (5) are $E_0 = 0.0035$ a.u. and $\omega = 0.16$ eV.

We analyze the solution of TDSE by means of the wavelet transform – the time-frequency technique, which allows us to track the process of harmonics formation in time, and which combines both the frequency and temporal resolution of the signal [17, 18, 19].

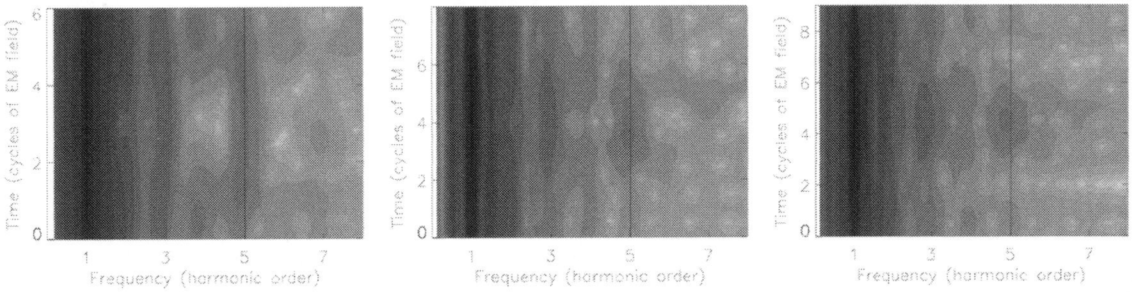

Figure 3. Wavelet time-frequency spectrum of Li for $\Delta = 0, 2T, 3T$.

In Figure 3 we show results of such analysis of the induced dipole momentum for the lithium atom and the field configuration described above. One can observe that, in agreement with the classical model of the process, with increasing Δ harmonics formation is confined to ever decreasing time interval in which the light polarization is linear.

4. Conclusion

We presented a procedure describing interaction of atoms with strong EM fields. The field-free atom is described within the Hartree-Fock approximation, including the non-local exchange interaction. This choice gives us more detailed account of the atomic system of interest, allowing to describe phenomena in which atomic structure plays important role, such as resonant enhancement of harmonics in the HHG spectrum.

As an illustration, we presented results of the optimization of the HHG cut-off value for the waveforms satisfying Eq. (4), and considered application of the polarization gating technique to the lithium atom.

[1] Agostini P, Fabre F, Mainfray G, Petiti G and Rahman N K 1979 *Phys. Rev. Lett.* **42** 1127
[2] Krause J L, Schafer K J and Kulander K C 1992 *Phys. Rev. A* **45** 4998
[3] Corkum P B 1993 *Phys. Rev. Lett.* **71** 1994
[4] Paulus G G, Becker W, Nicklich W and Walther H 1994 *J. Phys. B* **27** L703–L708
[5] Lewenstein M, Balcou P, Ivanov M Y, L'Huillier A and Corkum P B 1994 *Phys. Rev. A* **49** 2117
[6] Toma E S, Antoine P, de Bohan A and Muller H G 1999 *J. Phys. B* **32** 5843
[7] Ivanov I A and Kheifets A S 2008 *Phys. Rev. A* **78** 053406
[8] Bray I 1994 *Phys. Rev. A* **49** 1066–1082
[9] Chernysheva L V, Cherepkov N A and Radojevic V 1976 *Comp. Phys. Comm.* **11** 57
[10] Nurhuda M and Faisal F H M 1999 *Phys. Rev. A* **60** 3125–3133
[11] Goldberg A, Schey H M and Scwartz J L 1967 *Am. J. Phys.* **35** 177
[12] Radnor S B P, Chipperfield L E, Kinsler P and New G H C 2008 *Phys. Rev. A* **77** 033806
[13] Chipperfield L E, Robinson J S, Tisch J W G and Marangos J P 2009 *Phys. Rev. Lett.* **102** 063003
[14] Pfeifer T, Gallmann L, Abel M J, Nagel P M, Neumark D M and Leone S R 2006 *Phys. Rev. Lett.* **97** 163901
[15] Platonenko V T and Strelkov V V 1999 *J. Opt. Soc. Am. B* **16** 435
[16] Chang Z 2004 *Phys. Rev. A* **70** 043802
[17] Chu X and Chu S I 2001 *Phys. Rev. A* **64** 021403(R)
[18] Tong X M and Chu S I 2000 *Phys. Rev. A* **61** 021802(R)
[19] Tong X M and Chu S I 1999 *J. Phys. B* **32** 5593

Few-cycle intense laser interactions with complex atoms

Xiaoxu Guan[1] **and Klaus Bartschat**[2],

Department of Physics and Astronomy, Drake University, Des Moines, IA 50311, USA

E-mail: [1]xiaoxu.guan@drake.edu, [2]klaus.bartschat@drake.edu

Abstract. We present an *ab initio* and non-perturbative time-dependent approach to describe the response of a general atom to intense few-cycle laser pulses. After using a highly flexible B-Spline R-matrix method to generate field-free Hamiltonian and electric dipole matrices, the initial state is propagated in time using an efficient Arnoldi-Lanczos scheme. The method is illustrated with recent results for excitation and single ionization of Ne and Ar, as well as double ionization of He in a two-color pump-probe arrangement.

1. Introduction

As noted in the cover article for a recent special issue of *Journal of Physics B* [1], "we are [currently] witnessing a revolution in photon science, driven by the vision to time-resolve ultra-fast electronic motion in atoms, molecules, and solids..." Indeed, the ongoing development of ultra-short and ultra-intense light sources based on high-harmonic generation and free-electron lasers is providing new ways to generate optical pulses capable of probing dynamical processes that occur on attosecond time scales. These capabilities promise a revolution in our microscopic knowledge and understanding of matter [2].

Over the past three years, our group has been working on the development of a general *ab initio* theoretical approach to describe short-pulse intense laser interactions with atoms, which is applicable to complex targets beyond (quasi) two-electron systems. In recent papers [$3-5$], we outlined how field-free Hamiltonian and electric dipole matrices generated with the highly flexible B-Spline R-matrix (BSR) [6] suite of codes may be combined with an efficient Arnoldi-Lanczos time propagation scheme to calculate multi-photon excitation as well as well as single and even double ionization, albeit the latter so far has been restricted to the helium target. A computer code has also been made publicly available [7].

In this paper, we briefly summarize the computational method and then present a number of examples to demonstrate the feasibility of the approach. In addition to showing results for excitation and single ionization of Ne and Ar atoms by single laser pulses of variable length, intensity, and central photon energy, we investigate pump-probe processes in He involving two XUV laser pulses whose time delay is varied. This allows us to visualize the competition between direct and sequential double ionization as a function of the time delay.

Unless specified otherwise, atomic units (a.u.) are used throughout this manuscript.

2. Computational Approach

Since the method has been described in detail before [$3-5$], we only provide a brief summary here. We start with the time-dependent Schrödinger equation

$$i\frac{\partial}{\partial t}\Psi(\boldsymbol{r}_1,...,\boldsymbol{r}_N;t) = [H_0(\boldsymbol{r}_1,...,\boldsymbol{r}_N) + V(\boldsymbol{r}_1,...,\boldsymbol{r}_N;t)]\Psi(\boldsymbol{r}_1,...,\boldsymbol{r}_N;t) \tag{1}$$

for the N-electron wavefunction $\Psi(\boldsymbol{r}_1, \ldots, \boldsymbol{r}_N; t)$, where $H_0(\boldsymbol{r}_1, ..., \boldsymbol{r}_N)$ is the field-free Hamiltonian containing the kinetic energy of the N electrons, their potential energy in the field of the nucleus, and their mutual Coulomb repulsion, while

$$V(\boldsymbol{r}_1, \ldots, \boldsymbol{r}_N; t) = \sum_{i=1}^{N} \boldsymbol{E}(t) \cdot \boldsymbol{r}_i \tag{2}$$

represents the interaction of the electrons with the laser field $\boldsymbol{E}(t)$ in the dipole length form.

The tasks to be carried out in order to computationally solve this equation and to extract the physical information of interest are:

1) Generate a representation of the field-free Hamiltonian and its eigenstates; these include the initial bound state, other bound states, autoionizing states, as well as single-continuum and double-continuum states to represent electron scattering from the residual ion.
2) Generate the electric dipole matrices to represent the coupling to the laser field.
3) Propagate the initial bound state until some time after the laser field is turned off.
4) Extract the physically relevant information from the final state.

The solution of the TDSE requires an accurate and efficient generation of the Hamiltonian and electron–field interaction matrix elements. In order to achieve this goal, we approximate the time-dependent wavefunction as

$$\Psi(\boldsymbol{r}_1, ..., \boldsymbol{r}_N; t) \approx \sum_q C_q(t) \Phi_q(\boldsymbol{r}_1, ..., \boldsymbol{r}_N). \tag{3}$$

The $\Phi_q(\boldsymbol{r}_1, ..., \boldsymbol{r}_N)$ are a set of time-independent N-electron states formed from appropriately symmetrized products of atomic orbitals. They are expanded as

$$\Phi_q(\boldsymbol{r}_1, ..., \boldsymbol{r}_N) = \mathcal{A} \sum_{c,i,j} a_{ijcq} \Theta_c(x_1, \ldots, x_{N-2}; \hat{\boldsymbol{r}}_{N-1}\sigma_{N-1}; \hat{\boldsymbol{r}}_N \sigma_N) R_i(r_{N-1}) R_j(r_N). \tag{4}$$

Here $\mathcal{A}$ is the anti-symmetrization operator, the $\Theta_c(x_1, \ldots, x_{N-2}; \hat{\boldsymbol{r}}_{N-1}\sigma_{N-1}; \hat{\boldsymbol{r}}_N \sigma_N)$ are channel functions involving the space and spin coordinates (x_i) of $N-2$ core electrons coupled to the angular $(\hat{\boldsymbol{r}})$ and spin (σ) coordinates of the two outer electrons, $R_i(r)$ is a radial basis function, and the a_{ijcq} are expansion coefficients. Although resembling a close-coupling ansatz with two continuum electrons, the expansion (4) contains bound states and singly ionized states as well. In general, the atomic orbitals, $R_i(r)$, are not orthogonal to one another or to the orbitals used to describe the atomic core. If orthogonality constraints are imposed on these functions, additional terms would need to be added to the expansion to relax the constraints. This possibility still exists as an option in our computer code.

When the expansion (3) is inserted into the Schrödinger equation, we obtain

$$i\boldsymbol{S}\frac{\partial}{\partial t}\boldsymbol{C}(t) = [\boldsymbol{H}_0 + E(t)\boldsymbol{D}]\, \boldsymbol{C}(t), \tag{5}$$

where $\boldsymbol{S}$ is the overlap matrix of the basis functions, $\boldsymbol{H}_0$ and $\boldsymbol{D}$ are representations of the field-free Hamiltonian and the dipole coupling matrices, and $\boldsymbol{C}(t)$ is the time-dependent coefficient vector in (3).

The price to pay for the flexibility in the BSR approach, at least initially, is the representation of the field-free Hamiltonian and the dipole matrices in a non-orthogonal basis. To solve this problem, we use a transformation into the eigenbasis by solving the field-free generalized eigenvalue problem first. Details can be found in Guan *et al* [4]. In addition to simplifying the definition of the initial state and the extraction of the physically interesting information, this transformation makes it possible to cut unphysically high eigenvalues and the corresponding eigenvectors from the time propagation scheme, thereby improving its numerical stability and allowing us to use the standard Arnoldi-Lanczos method [8 – 10] propagate the initial state in time until the observables of interest can be extracted from the wavefunction.

These observables include the survival probability of the ground state, as well as the probabilities for single-electron excitation and ionization, ionization-excitation, and double ionization. Some of these probabilities are easy to extract via the coefficients of the time-propagated wavefunction in the eigenbasis of the problem, while others require projections to the one-electron or even two-electron continuum. Details can be found in refs. [3 – 5]. In some cases, it is possible to derive a "generalized cross section" from these probabilities, but care has to be taken in the definition of the "effective interaction time" [11, 12].

3. Example Results

3.1. Multiphoton single-ionization of neon

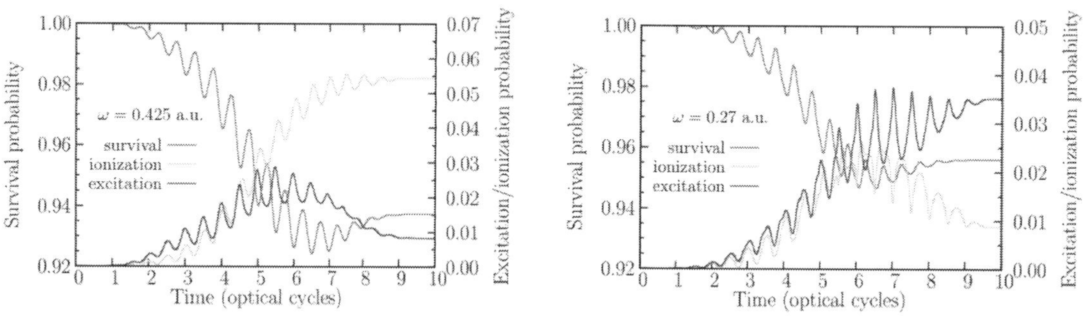

Figure 1. Ground-state survival (left scale) and total excitation and ionization probabilities (right scale) for Ne exposed to 10-cycle laser pulses of peak intensity 3.5×10^{14} W/cm^2 with a Gaussian envelope. The central laser frequencies are 0.425 a.u. (11.6 eV, left panel) and 0.27 a.u. (7.3 eV, right panel).

As our first example, Fig. 1 shows the response of the Ne atom to pulses with central photon energies of 11.6 and 7.3 eV, respectively. In these cases, at least two or three photons, respectively, need to be absorbed in order to ionize the system. This was our first attempt at such a problem, and hence we only used the ground state $(1s^2 2s^2 2p^5)^2P$ of Ne^+ as the target state for the "half collision" of the ejected electron with the residual ion. In order to ensure converged results for the above cases, we actually coupled LS symmetries up to a total orbital angular momentum $L_{max} = 6$ for the electron–ion collision system. Note that excitation rather than ionization appears as the dominating reaction process for $\omega = 0.27$ a.u. and the laser parameters chosen here.

3.2. Multiphoton single-ionization of argon

For the argon target, we performed a more sophisticated calculation [4], in which we coupled three states of the Ar^+ ion. In this project, we studied the effects of the finite pulse length and the intensity on various intermediate resonance states.

The left panel of Fig. 2 depicts the excitation probability at different photon energies. So-called Rabi oscillations occur when the photon energy matches the energy gap between the ground state and, in this case, the $(3p^5 4s)^1P$ excited state. This matching leads to oscillations in the excitation probability with a large amplitude and a long period. There are still oscillations when the photon energy is tuned away from the energy gap, but they have much smaller amplitudes and shorter periods.

As a result of coupling several ionic states, Rydberg-type resonances converging to different thresholds can be observed. However, both the finite length of the pulse and its intensity have an effect on the details of the observed structures. The right part of Fig. 3 shows the effect of the laser intensity on the generalized two-photon cross section. While both peak intensities, 10^{12} and 10^{13} W/cm^2, still lie in the perturbative regime, the height of the first resonance peak is significantly diminished for the more intense laser field. Similar results have been obtained by an other another time-dependent R-matrix approach, which is independently being developed by the Belfast group [14].

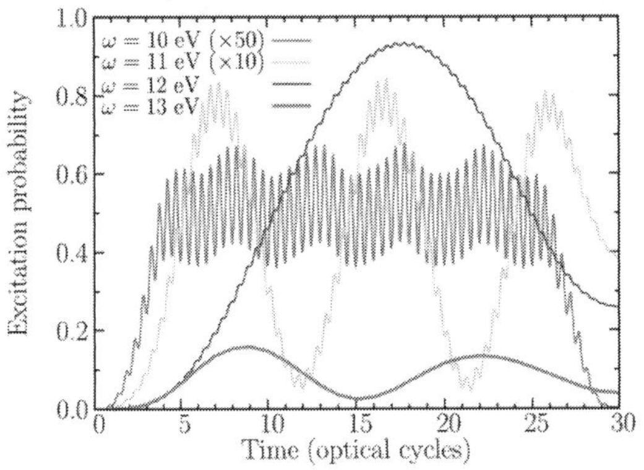

Figure 2. Excitation probability of argon for a 30-cycle laser pulse with peak intensity of 2×10^{13} W/cm^2 and photon energies of 10, 11, 12, and 13 eV. Note the different scales for the individual photon energies.

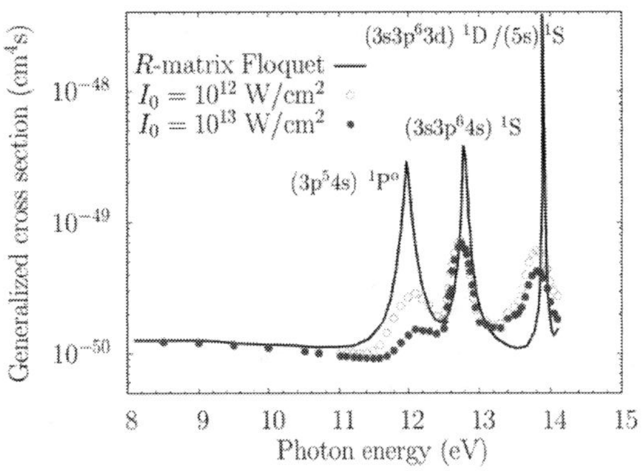

Figure 3. Effect of the laser peak intensity on the generalized two-photon cross section for a 30-cycle laser pulse interacting with argon. The Floquet-results are from McKenna and van der Hart [13].

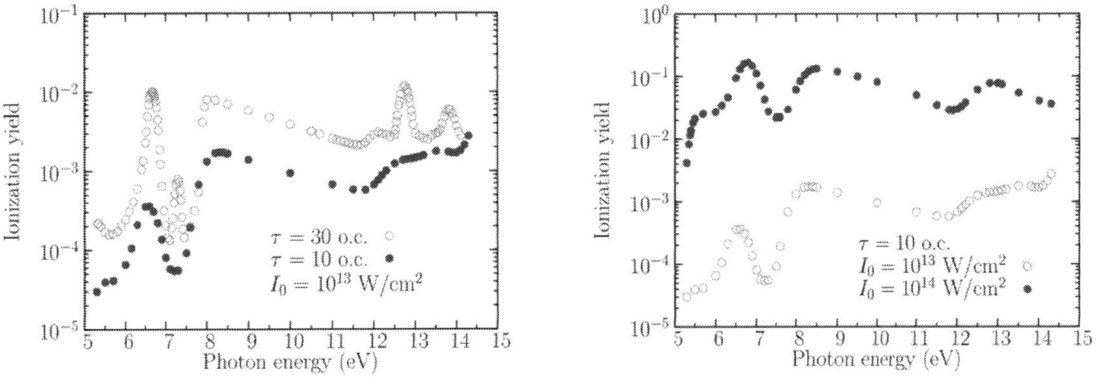

Figure 4. Ionization yield as a function of photon energy in 10-cyle and 30-cycle laser pulses of peak intensity 10^{13} W/cm^2 (left) and for 10-cycle laser pulses with peak intensities of 10^{13} W/cm^2 and 10^{14} W/cm^2 (right).

Figure 4 exhibits two examples for the single ionization rates in argon [4] obtained in few-cycle laser pulses of different lengths, intensities, and central wavelengths. There is clearly a non-trivial dependence on the various laser parameters, and once again resonances appear when the photon wavelength is varied. These predictions are currently awaiting experimental tests, but they show the richness of effects that can be expected in complex targets where inter-shell correlation effects play an important role.

3.3. Two-color two-photon double ionization of helium

The two-photon double ionization (DI) of the helium atom induced by intense short XUV laser pulses has received considerable attention from both theorists and experimentalists alike. Instead of listing a large number of references here, we note that much of the recent work was quoted in refs. [11, 12]. Even within the past few months, however, several additional papers appeared or are currently in press. In fact, even the calculation of the total cross section for this process is under heavy dispute, with ongoing debates about the need, or lack thereof, to account for the electron-electron interaction in the final state and the role of the direct vs. the sequential process [15]. For more details, we refer to the above references

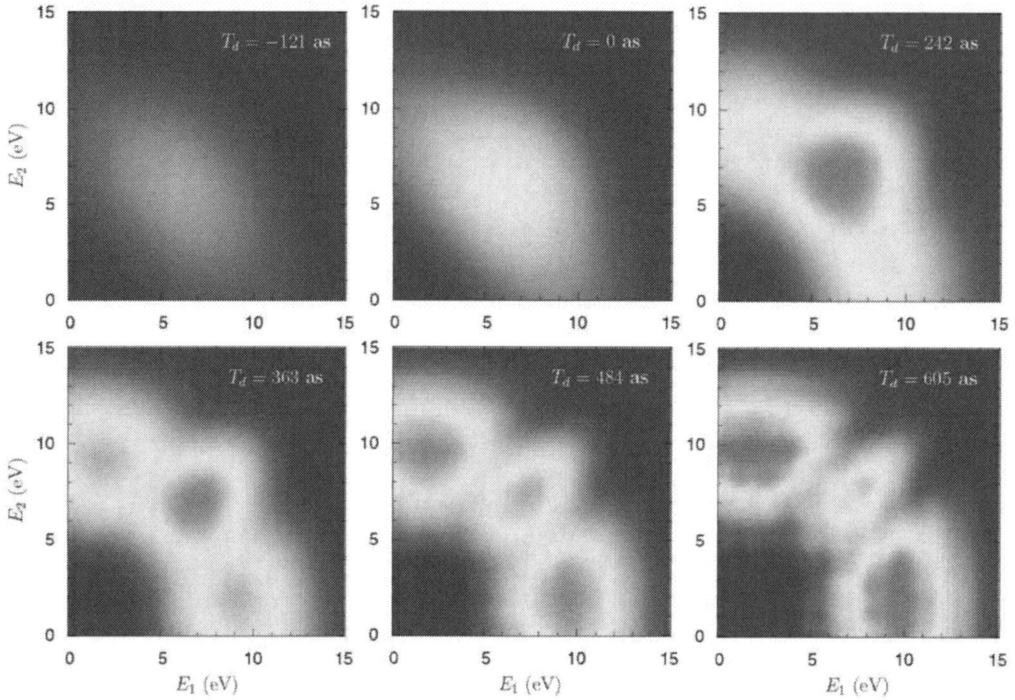

Figure 5. Energy distributions of the two escaping electrons in two-color laser pulses of 10 optical cycles each. The laser parameters are: $\omega_1 = 35.3\,\mathrm{eV}$ at a peak intensity of 10^{14} W/cm^2 and $\omega_2 = 57.1\,\mathrm{eV}$ at a peak intensity of 10^{13} W/cm^2. The delay between the two pulses is varied between -121 as and $+605$ as. See text for details.

and several papers in the special journal issue headed by ref. [1].

Here we consider the process of double photoionization by absorption of two photons at different central photon energies. In other words, the target helium atom is exposed to the irradiation by two laser pulses, of potentially different frequencies and with a controllable time delay. Specifically, the following two-color laser parameters were used for the results shown in Fig. 5: Pulse 1 has a central photon energy of $35.3\,\mathrm{eV}$ and a peak intensity of 10^{14} W/cm^2, while the corresponding parameters for pulse 2 are $57.1\,\mathrm{eV}$ and 10^{13} W/cm^2, respectively. The length of each pulse is 10 optical cycles with a Gaussian envelope, thus corresponding to pulse lengths of about 1.2 and 0.7 femtoseconds, respectively.

We are interested in the mechanism for the ejection of two electrons when the time delay between the two pulses is varied. We define this delay as the time distance between the peak intensities. Consequently, there is no overlap at all between the two pulses for a delay of about 0.95 femtoseconds, corresponding to a little less than 40 atomic units of time (1 a.u.$\approx$24 attoseconds).

Figure 5 shows the results for a variety of delays, ranging from about -120 attoseconds (i.e., the second photon with the higher energy comes first) to 600 attoseconds. Not surprisingly, the probability for double ionization is small in the first case, since the only chance for this process to happen is the two photons working together on the two electrons. Even when the two pulses come simultaneously, the probability for double ejection remains relatively small, and the peak for equal-energy sharing is a clear indication of the direct process. With increasing time delay, the peaks expected for the sequential process — one at 10.7 eV and the other at at 2.7 eV, corresponding to the 35.1 eV photon ionizing the neutral helium atom with an ionization potential of 24.6 eV and the other one ionizing the He^+(1s) ion with an ionization potential of 54.4 eV — start to grow, but the two processes still have about equal weight for a time delay as large as 400 attoseconds.

From this example, it is clear that the time delay plays a decisive role in determining how the two electrons are ejected by two-color XUV laser pulses. Depending on the details of the time delay, the electrons can be ejected in ways either similar to the sequential or the non-sequential process. Our findings qualitatively agree with those of Foumouo *et al* [16] for the two-color problem and Feist *et al* [17] in the single-color problem. They serve as an independent confirmation of their predictions, and also give us confidence in our newly developed computer code.

4. Summary and Outlook

We have presented a *general* method to calculate short-pulse intense laser interactions with complex atoms using a B-spline R-matrix approach in connection with an efficient Arnoldi-Lanczos time propagation scheme. Test calculations for several systems revealed good agreement with previous benchmark results obtained with different and entirely independent methods. Our application to two-photon double ionization of helium using two time-delayed XUV laser pulses confirmed that attosecond spectroscopy will provide a "microscope" to examine and also control the way electrons interact in atomic and molecular targets.

We are currently in the process of generating and then transforming the corresponding matrices for the two-photon double ionization problem of neon atoms. This will allow us to make a direct comparison with the recent experiments of Moshammer *et al* [18] carried out at the FLASH facility in Hamburg. While the additional complication of a residual core with non-zero angular momentum is substantial, the current method has been formulated in such a way that these calculations are effectively limited by the available hardware (i.e. supercomputer facilities) rather than special-purpose software.

Nevertheless, in order to move to systems like Ne and Ar, we need more work on parallelizing the code and substantial resources to handle the matrices that can quickly reach ranks of 50,000 – 100,000. To use the eigenbasis, we will need to solve a generalized eigenvalue problem *once* for each partial-wave symmetry, and these matrices are not sparse. However, in light of currently available computational resources, we are confident that we will be able to generate results for complex targets, including simple molecules, in the near future.

Acknowledgments

Much of this work was done in collaboration with C.J. Noble, B.I. Schneider, and O. Zatsarinny. We acknowledge support by the United States National Science Foundation under grants PHY-0757755 and PHY-0555226, and supercomputer resources through the Teragrid allocation TG-PHY090031.

References

[1] R. Moshammer and J. Ullrich, J. Phys. B **42** (2009) 134001.
[2] M. Uiberacker *et al*, Nature **446**, 627 (2007).
[3] X. Guan, O. Zatsarinny, K. Bartschat, B. I. Schneider, J. Feist, and C. J. Noble, Phys. Rev. A **76** (2007) 053411.
[4] X. Guan, C. J. Noble, O. Zatsarinny, K. Bartschat, and B. I. Schneider, Phys. Rev. A **77** (2008) 053402.
[5] X. Guan, O. Zatsarinny, C. J. Noble, K. Bartschat, and B. I. Schneider, J. Phys. B **42** (2009) 134015.
[6] O. Zatsarinny, Comp. Phys. Commun. **174** (2006) 273.
[7] X. Guan *et al*, Comp. Phys. Commun. (2009), in press; available on-line at http://dx.doi.org/10.1016/j.cpc.2009.03.005
[8] T. J. Park and J. C. Light, J. Chem. Phys. **85** (1986) 5870.
[9] E. S. Smyth, J. S. Parker, and K. T. Taylor, Comp. Phys. Commun. **114**, 1 (1998).
[10] B. I. Schneider and L. A. Collins, J. Non-Cryst. Solids **351** (2005) 1551.
[11] X. Guan, K. Bartschat, and B. I. Schneider, Phys. Rev. A **77** (2008) 043421.
[12] J. Feist *et al*, Phys. Rev. A **77** (2008) 043420.
[13] C. McKenna and H. W. van der Hart, J. Phys. B **37**, 457 (2004).
[14] M. A. Lysaght, H. W. van der Hart, and P. G. Burke, Phys. Rev. Lett. **101** (2008) 253001.
[15] P. Lambropoulos, L. A. A. Nikopoulos, M. G. Makris, and A. Mihelic, Phys. Rev. A **78** (2008) 055402.
[16] E. Foumouo, P. Antoine, H. Bachau, and B. Pireaux, New J. Phys. **10** (2008) 025017.
[17] J. Feist *et al*, ArXiv:0812.0373; Phys. Rev. Lett. (2009), in press.
[18] R. Moshammer *et al*, Phys. Rev. Lett. **98** (2007) 203001.

Double Photoionization of Strontium

Emma Sokell, Michael Grimm and Paul Sheridan

School of Physics, University College Dublin, Belfield, Dublin 4, Ireland.

emma.sokell@ucd.ie

Abstract. Resonant triple-differential cross-section (TDCS) measurements have been used to study the double photoionization process in strontium. Two sets of measurements were made at the photon energy of the 4p → 4d resonance. The coplanar geometry was used and the fixed analyser, positioned at -90° to the main axis of polarization of the photons, detected electrons with ~65% of the available excess energy. The mutual angle between the two electrons had a range just short of 90 → 270°. The TDCS exhibit unexpected lobes at a mutual angle of 180°. Comparison with other measurements made with the same geometry but with different sharings of the available energy indicate that these TDCS all show the unexpected lobe. Some possible explanations for the lobe are considered.

1. Introduction

Double photoionization (DPI), though forbidden by the one electron model of ionization, provides insight into electron correlation inside the atom. Helium, the DPI of which results in a bare nucleus and two electrons in the continuum, has been the focus of many investigations [1]. Alkaline-earth metals are also of interest since their electron configurations are such that there are two ns electrons in the outermost valence shell. These are the two electrons that are removed in the lowest energy DPI process and the remaining electrons may be considered as spectators, playing no direct role in the photoionization process. The focus of the present study is strontium although selected work on other alkaline-earths will be briefly reviewed.

One experimental method employed to study the DPI process is to measure the double to single photoionization ratio and use the single photoionization cross-section to determine the DPI cross-section. This method has been used to study beryllium [2,3] and more recently magnesium [4]. It has been argued that the photon energy dependence of the measured double to single photoionization ratios could be modeled by a suitably scaled ratio for helium [4]. This assumes that the dominant DPI mechanism is a two step one, the second step being that of internal electron impact ionization of the singly charged ion (created in the first step) by a single photoelectron, the impact happening as the initial electron is liberated. However, the scaling model, which worked for beryllium and lithium, did not work for magnesium [4]. The measured double to single ionization ratios for both magnesium and beryllium are in reasonably agreement with the convergent close coupling (CCC) calculations of Kheifets and Bray [5]. However it is important to note that the agreement is significantly poorer close to threshold for magnesium where the calculated ratio is larger than the experimental double to single ionization ratio, the calculated ratio for beryllium also exceeds the experimental one. It is possible that these discrepancies, at excess energies below 10 eV, are associated with the breakdown of the frozen core model adopted for the double photoionization calculations [5].

The DPI process can also be investigated more differentially by measuring triple differential cross-sections (TDCS), which determine the DPI cross-section with respect to the energy of one electron along with the angles of ejection of the two electrons for a particular excess energy. The only TDCS measurements for direct photoionization in alkaline-earth metals have been carried out on calcium [6]. The data differed markedly from similar data for helium, as do the corresponding CCC predictions [5]. However, again it is worth noting that there are discrepancies between the experimental data and predictions which have been ascribed to the fact that the measurements were made at a photon energy that is relatively close to the 3p → 3d giant resonance.

Due to the low density of target atoms that can be achieved with a metal vapour source and the extremely low cross-sections for the direct DPI process, further alkaline-earth TDCS measurements have been confined to resonant measurements in calcium and strontium [7-10]. Of particular relevance to the present article are the experimental TDCS recorded by West *et al* on the 4p → 4d resonance in strontium [9]. These measurements were taken in the coplanar geometry, the two electrons and the plane of polarization vector being in a plane perpendicular to the propagation vector of the incident photon beam. The fixed analyser was positioned at -90° to the major polarization axis of the linearly polarized light. As for the direct calcium TDCS, the strontium TDCS were significantly different from comparable helium measurements. It could be argued that this difference might be due to the resonant nature of the Sr TDCS. Although, Maulbetsch *et al* have shown that the concepts developed for direct PDI can be successfully used for resonant PDI, specifically TDCS data recorded for Ca in the region of the 3p → 3d resonance [11]. The Sr TDCS are especiallty noteworthy due to the presence of a lobe, non-zero, at a mutual emission angle for the two electrons of 180°. This feature violates the selection rules for a transition from a $^1S^e$ state, as is the case for DPI from the Sr $5s^2$ ground state [12]. In the present article further Sr TDCS measurements are considered. These TDCS data were taken at the photon energy of the 4p → 4d resonance, again with the fixed analyser at -90° to the polarization vector, with the aim of exploring the cause of the unexpected lobe.

2. Experimental details

Two experimental runs were carried out with the same spectrometer [10] using the coplanar geometry, the first at the McPherson beamline 3.2 at the Daresbury SRS and the second at the UE112-PGM1 beamline at BESSY II. The bandpass of the photons at the energy used to acquire the TDCS was 16 meV at BESSY and 25 meV at Daresbury. At BESSY the light was completely linearly polarized, whilst at Daresbury S_1 was measured to be 0.7 ± 0.02. S_1 describes the degree of linear polarization, parallel over perpendicular with respect to the major polarization direction [9]. The atomic strontium was produced in a cylindrical oven heated to 570°C and directed to the interaction region by 6 converging atomic beams. One of the two 150° spherical sector analysers, the fixed, was positioned at θ_1 = -90° to the polarization vector of the incident light. The second, rotatable analyser was able to move from approximately θ_2 = 0 to 180°, so that the mutual angle, θ_{12}, had a range just short of 90 → 270°. The analyser bandpass utilized at BESSY was 280 meV compared with 700 meV at Daresbury.

To acquire the TDCS presented here, the analysers were set to accept electrons that shared the available excess energy of 8.52 eV [10], such that the energy of the electron detected by the fixed analyser was 5.51 eV and the energy of the electron detected by the rotatable analyser was 3.01 eV. The rotatable analyser was positioned at 10° intervals across the available angular range and the number of coincidences was monitored by analysing the time difference between coincidence signals from the two analysers using a multi-channel analyser. The true coincidence rate was determined from the coincidence peak, after accounting for random coincidences, and these true coincidences were normalised to the counts in the fixed analyser to account for variations in the incident light flux or the metal vapour density.

3. Results and discussion

Figure 1 shows equivalent TDCS recorded at a) Daresbury and b) BESSY II. The photon energy in both cases was 25.26 eV, the energy of the 4p → 4d resonance [13]. In both cases the data have been averaged each side of θ_{12}, = 180° for presentation purposes. The two TDCS are similar in that there are the usual lobes either side of the mutual angle $\theta_{12} = 180^{\circ}$ and an additional lobe centred at 180°, which is forbidden by the selection rules [12]. Taking the quantization axis to be along the polarization vector of the photons and considering photo-excitation from an even parity S state, such as the ground state of Sr, with completely linearly polarized light, only odd parity states with M = 0 can result. The selection rules forbid electron emission perpendicular to the quantization axis for any M = 0 component of an odd parity state, so in the present geometry a node is anticipated in the TDCS at a mutual emission angle of 180°, for any sharing of the available excess energy. The TDCS shown in figure 1 are consistent with similar earlier data, recorded by West *et al* at the Daresbury SRS [9], also with the fixed analyser positioned at -90° to the polarization axis of the light. All of the available TDCS, recorded in this geometry, suggest that there is a lobe centred at 180°, for both equal and unequal sharing of the available excess energy by the two emitted electrons.

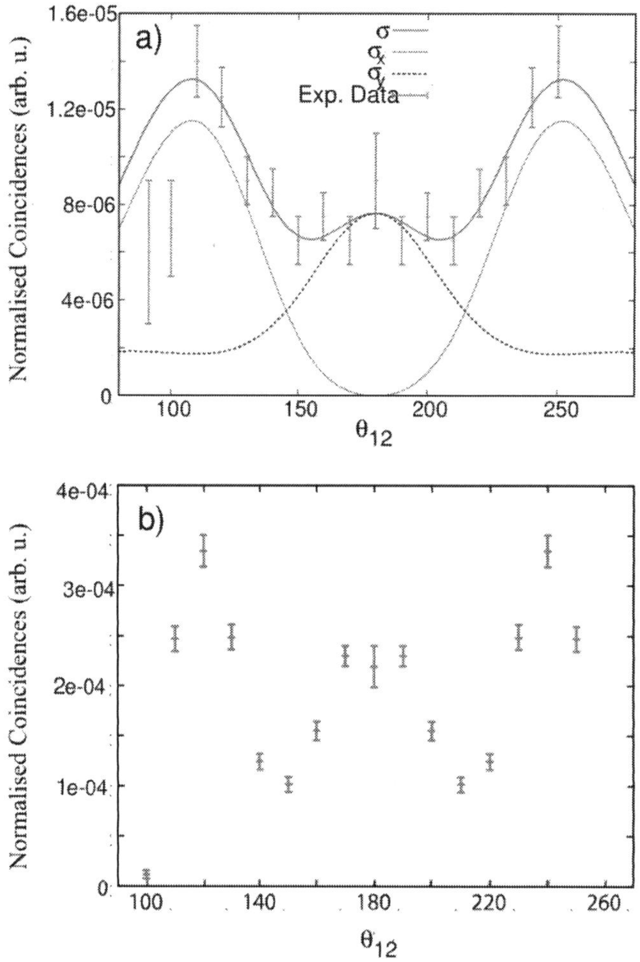

Figure 1. Sr TDCS recorded at a) the Daresbury SRS and b) BESSY II, at the energy of the 4p → 4d resonance. The fixed analyser detected 5.51 eV electrons emitted at an angle of -90° (with respect to the polarization direction of the light), whilst the rotatable analyser detected 3.01 eV electrons. θ_{12} indicates the mutual emission angle of the two electrons. The solid lines shown in panel a) are described in the text.

When comparing the TDCS it is important to consider the effect of the component of the incident photon beam that is polarized at 90° with respect to the main axis of polarization for all the data recorded at the Daresbury SRS, where the parameter $S_1 < 1$. Light polarized perpendicular to the main polarization axis will generally give rise to peaks in the TDCS at $\theta_{12} = 180°$. This point is illustrated by the lines included in figure 1a. The solid line is a fit, using the parameterization adopted previously [10], to the TDCS points, that is comprised of a contribution, labeled σ_x, corresponding to light polarized along the major axis and a smaller contribution, σ_y, resulting from light polarized at 90° to the main polarization axis. Although these two contributions could account for the unexpected lobe for unequal energy sharings, for equal energy sharing back to back emission is forbidden for either polarization direction [12] suggesting that another explanation is required.

Possible explanations for the unexpected lobe have been considered in detail previously [10]. Application of the selection rules [12] indicate that double photoionization from an intermediate state of even parity could give rise to the central lobe at $\theta_{12} = 180°$ and therefore implies that mechanisms that can access even parity, intermediate states should be considered. This is particularly relevant for the TDCS measured at BESSY II, figure 1b, where $S_1=1$, precluding any contributions from light polarized at 90° to the major axis of polarization. Perhaps the likeliest mechanism for population of even parity intermediate states that has been considered is the possibility that even parity states can be accessed through mixing of excited states coupled by stray electric fields. It is accepted that the ground state of Sr whilst predominately of the configuration $4s^24p^65s^2$ contains some components with configurations $4s^24p^65p^2$ and $4s^24p^64d^2$ [14]. Excitation from the ground state, including the effects of initial state configuration interaction and allowing for excitation of one or both valence electrons simultaneously with a 4p electron, results in a wealth of structure [14]. In the absence of external fields there is significant mixing between configurations with the same combined occupancy of the 5s, 4p and 4d sub-shells. It is therefore possible that stray fields could mix states with the same combined occupancy but different numbers of 5p electrons and result in intermediate states of even parity. DPI from these even parity states would give rise to a lobe at $\theta_{12} = 180°$.

However, there is evidence in the data of West *et al* [9] that stray fields are not the sole reason for the unexpected lobes. Among the TDCS recorded was one in which the excess energy was shared equally by the two electrons and the fixed analyser was position at $\theta_1 = 0°$ to the polarization axis. As the lobe at $\theta_{12} = 180°$ is prominent in the equal energy sharing TDCS recorded with the fixed analyser positioned at -90° to the polarization axis, it would be anticipated that there should be some evidence of it, arising from photons polarized perpendicular to the main polarization axis, in the TDCS recorded with the fixed analyser at 0°. There is no evidence of a lobe at $\theta_{12} = 180°$ in the $\theta_1 = 0°$ TDCS, suggesting that the unexpected lobe in the $\theta_1 = -90°$ TDCS is due, at least in part, to something other than mixing due to stray fields. One possibility, that requires further theoretical consideration, is that the resonant nature of the DPI process giving rise to the TDCS in question is significant.

4. Conclusion

Resonant triple-differential cross-section (TDCS) measurements have been performed at the energy of the $4p \rightarrow 4d$ resonance in atomic strontium. In particular TDCS in which the electron recorded in the fixed analyser, positioned at -90° to the main polarization axis, had an energy of 5.51 eV and the energy of the electron detected by the rotatable analyser was 3.01 eV have been considered. Measurements recorded at the Daresbury SRS and at BESSY II are similar and exhibit a lobe at a mutual emission angle of $\theta_{12} = 180°$ that is forbidden by the selection rules for the DPI process. Mechanisms to account for the unexpected feature have been considered. Although double ionization from even parity intermediate states, accessed through mixing caused by stray fields, is perhaps the most likely explanation there is at present no strong theoretical support for this mechanism. It is possible that TDCS measurements made in the presence of a well defined, external electric field could clarify this question. Additionally, mixing due to stray fields will not explain all of the available TDCS in Sr and indicate that theoretical modeling of the DPI would be useful. The present observations together with the discrepancies between calculation and experiment for less differential, double photoionization cross-section data suggest that there is more to be learned about the DPI process in alkaline-earth atoms.

Acknowledgments

This work was funded by Science Foundation Ireland, grant number SFI (04/BR/P0277). The experimental run at BESSY II was supported through the EC under BESSY IA-SFS Contract RII 3-CT-2004-506008, BESSY-ID.07.2.346A subsection. The authors also acknowledge the support of the Science and Technology Facilities Council for the Daresbury beamtime.

References

[1] Briggs J S and Schmidt V 2000 *J. Phys. B: At. Mol. Opt. Phys.* **33** R1-48.
[2] Wehlitz R and Whitfield S B 2001 *J. Phys. B: At. Mol. Opt. Phys.* **34** L719-725.
[3] Wehlitz R, Lukić D and Bluett J B 2005 *Phys. Rev. A.* **71** 012707.
[4] Wehlitz R, Juranić and Lukić D V 2008 *Phys. Rev. A.* **78** 033428.
[5] Kheifets A S and Bray I 2007 *Phys. Rev. A.* **75** 042703.
[6] Beyer H-J, West J B, Ross K J and De Fanis A 2000 *J. Phys. B: At. Mol. Opt. Phys.* **33** L767-771.
[7] Ross K J, West J B and Beyer H-J 1997 *J. Phys. B: At. Mol. Opt. Phys.* **30** L735-740.
[8] Ross K J, West J B, Beyer H-J and De Fanis A 1999 *J. Phys. B: At. Mol. Opt. Phys.* **32** 2927-2934.
[9] West J B, Ross K J, Beyer H-J, De Fanis A and Hamdy H 2001 *J. Phys. B: At. Mol. Opt. Phys.* **34** 4169-4181.
[10] Sheridan P, Grimm M and Sokell E 2008 *J. Phys. B: At. Mol. Opt. Phys.* **41** 165204.
[11] Maulbetsch F, Cooper I L and Dickinson A S 2000 *J. Phys. B: At. Mol. Opt. Phys.* **33** L119-125.
[12] Maulbetsch F and Briggs J S 1995 *J. Phys. B: At. Mol. Opt. Phys.* **28** 551-564.
[13] Jiménez-Mier J, Caldwell C D, Flemming M G, Whitfield S B and van der Meulen P 1993 *Phys. Rev. A.* **48** 442-451.
[14] Mansfield M W D and Newsom G H 1981 *Proc. R. Soc. Lond. A* **377** 431-448.

Double ionization of two-electron systems

L. U. Ancarani[1], C. Dal Cappello[1] and G. Gasaneo[2]

[1]Laboratoire de Physique Moléculaire et des Collisions, Université Paul Verlaine - Metz, 57078 Metz, France

[2]Departamento de Física, Universidad Nacional del Sur and Consejo Nacional de Investigaciones Científicas y Técnicas, 8000 Bahía Blanca, Buenos Aires, Argentina

E-mail: ancarani@univ-metz.fr

Abstract. We address various issues related to the double ionization by electron impact of two-electron systems. The emphasis will be put on the theoretical description of high incident energy $(e,3e)$ processes, for which the first Born approximation should be suitable. In the case of helium, absolute experimental data for fivefold differential cross sections are available in coplanar geometry. We will review and discuss the divergencies existing between the results obtained with different theoretical models, and those appearing when compared to the experiments in particular with respect to the absolute scale. We will then discuss some results obtained in a recently proposed out of plane geometry.

1. Introduction

The study of the double ionization of atoms by electron impact allows one to gain information on correlated systems. In a kinematically complete $(e,3e)$ experiment, the particles are detected in coincidence and a fivefold differential cross sections (FDCS) is deduced. In the case of two-electron atomic targets, like helium, one deals in the final state with a pure four-body Coulomb problem which, for high incident energies, can be reduced to a three-body problem from a theoretical point of view. However, even in this case, no exact wavefunction is known for either the scattering or the bound states. Hence, approximations are made, and $(e,3e)$ cross sections obtained with different theoretical descriptions of the initial and final states are generally not in agreement with each other. Moreover, when compared with high energy absolute experimental data on helium [1, 2], a rather confusing picture emerges. This has been the subject of many recent studies (see, e.g., [3, 4] and references therein), and will be further discussed in Section 2. We shall show how several models disagree with each other and with the experimental data: a particular attention will be put on the absolute scale of cross sections. In Section 3, we shall make some comments on the results obtained in a new out-of-plane geometry which was suggested as a tool to explore second Born effects. This contribution will end with some concluding remarks.

2. Coplanar geometry

Absolute $(e,3e)$ FDCS have been measured by Lahmam-Bennani *et al.* for high incident energy ($E_i = 5599$ eV) and ejected energies of (A) $E_1 = E_2 = 10$ eV [2] or (B) $E_1 = E_2 = 4$ eV [1] . In these coplanar experiments, the scattering angle was set at $\theta_0 = 0.45^0$, so that these conditions correspond to small momentum transfer $\mathbf{q}$ ($q = 0.24$ and $q = 0.22$ a.u., respectively) and thus to the dipolar regime. The absolute data allow one to make a detailed study – both in shape and magnitude – of the double ionization of helium. Cross sections were reported as a function of the angle of one of the ejected electrons θ_2, for 20 (A) and 16 (B) fixed values of θ_1.

The fully differential $(e,3e)$ cross section for the ejection of two electrons from a two-electron atom, is given theoretically within the first Born approximation (FBA) by

$$\frac{d^5\sigma}{d\Omega_0 d\Omega_1 d\Omega_2 dE_1 dE_2} = (2\pi)^4 \frac{k_0 k_1 k_2}{k_i} \left|T_{fi}(\mathbf{k}_0, \mathbf{k}_1, \mathbf{k}_2)\right|^2, \quad (1)$$

where $\mathbf{k}_i$ and $\mathbf{k}_0$ are the momenta of the incoming and outgoing projectile which define the scattering plane; $\mathbf{k}_1$ and $\mathbf{k}_2$ are the momenta of the ejected electrons after the collision (energy E_1, E_2); $d\Omega_0$, $d\Omega_1$ and $d\Omega_2$ denote, respectively, the solid angle elements for the scattered and the two ejected electrons. The transition matrix T_{fi} depends on the choice of the initial double bound state Ψ_i and the final double continuum Ψ_f^-; these wavefunctions which describe, respectively, the two electrons belonging to the target before and after their ejection, are not known exactly.

On the theoretical side, there have several published calculations which attempted to reproduce these experimental data: the Convergent Close Coupling (CCC) approach [1, 5]; the J–matrix method [6, 7]; a wave–packet evolution approach [8, 9]; the Green function expansion [10]; the distorted wave approaches with the "pure" 3C model [11, 12, 13, 14, 15, 16, 3, 4] or variants with effective charges [2, 17, 18] (it is worth underlying that all the 3C results, published by several groups and obtained with independent numerical codes, are consistent with each other); and the 6C model [11] (the latter takes into account the pairwise final state interactions between all four particles; as in the considered kinematical conditions the 6C model results do not differ much from those of the 3C [11], we shall no longer discuss it). Since the choice of effective charges is arbitrary, and the improvement with respect to the pure 3C model is highly dependent on the kinematical conditions, we shall not consider here any of its variants.

Since the experimental energy of the incoming projectile is high and the momentum transfer is small, the comparison between the theoretical calculations and the measured data can be performed within the frame of the FBA in the interaction of the projectile with the target atom. Indeed, explicit second Born calculations showed that little difference (about 10%) is observed with either the CCC approach [19] or the 3C model [20, 21]. The validity of the FBA, is further supported by the comparison of 3C calculations for electron and positron impact [11] (in contrast, differences between the two projectiles were observed in a Green function expansion combined with a very simple Ψ_i wavefunction [10], suggesting that the Born limit is not reached).

In the present contribution, we shall restrict the analysis to FBA results. Even within this approximation, the FDCSs obtained with different theoretical descriptions of the initial and final states are not in agreement with each other, and yield a rather confusing picture which is the subject of a lively debate amongst theorists.

Consider first the set, named (A) hereafter, of experimental data for two electrons ejected at $E_1 = E_2 = 10$ eV. Based on the published material, we may state that practically all theoretical models manage to reproduce overall the FDCS shapes; some minor details of the 20 situations may differ from one approach to the other. On the other hand, large differences appear as the magnitude is concerned and will be now investigated. To facilitate the task, we introduce the following experimental to theoretical FDCS ratio

$$R_{\theta_1}(E_1, E_2) = \frac{FDCS[\text{Experimental}]}{FDCS[\text{Theoretical}]} \quad (2)$$

evaluated at one of the cross sections maxima (see choice below), for a given set of energies (E_1, E_2) and geometrical configuration (θ_1). Ratios found for several models are presented in Table 1 with the intention of providing an overview of the magnitudes (different, but relatively similar, ratios can be deduced for other angles θ_1).

Let us start with those theoretical approaches which are fundamentally "numerical", and which should – within their convergence limits – be "exact". To compare with the measured data, all the FDCS obtained with the CCC approach were rescaled – at the theoretical maximum at $\theta_2 = 263^0$ for $\theta_1 = 41^0$ – by a factor 3.2 [1], hence $R_{41}(10, 10) = 3.2$. This important magnitude disagreement surprised Kheifets and Bray [5], as the same model gave excellent gauge agreement for $(\gamma, 2e)$ processes with two electrons escaping with 10 eV each. Recently, a purely numerical calculation based on a wave-packet evolution approach [8] yielded $(e, 3e)$ cross sections which are close, in both shape and magnitude, to those found with the CCC approach. However, a magnitude difference of about 20% is observed [9], hence $R_{41}(10, 10) = 2.6$. While these two independent "numerical" approaches support each other, and agree also when applied to $(\gamma, 2e)$ processes, they do not reproduce the absolute scale of the experimental $(e, 3e)$ data. It is worth mentioning that these two calculations were performed using different type of initial states: the CCC used a 20-parameters Hylleraas helium function while the second [8] derived its own wavefunction using a relaxation method. Calculations with the *J*-matrix approach to Faddeev-Merkuriev differential equations [6, 7] yield a reasonable agreement in $(e, 3e)$ cross sections magnitude, but to a lesser extent in their shapes. It should be noted that previous published calculations [6], where the pseudostates method was employed, showed important magnitude disagreements with experimental data, similarly to the CCC approach. The latest, presumably more converged, calculations [22] give a ratio $R_{41}(10, 10) = 1.2$.

Consider now more "analytical" descriptions of the double continuum. These include the Green's function expansion [10] which corresponds to the lowest order iteration of an incremental approach to the four-body Coulomb problem. In combination with an independent particle description of the helium bound state, the published results indicate that FDCS magnitudes are reproduced for five θ_1 values, i.e. $R_{\theta_1}(10, 10) \simeq 1$; the cross section for $\theta_1 = 41^0$ was not given. The combination of the "pure" 3C double continuum with different bound wavefunctions yields results which depend on the level of correlation included in the initial state. When highly correlated ("advanced") Ψ_i are used for helium [12, 13, 21, 14], a disagreement in magnitude is found, $R_{41}(10, 10) \simeq 0.5$. On the other hand, overall agreement in magnitude is found [11, 12, 15, 3] when using "simple" Ψ_i, like that proposed by Pluvinage (which solves or "diagonalizes" the three two-body Coulomb potentials) or others with a similar analytical structure; at $\theta_1 = 41^0$ we have $R_{41}(10, 10) = 0.9$ i.e. a theoretical cross section about 1.8 times smaller than with advanced Ψ_i.

In summary, when compared to the "numerical" models, the 3C model and the Green's function expansion, combined with simple helium wavefunctions, lead to the best level of agreement, on the absolute scale, with the experimental set (A). The agreement of the 3C model, however, is fortuitous [3] as will be briefly explained below.

Table 1. Ratio $R_{\theta_1}(E_1, E_2)$ defined by (2) obtained with different theoretical models.

	$R_{41}(10, 10)$	$R_{45}(4, 4)$
CCC [1]	3.2	14
Time Dependent [8, 9]	2.6	8.5
J-matrix [6, 7, 22]	1.2	7
3C (advanced Ψ_i) [Present] and [12, 4]	0.5	30
3C (simple Ψ_i) [Present] and [11, 12]	0.9	46

Let us now turn to the second set of data, named (B) hereafter, at $E_1 = E_2 = 4$ eV ejected energies. Contrary to the $E_1 = E_2 = 10$ eV case, the theoretical models do not reproduce so well the shapes of the 16 measured FDCS. Moreover, large magnitude differences appear again, and yield a picture which differs with the one described above.

The CCC results were rescaled – at the experimental maximum at $\theta_2 = 279^0$ for $\theta_1 = 45^0$ – by a factor 14 [1], hence $R_{45}(4,4) = 14$. At the same angle, the wave–packet evolution approach [9] reproduce again the CCC shapes but with $R_{45}(4,4) = 8.5$. The agreement (within 20%) observed for the (A) set does not hold for the (B) case, since the magnitude differs by up to 50%. On the other hand, the *J*-matrix method gives a similar FDCS shape but with $R_{45}(4,4) = 7$ [22]. In summary, the three "numerical" approaches, which in principle are exact solutions for the three-body problem, do not give the same conclusions as in the (A) case; they disagree on the absolute scale not only with the experimental data, but also between each other (moreover the relative ratios are different for sets (A) and (B)).

Turning to "analytical" descriptions, some results with the Green's function expansion were presented in [10] but with an unspecified scaling factor. As the 3C double electron continuum model is expected to be very poor for ejected energies as low as 4 eV, FDCS have not been published previously. Indeed, very important magnitude disagreements are found: $R_{45}(4,4) = 30$ with advanced helium Ψ_i and $R_{45}(4,4) = 46$ with simple Ψ_i (present calculations). Note that, again, a ratio of about 1.5 is observed between these two results: while this ratio played in favor in case (A), here it implies a worse disagreement with the measurements. Hence, the 3C magnitude agreement obtained in case (A) has to be considered as fortuitous. This conclusion could be reached after a series of investigations [12, 16, 21, 3, 4] which will now be summarized.

Let us recall that the 3C model consists in describing the double continuum by a symmetrized and orthogonalized version of the pure C3 (or BBK) function [23]. While this function diagonalizes all three two-body Coulomb interactions and has the correct asymptotic behavior when all interparticle distances are large, it describes poorly the behavior at intermediate distances or when one particle is far away from the other two. Previous applications to $(e,2e)$, $(\gamma,2e)$ or $(e,3e)$ processes have shown that the 3C model is able to reproduce reasonably well most of the shapes of the observed differential cross sections, but rather badly their magnitude. Curiously, when combined with simple Ψ_i, the 3C model leads to a very good agreement, on the absolute scale, for the $(e,3e)$ experimental data at $E_1 = E_2 = 10$ eV.

The particular choice used to describe the helium initial state Ψ_i has been thoroughly studied in combination with the 3C double continuum. In [11], the theory-experiment magnitude agreement was directly related to the quality of Pluvinage's wavefunction, in particular because it diagonalizes the Hamiltonian in all three Coulomb interactions. In [12, 16], it was shown that this property is not the deciding factor. The agreement was otherwise attributed [13] to the fact that the Pluvinage's wavefunction treat the interactions in the initial and final channels in a "balanced" way (on equal footing). In order to check this, we have studied [3] the interplay of the 3C final state with several initial bound state descriptions, not only for $(e,3e)$ but also for $(\gamma,2e)$ processes. The approximate initial (Pluvinage) and final (3C) wavefunctions appear to be "balanced" for $(e,3e)$ – but not for $(\gamma,2e)$ – processes since large $(\gamma,2e)$ gauge discrepancies are observed. Moreover, Pluvinage wavefunction is not the doubly bound analog of the C3 [24].

As the behavior of Ψ_i near the two-body coalescence points is concerned (Kato cusp conditions [25]), it has been demonstrated, unambiguously, that it has no influence in the experimental $(e,3e)$ conditions under scrutiny, with helium wavefunctions of both moderate (with several asymptotic behaviors) [12, 3] and very good quality [14, 4].

In view of the present analysis no theoretical model, whether of "numerical" or "analytical" nature, is able to reproduce the absolute scale of all presently available $(e,3e)$ measurements [1, 2]. Moreover, no agreement is found between numerically exact approaches like the CCC, wave-packet evolution and *J*-matrix, placing doubts about the approaches themselves since, within the FBA, agreement between all of them is expected. On the more "analytical" side, the 3C model is able to reproduce reasonably the FDCS shapes. However, whatever the chosen initial state trial wavefunction – diagonal or not, and satisfying or not Kato cusp conditions – the 3C model is not suitable to describe the absolute scale of the considered $(e, 3e)$ processes. The magnitude agreement obtained for set (A) with simple functions should be considered as fortuitous; the combination of simple functions can accidentally yield good absolute results by a lucky compensation of deficiencies.

In spite of these considerations on the absolute scale, we have calculated [4], for the first time, FDCSs for helium-like ions within the 3C model in combination with highly correlated initial states which, by construction, satisfy exactly two-body cusp conditions. No experimental data exist, but under similar kinematical and geometrical conditions, the H^- cross sections are predicted to be much larger than for helium, while for helium-like positive ions the magnitudes decrease fast with increasing nuclear charges. These results were to be expected, and have been observed in previous publications with, however, only poor models of Ψ_i and/or Ψ_f^- (see references in [4]). Moreover, thanks to the analytical character of the 3C final state and by adequately scaling the incident and the ejected energies, an approximate scaling law for $(e, 3e)$ FDCS has been mathematically identified along the helium isoelectronic sequence [4]. The scaling law is rather well verified, and can be used to easily predict cross sections which are extremely difficult to measure.

3. Out of plane geometry

Let us now turn to the new out of plane arrangement, recently proposed by Choubisa and Sud [26], which supposedly would allow one to observe second Born effects. In this quasi-binary geometry the two electrons are ejected with the same energy out of the scattering plane, and are such that the momentum of the centre of mass is constant and lies in the scattering plane.

In [26] the authors obtained, within a "3C approach", a series of cross sections for helium which are peculiar. Indeed, in coplanar geometry their results, in first and second Born approximations, are at odds with all other 3C previously published calculations obtained with independent numerical codes [11, 12, 15, 21, 14]: the FDCS differ substantially in both shapes and magnitudes, and moreover show a much larger sensitivity to the choice of the initial state wavefunction. As demonstrated in [27], their FBA calculations do not correspond to the 3C model but an approximate version of it, as later acknowledged by the authors [28]. As it is now known, the model they used (named 2CG in [27]) in which the electron-electron interaction is described only by a Gamow factor, is a poor description of the double continuum. It yields large differences when compared to the 3C model, and cannot reproduce the proper shapes and order of magnitude (it largely underestimates the cross sections) of experimental data. We should also point out that there is in Ref. [26] a magnitude problem related to the initial state normalization which brings in further confusion. As a consequence, Choubisa and Sud [26] observed also important second Born effects which have not been found before in the same kinematics with the 3C model [21] (or with the CCC approach [19]). In summary, in [27] it was made clear that the first (and second) Born calculations presented in [26] (i) do not correspond to a 3C but a 2CG calculation; (ii) show a systematic magnitude problem.

The investigations in the new out of plane geometry [26], whether in first or second Born approximations, are equally flawed. Again, their results are 2CG calculations, and the geometry effects observed were not confirmed within the 3C model. Hence, the new arrangement does not seem, unfortunately, to provide a useful tool to investigate geometrical or second order effects.

4. Concluding remarks

None of the theoretical models discussed here are satisfactory for both $(e, 3e)$ and $(\gamma, 2e)$ processes, so the study of differential double ionization processes cannot be considered as resolved from a theoretical point of view. Further studies are clearly needed to understand the interplay of initial and final states. Numerical approaches, like the CCC or the wave-packet evolution approach, while working well for $(\gamma, 2e)$ processes do not seem to work as well for the $(e, 3e)$ processes, in particular as the absolute scale is concerned. If agreement between these numerically exact approaches were found, and still in disagreement with the experimental data, then one of the two (or the two) statements would emerge: either higher Born orders are necessary or the experimental data are not on the right absolute scale. However, agreement between theories is not observed, bringing more confusion to the subject. The description of the final state by the 3C model manages to reproduce rather well the shapes; however severe magnitude problems appear for low energy ejected electrons. The agreement for two electrons ejected at 10 eV each, in combination with simple initial state wavefunctions, is fortuitous.

In view of the present analysis of $(e, 3e)$ processes, and in order to elucidate the absolute scale divergencies, it would be very useful to have new and independent absolute experimental data (i) in the same kinematical conditions discussed here in order to confirm the measurements of [1, 2]; (ii) for asymmetric ejected energies $E_1 \neq E_2$ since this would decrease electron-electron correlation effects in the final channel; and (iii) at higher ejected energies $E_1 = E_2$ as proposed in [3] (this would allow us to (a) confirm the fortuitous character of the 10 eV agreement observed with the 3C model combined with simple functions Ψ_i; (b) to see from which value $E_1 = E_2$ the 3C model is adequate to describe the double electron continuum).

Acknowledgments We are grateful to Azzedine Lahmam-Bennani for providing experimental data in tabulated form, and to Vladislav Serov and Yuri Popov for providing published and unpublished data. G. G. would like to acknowledge the support provided by the Région Lorraine and by PGI 24/F038 of the Universidad Nacional del Sur.

References

[1] Kheifets A, Bray I, Lahmam-Bennani, Duguet A and Taouil I 1999 *J. Phys. B* **32** 5047
[2] Lahmam-Bennani A *et al.* 1999 *Phys. Rev. A* **59** 3548
[3] Ancarani L U, Gasaneo G, Colavecchia F D and Dal Cappello C 2008 *Phys. Rev. A.* **77** 062712
[4] Ancarani L U, Dal Cappello C, Charpentier I, Rodriguez K V and Gasaneo G 2008 *Phys. Rev. A* **78** 062709
[5] Kheifets A S and Bray I, 2004 *Phys. Rev. A* **69** 050701(R)
[6] Knyr V A, Nasyrov V V and Popov Yu V 2003 *Correlation and Polarization in Photonic, Electronic, and Atomic Collisions* (AIP Conf. Proc. Vol. 697) ed G F Hanne *et al.* (New York: AIP, Melville) p 76
[7] Zaytsev S A, Knyr V A and Popov Yu V 2007 *Phys. At. Nucl.* **70** 676
[8] Serov V V, Derbov V L, Joulakian B B and Vinitsky S I 2007 *Phys. Rev. A* **75** 012715
[9] Serov V V, private communication
[10] Berakdar J 2000 *Phys. Rev. Lett.* **85** 4036
[11] Jones S and Madison D H 2003 *Phys. Rev. Lett.* **91**, 73201
[12] Ancarani L U, Montagnese T and Dal Cappello C 2004 *Phys. Rev. A* **70** 12711
[13] Jones S, Macek J H and Madison D H 2004 *Phys. Rev. A* **70** 12712
[14] Chuluunbaatar O *et al.* 2006 *Phys. Rev. A* **74** 14703
[15] Colavecchia F D, Gasaneo G and Rodriguez K V 2007 *J. Electron Spectrosc. Relat. Phenom.* **161** 73
[16] Ancarani L U and Dal Cappello C 2007 *J. Electron Spectrosc. Relat. Phenom.* **161** 22
[17] Gasaneo G, Otranto S and Rodriguez K V 2006 *Proc. XXIV Int. Conf. on Photonic, Electronic and Atomic Collisions* (Singapore: World Scientific) p 360
[18] Ancarani L U, Dal Cappello C and Montagnese T 2006 *Ionization, Correlation, and Polarization in Atomic Collisions* (AIP Conf. Proc. Vol. 811) ed A Lahmam-Bennani *et al.* (New York: AIP, Melville) p 1
[19] Kheifets A S 2004 *Phys. Rev. A* **69** 032712
[20] Grin M, Dal Cappello C, El Mkhanter R and Rasch J 2000 *J. Phys. B* **33** 131
[21] Ancarani L U, Montagnese T and Dal Cappello C 2005 *Electron and Photon Impact Ionization and Related Topics* (IOP Conf. Proc. Vol 183) ed B Piraux (London: IOP) p 21
[22] Popov Yu V, private communication
[23] Joulakian B, Dal Cappello C and Brauner M 1992 *J. Phys. B* **25** 2863
[24] Ancarani L U and Gasaneo G 2007 *Phys. Rev. A* **75** 032706
[25] Kato T 1957 *Commun. Pure Appl. Math.* **10** 151
[26] Choubisa R and Sud K K 2008 *J. Phys. B* **41** 035202.
[27] Dal Cappello C and Ancarani L U 2008 *J. Phys. B* **41** 208001
[28] Choubisa R and Sud K K 2008 *J. Phys. B* **41** 208002

Journal of Physics: Conference Series **212** (2010) 012026 doi:10.1088/1742-6596/212/1/012026

Multiply excited molecules in electron collisions as probed by energy and angle-resolved coincidence measurements between a scattered electron and a photon from a fragment atom

Takeshi Odagiri and Noriyuki Kouchi

Department of Chemistry, Tokyo Institute of Technology, O-okayama 2-12-1-W4-4, Meguro-ku, Tokyo 152-8551, Japan

Abstract. Doubly excited states of CH_4 in electron collisions were investigated for the first time by measuring electron energy-loss spectra tagged with the emission of the Lyman-α photon. It turned out that the collisional cross section resulting in the emission of the Lyman-α photon originating from the doubly excited states are larger than or comparable to those originating from the nearby singly excited states and the electron interaction with molecules promotes the double excitation against the single excitation as compared with the photon interaction.

1. Introduction

Multiply excited states of molecules are highly correlated systems involving light and heavy particles strongly interacting with each other and thus revealing the dynamics of them in electron collisions is one of important and challenging subjects in physics and chemistry. However, it has been difficult to observe the multiply excited molecules until we established a new experimental method based on the coincidence detection between a scattered electron and a photon emitted from an excited neutral fragment, *i.e. the coincident electron-energy-loss spectroscopy* [1]. The key to observing the multiply excited states embedded in ionization continua is measuring the cross sections free from ionization. In the present paper, the recent investigations on the the dynamics of multiply excited states of CH_4 in electron collisions [2–4] are presented. The density of the dipole oscillator strength for the emission of the Lyman-α fluorescence in the photoexcitation of CH_4 was measured as well since the comparative study between electron and photon interactions gives us a deeper understanding of the multiple excitation [2, 5].

These studies strongly motivated by our previous measurements of the cross sections for the emission of dispersed fluorescence from neutral atoms and radicals produced in the photoexcitation of CH_4 NH_3 and H_2O [6–8]. We found that the dipole oscillator strengths for the emission of the Balmer fluorescence originating from the doubly excited states are larger than or comparable to those from the nearby singly excited states, which seems to indicate the breakdown of the independent electron model.

2. Experiment

The experimental apparatus is composed of an electron energy loss spectroscopy system for static gas target, a Lyman-α photon detector and a delayed coincidence system. Electrons scattered into a small solid angle were energy-analyzed by a hemispherical electron energy analyzer and detected by a channel electron multiplier. True coincidence rates were normalized for the target gas density, incident electron beam current and geometric factor to obtain differential cross sections for electron scattering accompanied by the emission of the Lyman-α photon, $\mathrm{d}^3\sigma/\mathrm{d}E\mathrm{d}\Omega_e\mathrm{d}\Omega_{ph}$, where $\mathrm{d}\Omega_e$ and $\mathrm{d}\Omega_{ph}$ are elements of solid angles for the scattered electron and emitted photon, as a function of energy loss E, *i.e.*, the coincident electron-energy-loss spectrum. Details of the experiments were described in [2] and [5].

For the measurement of the scattering angle dependence of the coincident electron energy loss spectra, we developed a new apparatus using an imaging detector, *i.e.*, we employed an electron energy analyzer equipped with a position sensitive detector [3].The fast-slow coincidence technique between energy-dispersed electrons and Lyman-α photons enable simultaneous accumulation of the coincidence time spectra at different energy losses and hence the time needed for measuring one coincident electron-energy-loss spectrum was greatly reduced as compared with that by the apparatus mentioned above. To put the coincident energy loss spectra measured at different scattering angles on the same vertical scale, we calculated the geometrical factor as a function of electron energy loss and scattering angle based upon the simulation of the trajectories of scattered electrons and emitted photons.

Measurement of the density of the dipole oscillator strength for the emission of the Lyman-α photon in the photoexcitation of CH_4 as a function of incident photon energy were carried out at the BL-20A of the Photon Factory, KEK. Linearly polarised light from a 3m normal incidence monochromator was introduced into a gas cell equipped with the Lyman-α photon detector which is the same as that used in the electron collision experiment. The details were mentioned in [2] and [5]. In this paper we use the same symbol E for the energy loss and incident photon energy because both quantities give the excitation energy of the target molecule.

3. Result and discussion

The coincident electron-energy-loss spectrum of CH_4 measured at incident electron energy of 80 eV and electron scattering angle of 10° (hereafter abbreviated as "80eV/10°") is shown in fig. 1(b) together with the ordinary electron-energy-loss spectrum of CH_4 measured at 80eV/10° in (a). It is noted that superexcited states of CH_4 are clearly visible in fig. 1(b) owing to the coincidence detection of the scattered

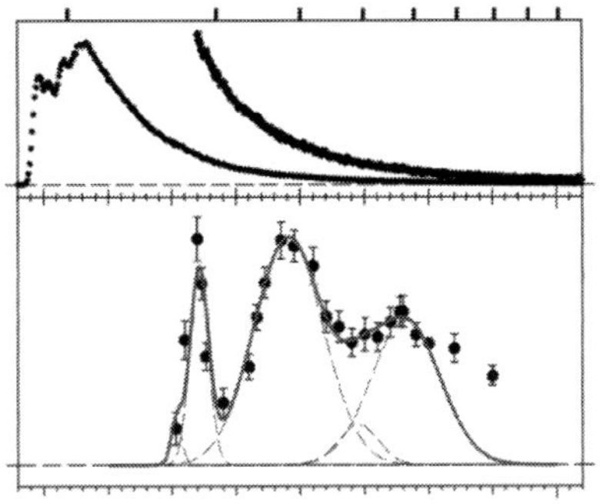

Figure 1. The electron energy-loss spectra of CH_4 measured at 80 eV incident electron energy and 10° electron scattering angle [2]: (a) the ordinary energy-loss spectrum, (b) the spectrum tagged with the emission of the Lyman-α photons, *i.e.* the coincident energy-loss spectrum. The curves in (b) show the results of the fits (see text). The square of the momentum transfer, K^2, is indicated on the upper axis.

electron and the Lyman-α photon, while those are buried in the continuous cross section curve due to direct ionizations in the ordinary energy-loss spectrum in fig. 1(a).

The superexcited states of CH_4 take also part in the emission of the Lyman-α photon following the photoexcitation. In fig. 2(b) the density of the dipole oscillator strength (DOS) for the emission of the Lyman-α photon, $(df/dE)_{Ly-\alpha}$, is shown as a function of incident photon energy E. The DOS curve was decomposed into five contributions from superexcited states at 20.3, 22.1, 29.5, 34.7 and 38.6 eV by a fit based on the multi-dimensional reflection approximation [9] and a semiclassical treatment of the decay dynamics of the superexcited states involved. The peaks at 20.3 and 22.1 eV were well investigated by various experimental methods and assigned to the singly excited $(2a_1)^{-1}(3pt_2)$ and $(2a_1)^{-1}(4pt_2)$ superexcited states [6]. It followed from the vertical ionization potentials for the $(1t_2)^{-1}$ (14.5 eV), $(2a_1)^{-1}$ (23.0 eV) and $(1a_1)^{-1}$ (290.707 eV) states of CH_4^+ that the other three peaks are attributed to nothing other than the doubly excited states. Kato *et al* [6] found three doubly excited states of CH_4, D1, D2 and D3, in the densities of the dipole oscillator strength for the emission of Balmer series and CH(A, B$\rightarrow$X) fluorescence as a function of incident photon energy in the photoexcitation of CH_4. On account of the energy positions and peak widths the peaks at 29.5 and 34.7 eV in fig. 2(b) seem to be attributed to the doubly excited D2 and D3 states, respectively. The 38.6 eV peak seems to originate from the newly found doubly excited state, which was labelled D4. It was found that the DOS originating from the doubly excited D2 - D4 states are larger than or comparable to those from the nearby singly excited $(2a_1)^{-1}(npt_2)$ (n=3,4) states, which seems to indicate the breakdown of the independent electron model.

In fig. 2(b) the energy positions of the four correlation states of CH_4^+ found in the X-ray photoelectron spectrum of CH_4 [10] are shown by vertical bars. It turned out from the viewpoint of energy that the doubly excited D2 - D4 states are built on the correlation states of CH_4^+ since a doubly excited state, *i.e.* two-particle/two-hole state, is expected to lie in a lower energy-range than the energy of the corresponding correlation state, *i.e.* one-particle/two-hole state, as an ion-core. In conclusion, the

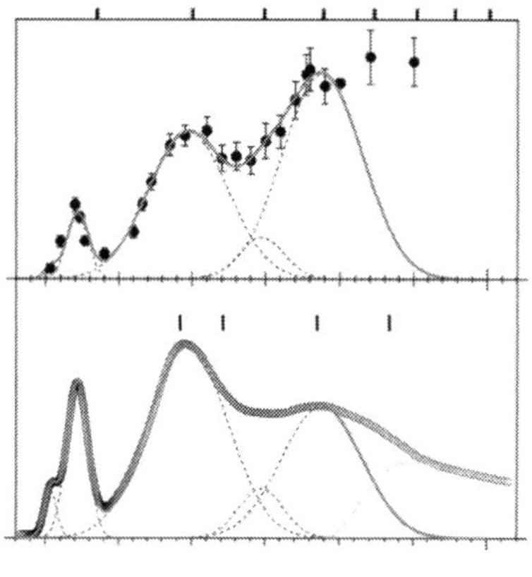

Figure 2. (a) The density of the effective generalized oscillator strength for the emission of the Lyman-α photon in CH_4 at 80 eV incident electron energy and 10° electron scattering angle, $(\mathrm{d}F/\mathrm{d}E)_{Ly-\alpha}$, as a function of electron energy-loss E and (b) the density of the dipole oscillator strength for the emission of the Lyman-α photon in CH_4, $(\mathrm{d}f/\mathrm{d}E)_{Ly-\alpha}$, as a function of incident photon energy E. The curves show the results of the fits (see text). The vertical bars in (b) represent the energies of the correlation states of CH_4^+ [10].

doubly excited D2, D3 and D4 states are built on the $(1t_2)^{-2}(3a_1)$, $(2a_1)^{-1}(1t_2)^{-1}(2t_2)$ and $(2a_1)^{-2}(3a_1)$ states of CH_4^+, respectively.

The coincident electron energy-loss spectrum was decomposed into the five contributions from the superexcited states by the similar fitting procedure as is shown in fig. 1(b). The superexcited states are the same as those in fig. 2(b). It is interesting that the cross sections originating from the doubly excited states are larger than or comparable to those originating from the nearby singly excited $(2a_1)^{-1}(npt_2)$ (n=3,4) states. We note that forbidden superexcited states make almost no contribution in the coincident electron energy-loss spectrum at 80eV/10°, while optically forbidden doubly excited states make a clear contribution at 100eV/8° electron collision with NH_3 [5].

Formation of the doubly excited D2 - D4 states of CH_4 by photon and electron interactions is discussed in terms of the oscillator strengths as follows. The cross section originating from the electronic state 'α', $\mathrm{d}^3\sigma_\alpha/\mathrm{d}E\mathrm{d}\Omega_e\mathrm{d}\Omega_{ph}$, is related to density of the effective generalized oscillator strength (effective GOS) for the emission of the Lyman-α photons originating from 'α', $(\mathrm{d}F_\alpha(K,k_i)/\mathrm{d}E)_{Ly-\alpha}$, by

$$\frac{\mathrm{d}^3\sigma_\alpha}{\mathrm{d}E\mathrm{d}\Omega_\mathrm{e}\mathrm{d}\Omega_\mathrm{ph}} = \frac{k_\mathrm{s}}{k_\mathrm{i}}\frac{4R_y}{E}\frac{1}{K^2}\frac{1}{4\pi}\left(\frac{\mathrm{d}F_\alpha(K,k_\mathrm{i})}{\mathrm{d}E}\right)_{Ly\text{-}\alpha}, \tag{1}$$

where k_i and k_s are the magnitudes of the wavenumber vectors of the incident and scattered electrons, respectively, R_y the Rydberg energy and K the momentum transfer [11]. According to the limit theorem [12, 13],

$$\left(\frac{\mathrm{d}F_\alpha(K,k_\mathrm{i})}{\mathrm{d}E}\right)_{Ly\text{-}\alpha} \longrightarrow \left(\frac{\mathrm{d}f_\alpha}{\mathrm{d}E}\right)_{Ly\text{-}\alpha} \tag{2}$$

as the momentum transfer $K \to 0$. Fig. 2(a) shows $(\mathrm{d}F(K,k_\mathrm{i})/\mathrm{d}E)_{Ly\text{-}\alpha} = \sum_\alpha \mathrm{d}F_\alpha(K,k_\mathrm{i})/\mathrm{d}E)_{Ly\text{-}\alpha}$ as a function of energy loss E. It was concluded from figures 2(a) and (b) that the electron interaction with CH_4 at 80eV/10° involves the mechanism that enhances the double excitation against the single excitation as compared with the photon interaction. The photon interaction is expressed by a sum of single-electron dipole operators. The same conclusion was also obtained in the study on the electron (100eV/8°) and photon interactions with NH_3 [5].

To investigate further the electron-CH_4 collision to form the doubly excited states, the coincident electron energy loss spectra were measured at 80 eV incident electron energy and electron scattering angles of 4°, 12°, 24°, 36° and 48° [4]. The spectral shape changes drastically with increasing the electron scattering angle. All the coincident energy-loss spectra were decomposed into the contributions from the same singly and doubly excited states as those in fig. 2(a) and no discernible contribution of optically forbidden doubly excited states was observed up to 48°. It is remarkable that the fraction of the doubly excited D4 state increases and that of the doubly excited D2 state decreased with increasing the electron scattering angle. The details will be given in [4].

4. Conclusions

We measured electron energy-loss spectra of CH_4 tagged with the emission of the Lyman-α photon, *i.e.* the coincident electron-energy-loss spectra, and the density of the dipole oscillator strength for the emission of the Lyman-α photon to investigate the double excitation in electron and photon interactions. The doubly excited states of CH_4 contributing to the emission of the Lyman-α photon were revealed for the first time. The density of the dipole oscillator strength seems to show the breakdown of the independent electron model: the dipole oscillator strengths originating from the doubly excite states are larger than or comparable to those originating from the nearby singly excited states. It turned out that the cross sections of the electron interaction originating from the doubly excited states are also larger than or comparable to those originating from the nearby singly excited states and the electron interaction promotes the double excitation against the single excitation as compared with the photon interaction.

References

[1] Odagiri T and Kouchi N 2004 *Phys. Scr.* 183
[2] Fukuzawa H, Odagiri T, Nakazato T, Murata M, Miyagi H and Kouchi N 2005 *J. Phys. B: At. Mol. Opt. Phys.* **38** 565
[3] Nakazato T, Odagiri T, Fukuzawa H, Miyagi H, Ishikawa L, Kitajima M and Kouchi N 2007 *J. Phys. B: At. Mol. Opt. Phys.* **40** 2459
[4] Yachi K, Odagiri T, Ishikawa L, Nakazato T, Ohno N, Tsuchida T and Kouchi N 2009 To be published
[5] Ishikawa L, Odagiri T, Yachi K, Nakazato T, Kurokawa M, Kitajima M and Kouchi N 2008 *J. Phys. B: At. Mol. Opt. Phys.* **41** 195204
[6] Kato M, Kameta K, Odagiri T, Kouchi N and Hatano Y 2002 *J. Phys. B: At. Mol. Opt. Phys.* **35** 4383
[7] Kato M, Odagiri T, Kameta K, Kouchi N and Hatano Y 2003 *J. Phys. B: At. Mol. Opt. Phys.* **36** 3541
[8] Kato M, Odagiri T, Kodama K, Murata M, Kameta K and Kouchi N 2004 *J. Phys. B: At. Mol. Opt. Phys.* **37** 3127

[9] Schinke R 1993 *Cambridge Monographs on Atomic, Molecular and Chemical Physics 1. Photodissociation Dynamics* (Cambridge: Cambridge University Press) p 115

[10] Gothe M C, Wannberg B, Karlsson L, Svensson L, Baltzer P, Chau F T and Adam M Y 1991 *J. Chem. Phys.* **94** 2536

[11] Odagiri T, Uemura N, Koyama K, Ukai M, Kouchi N and Hatano Y 1996 *J. Phys. B: At. Mol. Opt. Phys.* **29** 1829

[12] Lassettre E N, Skerbele A and Dillon M A 1969 *J. Chem. Phys.* **50** 1829

[13] McDaniel E W 1989 *Atomic Collisions, Electron and Photon Projectiles* (New York: Wiley) p 319

The dynamics of the ionization of atoms exposed to strong low-frequency fields

Javier Madroñero[1,2] and Bernard Piraux[1]

[1] Laboratoire de Physique Atomique, Moléculaire et Optique (PAMO), Université catholique de Louvain, 2, chemin du Cyclotron, 1348 Louvain-la-Neuve, Belgium

[2] Physik Department, Technische Universität München, James-Franck-Straße, 85747 Garching, Germany

E-mail: Javier.Madronero@ph.tum.de

Abstract. We present a theoretical study of strong-low-frequency-field ionization of atoms in the intermediate regime where the Keldysh parameter is close to one. For that purpose we have developed a novel method based on an *explicit* algorithm for the propagation of the time-dependent Schrödinger equation which is known to be very stiff. This method overcomes the stiffness by taking into account the eigenfrequencies of the interacting system. With the help of this approach we are able in particular to investigate the ionization of the hydrogen atom by ultra-short (25 fs) 790 nm laser pulses at intensities 0.5–6 $\times 10^{14}$ W/cm^2.

1. Introduction

Atomic photoionization under intense laser radiation has been intensively investigated. It is widely perceived that the ionization process occurs through multiphoton absorption or tunnelling [1–4]: in the low-intensity/short-wavelength limit, Above-Threshold Ionization (ATI) is the absorption of photons beyond the minimum required for ionization. In the high-intensity/long-wavelength limit, Keldysh theory suggests that the valence electron escapes by tunnelling through the barrier created by the time-dependent bending of the Coulomb potential by the field followed by an oscillating motion of the free electron induced by the field of frequency ω and intensity I. These two limits are linked through the dimensionless parameter $\gamma = \sqrt{I_p/2U_p}$, where I_p is the ionization potential and $U_p = I^2/4\omega^2$ is the pondemorotive energy.

At intensities of the order of $10^{14} - 10^{15}$ W/cm^2 and infrared frequencies, there is no clear separation between the regime where multiphoton processes dominate the ionization of atoms exposed to such fields and the strong field limit dominated by tunnel ionization. In fact, there is a subtle interplay between the two mechanisms as indicated by the fact that the Keldysh parameter γ is close to one. From the theoretical point of view, de Bohan *et al.* [5] have shown, in the case of the interaction of atomic hydrogen with such fields, that the low-energy part of the ATI spectrum does not result from tunnel ionization but rather involves some multiphoton transitions. This explains why the tunnelling formula for the total ionization rate gives results that disagree with those obtained by solving the Time-Dependent Schrödinger Equation (TDSE). From the experimental side, this was confirmed by Rudenko *et al.* [6] who found that the ionization of rare atoms in this regime shows evidence of resonance structures in the electron ATI spectrum which are thought to be characteristic of the multiphoton regime. By contrast, another experiment [7] in the same regime shows that the ionization rate follows adiabatically the oscillations of the field. Understanding the interplay of such mechanisms poses a challenge for both experiment and theory. From the theoretical point of view, one of the main difficulties to deal with this regime is the solution of the TDSE, which requires an accurate treatment of the very high-order angular momentum coupling in this regime of low frequency, rather long pulse duration and high field intensity. Due to this and the infinite range of the coulombic interactions, this requires either very large grids or big bases. In both cases, one has to solve large systems of coupled first order differential equations that are well known to be stiff [8]. By this, it is meant that the time step decreases rapidly with the increasing size of the system. The origin of the stiffness is clear: by increasing the size of the grid or the basis, the diagonalization of the atomic Hamiltonian generates large positive energy eigenvalues. It is precisely the highest eigenvalue that controles the time step while leading to highly oscillating solutions. In order to overcome this problem, one can use implicit time propagation methods. This requires to solve very large systems of algebraic equations at each time step. A typical example of such an implicit scheme is the Crank-Nicholson algorithm which is used in grid methods based on finite differences [9]. This propagation method is tractable when the system is banded but as soon as the complexity of the atomic system increases, the bandwidth of the system increases rapidly together with the computation time. An alternative way of overcoming the stiffness of the problem is to propagate in the atomic basis where the atomic Hamiltonian is diagonal. Time propagating in the atomic basis has three advantages: first, it is possible to eliminate from the propagation, very high energy eigenstates that play a minor role in the dynamics. Second, it allows one to work in the interaction picture where the free evolution of the atomic system is somehow subtracted. And finally, explicit methods which involve only matrix-vector products are numerically stable in this case. However, this method requires the diagonalization of large matrices which is a computationally very demanding problem.

In this contribution, we present a theoretical study of strong-field ionization of atoms in the intermediate regime where the Keldysh parameter is close to 1. We consider the interaction of atomic hydrogen with a 790 nm laser pulse which duration is of the order of 25 femtoseconds and the intensity ranges from 5×10^{13} to 6×10^{14} W/cm^2. The propagation of the spectral representation of the TDSE is accomplished through a very efficient novel *explicit method* which is briefly described in Sec. 2. Our results are presented in Sec. 3.

2. Time propagation algorithm

In this section, a brief review of the method proposed in [10, 11] is given. Let us start with the m-dimensional *stiff* first order differential equation

$$\boldsymbol{y}' = \boldsymbol{f}(x, \boldsymbol{y}), \qquad \boldsymbol{y} = (y_1, y_2, \dots, y_m), \tag{1}$$

where $\boldsymbol{f}(x, y)$ is in general a complex m-dimensional function.

The stiffness of the equation leads to a solution $\boldsymbol{y}(x)$ which is an oscillating function. In a given interval (x_n, x_{n+1}), $x_{n+1} = x_n + h$, with h a small number, $\boldsymbol{y}(x)$ is approximated by the function

$$\boldsymbol{F}(x) = (I - e^{\Omega_1 x})\boldsymbol{a} - (I - e^{-\Omega_2 x})\boldsymbol{b} + \boldsymbol{c}, \tag{2}$$

with I the identity matrix, $\Omega_i = \mathrm{diag}(\omega_1^{(i)}, \dots, \omega_m^{(i)})$, $i = 1, 2$, and $\boldsymbol{a}, \boldsymbol{b}, \boldsymbol{c}$ constant vectors. The complex numbers $\omega_1^{(i)}, \dots, \omega_m^{(i)}$, $i = 1, 2$ are called *stiffness parameters*. Assuming that $\boldsymbol{F}(x)$ coincides with $\boldsymbol{y}(x)$ at x_n and x_{n+1}, that $\boldsymbol{F}'(x)$ coincides with $\boldsymbol{f}(x, \boldsymbol{y})$ at x_n, and that $\boldsymbol{F}''(x)$ coincides with $\boldsymbol{f}'(x, \boldsymbol{y})$ at x_n, the solution $\boldsymbol{y}_{n+1} = \boldsymbol{y}(x_{n+1})$ at x_{n+1} can be expressed recursively in terms of $\boldsymbol{y}_n = \boldsymbol{y}(x_n)$, $\boldsymbol{f}_n = \boldsymbol{f}(x_n, \boldsymbol{y}_n)$ and $\boldsymbol{f}_n^{(1)} = d\boldsymbol{f}/dx|_{x=x_n}$ according to

$$\boldsymbol{y}_{n+1} = \boldsymbol{y}_n + R\boldsymbol{f}_n + S\boldsymbol{f}_n^{(1)}. \tag{3}$$

R and S are diagonal matrices which can be written in terms of the stiffness parameters:

$$R = \Omega_2 \Phi - \Omega_1 \Psi, \qquad S = \Phi + \Psi, \tag{4}$$

where Φ and Ψ are diagonal matrices whose nonzero entries are

$$\Phi_i = \frac{e^{\omega_i^{(1)} h} - 1}{\omega_i^{(1)}(\omega_i^{(1)} + \omega_i^{(2)})} \quad \text{and} \quad \Psi_i = \frac{e^{-\omega_i^{(2)} h} - 1}{\omega_i^{(2)}(\omega_i^{(1)} + \omega_i^{(2)})}. \tag{5}$$

The recursive relation (3) depends on the so far unknown stiffness matrices Ω_1 and Ω_2. However, these matrices can be written in terms of the function $\boldsymbol{f}(x_n, \boldsymbol{y}_n)$ and its derivatives up to 3rd order at x_n. This is obtained after substituting the Taylor expansion of $\boldsymbol{y}_{n+1} = \boldsymbol{y}(x_n + h)$, and the Maclaurin series of $\exp(\Omega_1 h)$ and $\exp(-\Omega_2 h)$ in (3) and equating the coefficients of h^k, $k = 0, \dots, 4$. The components of the stiffness matrices obtained after solving these equations read [10]

$$\omega_i^{(1)} = \frac{1}{2}\left[-D_i + \sqrt{D_i^2 + 4E_i}\right] \quad \text{and} \quad \omega_i^{(2)} = \omega_i^{(1)} + D_i, \tag{6}$$

where D_i and E_i, $i = 1, \dots, m$ are given in terms of the respective components $f_{in}^{(k)}$ of the derivatives $\boldsymbol{f}_n^{(k)}$, $k = 0, 1, 2, 3$, of $\boldsymbol{f}(x, \boldsymbol{y})$ at $x = x_n$ by

$$D_i = \frac{f_{in}^{(0)} f_{in}^{(3)} - f_{in}^{(1)} f_{in}^{(2)}}{f_{in}^{(1)} f_{in}^{(1)} - f_{in}^{(0)} f_{in}^{(2)}}, \quad \text{and} \quad E_i = \frac{f_{in}^{(1)} f_{in}^{(3)} - f_{in}^{(2)} f_{in}^{(2)}}{f_{in}^{(1)} f_{in}^{(1)} - f_{in}^{(0)} f_{in}^{(2)}}, \quad i = 1, \dots, m. \tag{7}$$

If the denominator of the previous expressions is zero, special care has to be taken into account [12].

The implementation of the recursion (3) is now rather simple. It requires the calculation of the function $\boldsymbol{f}_n$ and its derivative $\boldsymbol{f}_n^{(1)}$ at each value of x_n. For the stiffness matrices Ω_1 and Ω_2, and thus also for the matrices R and S, the derivatives $\boldsymbol{f}_n^{(2)}$ and $\boldsymbol{f}_n^{(3)}$ are also needed. Ω_1 and Ω_2 have to be calculated in principle at each integration step, since they characterize the local frequencies of the solution $\boldsymbol{y}(x)$ [12]. In addition, the truncation error [10, 12]

$$T_{n+1} = \frac{h^5}{5!}\frac{1}{\omega_1 + \omega_2}\left[(\omega_1 + \omega_2) f_n^{(4)} + (\omega_2^4 - \omega_1^4) f_n^{(1)} - (\omega_1^4 \omega_2 + \omega_1 \omega_2^4) f_n^{(0)}\right] + \mathcal{O}(h^6) \tag{8}$$

can be used to control the size of the integration step, e.g., by imposing a boundary criterion for $|T_n|$. For this also the derivative $\boldsymbol{f}_n^{(4)}$ must be provided.

3. Resonance structures in the tunnelling regime of ATI of hydrogen atoms

In this section we consider hydrogen atoms exposed to intense low frequency femtosecond laser pulses. We consider linearly polarized light, and apply the dipole approximation. In the velocity gauge the interaction operator reads

$$V(t) = \boldsymbol{A}(t) \cdot \boldsymbol{p}, \tag{9}$$

where the vector potential takes the form $\boldsymbol{A}(t) = A_0(t)\sin(\omega t)\boldsymbol{e}_z$, for $-\tau/2 \le t \le \tau/2$, and zero outside this time interval. $A_0(t) = A_0 \cos^2(\pi t/\tau)$ is the envelope containing n_c optical cycles, ω is the pulse angular frequency, $\tau = 2\pi n_c/\omega$ is the pulse duration and $\boldsymbol{e}_z$ is the unit vector along the z-axis in the laboratory frame. The Hamiltonian of the driven atom is thus

$$H(t) = H_0 + V(t), \tag{10}$$

Journal of Physics: Conference Series **212** (2010) 012027 doi:10.1088/1742-6596/212/1/012027

with

$$H_0 = \frac{p^2}{2} - \frac{Z}{r}. \tag{11}$$

We use a spectral representation of the Hamiltonian in terms of spherical harmonics and Coulomb Sturmian functions defined by

$$S_{n\ell}^{(\alpha)}(r) = \sqrt{\frac{(n-\ell-1)!}{2(n+\ell)!}} \left(\frac{2r}{\alpha}\right)^{\ell+1} L_{n-\ell-1}^{2\ell+1}\left(\frac{2r}{\alpha}\right) e^{-r/\alpha}, \tag{12}$$

where $L_n^{(\beta)}(x)$ are the Laguerre polynomials and α is a real number called the dilation parameter.

Substitution of the expansion

$$\psi(\boldsymbol{r}, t) = \sum_{n\ell} c_{n\ell}(t) \frac{S_{n\ell}^{(\alpha)}(r)}{r} Y_{\ell m}(\theta, \varphi) \tag{13}$$

of the wave packet $\psi(\boldsymbol{r}, t)$ in the TDSE leads to the following matrix representation:

$$iS\frac{d\boldsymbol{y}}{dt} = \mathcal{A}\boldsymbol{y} + g(t)\mathcal{V} \tag{14}$$

where S is the overlap matrix, $\mathcal{A}$ is the matrix representation of the unperturbed Hamiltonian of the system, $g(t) = A_0 \cos^2(\pi t/\tau)\sin(\omega t)$ is the scalar time-dependent part of the field interaction (9) and $\mathcal{V}$ is the matrix representation of the dipole operator p_z. $\boldsymbol{y}$ is the vector of the coefficients $c_{n\ell}(t)$.

In order to solve numerically the TDSE, the basis (12) has to be truncated: $n_{\max}$ is the number of Sturmian functions per angular configuration for each of the values of $\ell = 0, \ldots, \ell_{\max}$.

We consider the ionization process from the ground state of atomic hydrogen exposed to a laser pulse of 790 nm ($\omega = 0.057$ a.u.), which is chosen to match the Ti:sapphire lasing frequency, and peak intensities: from 5×10^{13} to 6×10^{14} W/cm^2. The Keldysh parameter γ for these choices ranges from $\gamma = 1.510$ to $\gamma = 0.435$. We consider a pulse of $n_c = 10$ optical cycles which corresponds to a pulse duration of about 25 fs (1102.3 a.u.). The laser parameters are chosen in this way in order to match those of Rudenko's experiment [6]. The minimum number of photons needed for ionization ranges from 10 to 31, which implies a high-order angular momentum coupling. Therefore, an accurate solution of the TDSE requires a rather big basis containing a large number of Sturmian functions (up to $n_{\max} = 800$) and large number of angular configurations (up to $\ell_{\max} = 70$).

Fig. 1 displays the energy spectrum for $I = 5 \times 10^{13}$ (a), 10^{14} (b), 4×10^{14} (c), and 6×10^{14} W/cm^2 (d) obtained after propagation of the TDSE with the help of the method described above. The accuracy and efficiency of this method has been confronted with a diagonally implicit Runge-Kutta propagator [12]. Our approach is typically one order of magnitude faster than the implicit Runge-Kutta method. The results for $I = 5 \times 10^{13}$ and 10^{14} W/cm^2 were obtained with $\ell_{\max} = 50$, and $n_{\max} = 300$ and 400, respectively. For higher field intensities some features of the energy spectrum at low energy can be already seen by coupling at least 40 angular momenta with $n_{\max} = 400$, however the resolution of structures at higher energies require larger bases (Fig. 1(c), (d)).

For a field intensity $I = 5 \times 10^{13}$ W/cm^2 the ionization process is a multiphoton process as expected. The peaks in the spectrum are precisely located at positions given by

$$E_n = n\omega - U_p - I_p, \tag{15}$$

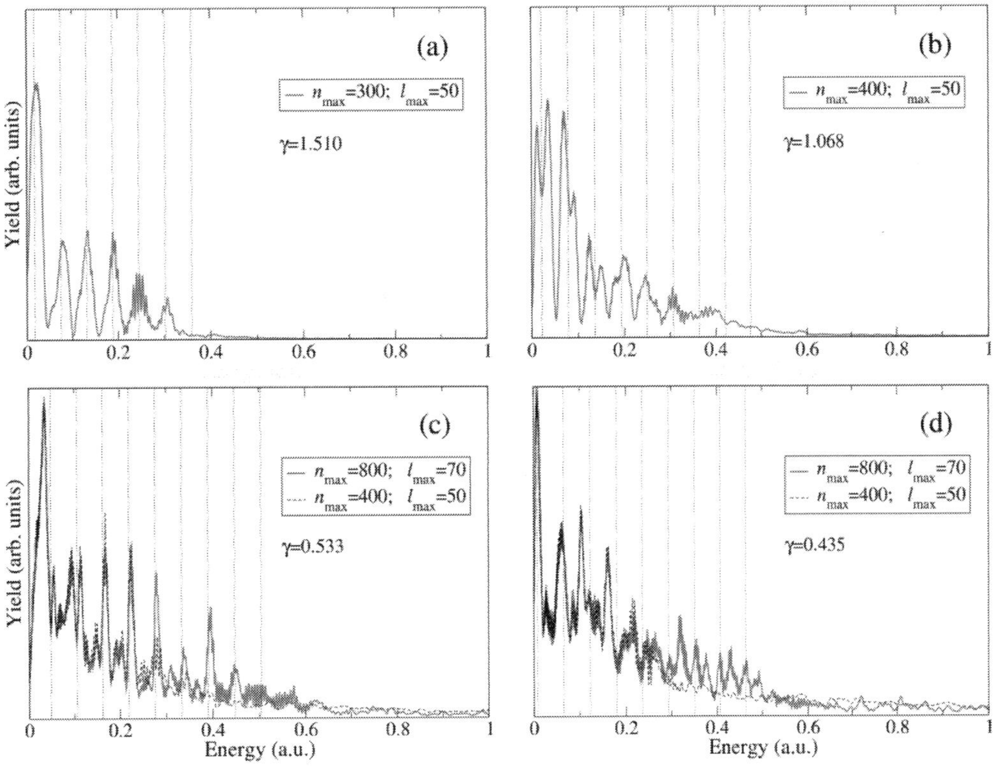

Figure 1. Electron energy distributions for single ionization of the hydrogen atom by 25 fs pulses at $I = 5 \times 10^{13}$ (a), 10^{14} (b), 4×10^{14} (c), and $6 \times 10^{14}\,\mathrm{W/cm^2}$ (d). The dotted vertical lines denote the expected position of the peaks according to Eq. (15).

which are highlighted in Fig. 1 by the vertical dotted lines. As the intensity increases new structures appear at the low-energy part of the spectrum. At $I = 10^{14}\,\mathrm{W/cm^2}$ the peaks associated to $n = 1, 2$ and 3 are splitted (Fig. 1(b)). At higher intensities the spectrum consists of peaks which are rather irregularly distributed, specially in the case of $I = 6 \times 10^{14}\,\mathrm{W/cm^2}$.

Understanding the origin of these structures is not the purpose of the present contribution. It requires further investigation and will be presented elsewhere. Indeed, we are also exploring this problem under two other approaches: a Floquet formalism combined with complex rotation which allows us to access the resonances induced by the field [13] and a model calculation based on series of separable potentials in momentum space, each of them supporting a bound state of the hydrogen atom [14].

4. Conclusions

We presented a study of the ionization process of hydrogen atoms by intense fields in a regime where the Keldysh parameter is close to 1. For this purpose we developed a method for the propagation of the TDSE which includes an efficient implementation of an explicit algorithm for the solution of stiff differential equations. Some structures in the ATI energy spectrum were found which can be only partially associated to multiphoton ionization peaks. The nature of the remaining structures still remains unclear and deserves further investigation.

Our calculations show a stupendous efficiency compared to diagonally implicit propagation algorithms. This opens new perspectives for the treatment of more complicate ionization processes like two-photon ionization of helium [15–23], where the non-linearity of this process poses tremendous experimental difficulties that so far prevent the measurement of accurate total and differential cross sections [20, 21].

Acknowledgments
The authors are indebted to Robin Shakeshaft for bringing [11] to their attention. J. M. thanks support by the FNRS ("Bourse postdoctorale FNRS dans le cadre de la convention IISN No. 4.4.503.02").

References

[1] Keldysh L V 1965 *Sov. Phys. JETP* **20** 1307
[2] Faisal F H M 1973 *J. Phys. B* **6** L89
[3] Reiss H R 1980 *Phys. Rev. A* **22** 1786
[4] Delone N B and Krainov V P 1991 *J. Opt. Soc. Am. B* **8** 1207
[5] de Bohan A, Piraux B, Ponce L, Taïeb R, Véniard V and Maquet A 2002 *Phys. Rev. Lett.* **89** 113002
[6] Rudenko A, Zrost K, Schröter C D, de Jesus V L B, Feuerstein B, Moshammer R and Ullrich J 2004 *J. Phys. B* **37** L407
[7] Uiberacker M *et al* 2007 *Nature* **446** 627
[8] Hairer E and Wanner G 1991 *Solving Ordinary Differential Equations II: Stiff and Differential-Algebraic Problems* (Springer-Verlag, Berlin)
[9] Cerjan C and Kulander K C 1991 **63** 529
[10] Fatunla S O 1980 *Math. Comp.* **34** 373
[11] Fatunla S O 1978 *Math. Comp.* **32** 1
[12] Madroñero J and Piraux B *submitted to Phys. Rev. A*
[13] Buchleitner A, Delande D and Gay J 1994 *J. Opt. Soc. Am. B* **12** 505
[14] Tetchou-Nganso H M, Giraud S, Piraux B, Popov Y V and Kwato-Njock M 2007 *J. Elect. Spect. Rel. Phen.* **161** 178
[15] Feng L and van der Hart H W 2003 *J. Phys. B* **36** L1
[16] Shakeshaft R 2007 *Phys. Rev. A* **76** 063405
[17] Nikolopoulos L A A and Lambropoulos P 2007 *J. Phys. B* **40** 1347
[18] Foumouo E, Antoine P, Piraux B, Malegat L, Bachau H and Shakeshaft R 2008 *J. Phys. B* **41** 051001
[19] Kheifets A S and Ivanov I A 2006 *J. Phys. B* **39** 1731
[20] Rudenko A *et al* 2008 *Phys. Rev. A* **101** 073003
[21] Antoine P, Foumouo E, Piraux B, Shimizu T, Hasegawa H, Nabekawa Y and Midorikawa K 2008 *Phys. Rev. A* **78** 023415
[22] Horner D A, Morales F, Rescigno T N, Martín F and McCurdy C W 2007 *Phys. Rev. A* **76** 030701(R)
[23] Palacios A, Rescigno T N and McCurdy C W 2009 *Phys. Rev. A* **79** 033402

Polarization and correlation aspects of resonant coherent excitation of fast highly charged ions in crystals

V V Balashov

Skobeltsyn Institute of Nuclear Physics, Moscow State University, Moscow, Russia

E-mail: balvse@anna19.sinp.msu.ru (V V Balashov)

Abstract. We report selected results of our latest calculations on polarization properties of the in-crystal resonating electric field and their manifestation in the yield, angular distribution and polarization parameters of the characteristic X-ray radiation from resonant coherent excited relativistic highly charged ions under planar channeling.

1. Introduction

Unambiguous observation of resonant coherent excitation (RCE) of channeled ions in crystals made by S.Datz group three decades ago [1] and, now, the comprehensive experiments performed by the Tokyo RCE collaboration using relativistic highly charged ion beams (see their latest publication [2] and references therein) revert us to fundamental predictions made in 1965 by V.V.Okorokov [3] and to his pioneer experimental and theoretical studies on this phenomenon. The Okorokov effect appears when one of the harmonics of the spatially periodic electric field of the crystal acts on the moving ion with a resonance frequency corresponding to the ion excitation energy. The anisotropic character of the field manifests itself in the produced alignment (angular momentum polarization) of the excited ions and, hence, in possible angular anisotropy, linear polarization and, under special conditions, circular polarization of their characteristic X-ray radiation.

Our theoretical studies link together the origin and the wide set of observables of the RCE process. Treating the propagating ion as an open quantum system involved into both coherent and incoherent (relaxation) interactions with its environs, they are based of a unified concept [4] of the generalized Master equation description of the time evolution of the excitation and decay properties of the propagating ions together with the well-known density matrix (statistical tensors) technique to deal with polarization and correlation phenomena in atomic collisions [5]. Our recent calculations suggest a quantitative theoretical analysis of such fundamental RCE experiments by the Tokyo group as observation of the axial symmetry violation in the angular distribution of the X-rays from planar channeled resonant coherently excited relativistic Fe^{24+}ions in Si crystal [6,7], first double-resonant excitation of ionic autoionizing states [8,9], a comprehensive analysis of recent observation of the Autler-Townes doublet in resonant coherent excitation of the helium-like ions in crystals beyond the channeling conditions [2,10,11] and, also in the non-channeling conditions, fine structure of resonances in relativistic hydrogen-like ions [11,12]. At the same time a number of new aspects of the RCE studies has been considered and modelled by computer simulation for corresponding future experiments. They concern the yield and angular anisotropy of the Auger electrons produced by resonant coherent excitation of relativistic highly charged ions in channeling and non-channeling conditions [9,11], the metastable ion production in the RCE process [13, 14], the trajectory resolved X-ray measurements in this process in the ion – X-ray coincidence experiments [14] and others.

A compact presentation of our approach together with main results of its application was given at the conference HCI2008 [14]. Supplementing it by our new calculations, we concentrate here on one special aspect of the problem strictly corresponding to the subject (of the second part) of the Symposium to draw a line from polarization properties of the resonating field of the crystal acting on the ion to the angular anisotropy and polarization properties of the characteristic X-ray radiation in the RCE process including possible perspectives of its applications.

2. From spectral and geometrical properties of the resonating electric field of the crystal to the Stokes parameters of the RCE ion characteristic X-ray radiation

In paper [15] on the RCE process with planar channeled ions, we showed that taking a special coordinate axes and, also, using the dipole approximation to represent the crystal electric field in the ion rest frame one reaches considerable advantages to parameterize the excited ion density matrix and, hence, to consider the angular distribution of the photons in both cases of trajectory resolved and non-resolved RCE measurements. On this way, we revealed that the field in the ion rest frame is elliptically polarized but its opposite circulation in the upper and the lower parts of the channel compensate the produced vector polarization of the whole ensemble of the excited ion when they enter the channel symmetrically according to standard conditions of the RCE experiments. Therefore, according to general symmetry rules, the observation of circularly polarized photons in all these RCE experiments is strictly forbidden.

This is not so as concerns tensor polarization (alignment) of the excited ions and, hence, angular anisotropy and linear polarization of the radiation where both parameters are determined by the same set of statistical tensors of rank 2 of the excited ion. In Fig.1 we reproduce our calculation [7] for the first unambiguous observation of the alignment effect in the RCE process made by the Tokyo group [6] with 423 Mev/u Fe^{24+} ions in the $(2\bar{2}0)$ planar channel in Si crystal. Contrary to the case of RCE of light channeled ions of relatively low energies where the Okorokov effect is concentrated, due to strong relaxation processes near the channel walls, only in a very narrow central part of the channel [16], here the corresponding *coherence zone* turns out to be on the level of about 90%.

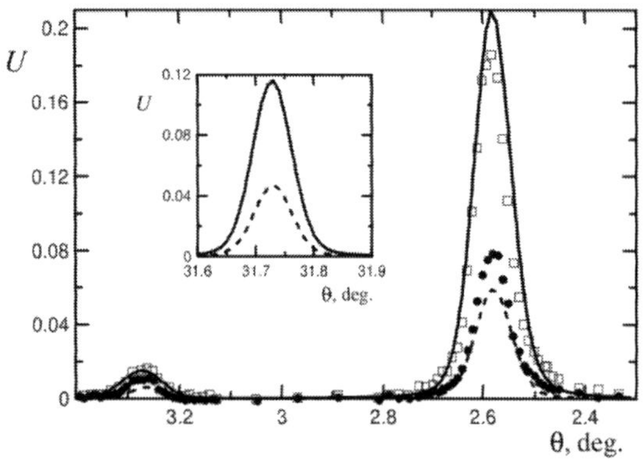

Figure 1. Differential yield of the X-ray radiation as a function of the crystal rotation angle θ at $(k,l) = (2,-1)$ for 423 MeV/u Fe^{24+} ion in 21 μm Si crystal, measured in [6] and calculated in [7] for two directions: in the channeling plane (white squares — experiment, solid line — calculation) and perpendicular to the channel (black circles — experiment, dashed line — calculation). Inset shows calculation results in the same conditions for the major peak of $(k,l) = (1,3)$ resonance.

2.1. Our calculation procedure (briefly)

Scalar potential $\varphi(\vec{r})$ of the electric field of the crystal is composed in the laboratory frame into its Fourier (kln) harmonics

$$\varphi(\vec{r}) = \sum_{kln} \Phi_{kln} e^{i\vec{G}_{kln}\vec{r}}, \tag{1}$$

with wave vectors $\vec{G}_{kln}$ being reciprocal lattice vectors. After a set of successive coordinate frame transformations this potential is transformed into the the scalar and vector electromagnetic potentials in the ion rest frame:

$$\begin{aligned} \varphi'(\vec{r}', t') &= \gamma \sum_{kln} e^{ivt'(\vec{G}'_{kln})_x} \Phi_{kln} e^{\pi in(z_{\rm ion}/d+1/2)} e^{i\vec{G}'_{kln}\vec{r}'}, \\ \vec{A}'(\vec{r}', t') &= -\vec{e}_x \frac{\gamma v}{c} \sum_{kln} e^{ivt'(\vec{G}'_{kln})_x} \Phi_{kln} e^{\pi in(z_{\rm ion}/d+1/2)} e^{i\vec{G}'_{kln}\vec{r}'}. \end{aligned} \tag{2}$$

with vectors $\vec{G}'_{kln} = \vec{G}_{kln} + (\gamma - 1)(\vec{G}_{kln})_x\vec{e}_x$ standing for the reciprocal lattice vectors in the ion rest frame, which x-component is γ times larger, due to relativistic lattice contraction, than x-component of $\vec{G}_{kln}$. Hamiltonian H_0 of the free ion and the operator

$$V = -e\varphi' + \frac{e}{2mc}(\vec{p}\vec{A}' + \vec{A}'\vec{p}) \tag{3}$$

of its interaction with the potentials $\varphi'(\vec{r}', t')$ and $\vec{A}'(\vec{r}', t')$ form the total Hamiltonian of the channeled ion $H = H_0 + V$ entering the generalized Master equation

$$i\frac{\partial \rho}{\partial t} = [H, \rho] + R\rho, \tag{4}$$

which describes time evolution of the density matrix ρ of the ion. The spin-orbit interaction is included in H_0. Time dependence of the field acting on the ion, expressed by the first exponents in (2), leads to the Okorokov resonance condition between the field frequency and the ion transition energy $E_{\rm trans}$:

$$\frac{2\pi\gamma v}{a}(k\sqrt{2}\cos\theta + l\sin\theta) = E_{\rm trans}. \tag{5}$$

Indices (k, l) represent a family of the reciprocal lattice vectors $\vec{G}_{kln} = \vec{G}_{kl0} + \vec{G}_{00n}$ with various n values. All they have the same $\vec{G}_{kl0}$ component lying in the channel plane but different perpendicular to the channel $\vec{G}_{00n}$ components. Solving equation (4) numerically in a chosen basic of states of a free ion, we project density matrix $\rho(t)$ on the subspace of the ion states emitting photons (say, states 2^1P_1 in case of RCE of helium-like ions) and transform this submatrix $\rho^{(\gamma)}(t)$ into its statistical tensors to use standard methods of calculating angular distribution and polarization of the radiation.

2.2. Elliptically polarized electric field of the crystal in the ion rest frame

Following [15] consider the resonant (k,l) term

$$\vec{E}'_{kl}(\vec{r}', t') = -i\gamma e^{ivt'(\vec{G}'_{kl0})_x} \sum_n \vec{G}''_{kln} \Phi_{kln} e^{\pi in(z_{\rm ion}/d+1/2)} e^{i\vec{G}'_{kln}\vec{r}'} \tag{6}$$

of the vector of the electric field strength $\vec{E}'(\vec{r}', t') = -\nabla_{\vec{r}'}\varphi' - (1/c)(\partial\vec{A}'/\partial t')$ acting upon the ion in its rest frame. It is a superposition of the components directed along auxiliary vectors

$$\vec{G}''_{kln} = \vec{G}_{kln} - \frac{\gamma - 1}{\gamma}(\vec{G}_{kln})_x\vec{e}_x, \tag{7}$$

where their y- and z-components are the same as those of corresponding $\vec{G}_{kln}$ vectors but x-components are γ times smaller. All vectors $\vec{G}''_{kln}$ with fixed (k, l) lie in the same plane which is perpendicular to the channel plane and crosses it on the line directed along the vector $\vec{G}''_{kl0}$.

To reveal the elliptic character of the oscillating electric field of the crystal, take a special ion rest frame (X,Y,Z) with X axis directed towards $\vec{G}''_{kl0}$ and Z axis is perpendicular to the channel plane and consider a general case of resonant coherent excitation of $1s^2 : {}^1S_0 \rightarrow 1s2p : {}^1P_1$ transition in helium-like ions in $(2\bar{2}0)$ planar channel. Then, using the dipole approximation, reduce operator (3) to scalar product of the ion electric dipole moment operator $\vec{d}_{\rm ion}$ and the strength of the field $\vec{E}'_{kl}(0, t')$ at the ion nucleus: $V^{(\rm dip)}(t') = -\vec{d}_{\rm ion}\vec{E}'_{kl}(0, t')$.

In [15], we showed that the phase shift between time oscillations of the $[\vec{E}'_{kl}(0, t')]_X$ and $[\vec{E}'_{kl}(0, t')]_Z$ Cartesian components of the field $\vec{E}'_{kl}(0, t')$ in the ion rest frame (X, Y, Z) is $\pm\pi/2$ at any position $z_{\rm ion}$ of the ion in the channel. The ellipse of this rotating field lies in (X, Z) plane and its axes are directed along X and Z coordinate axes (Fig.2). The specific character

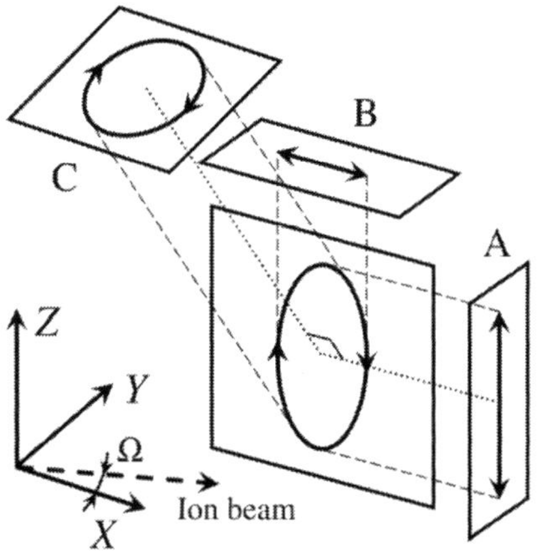

Figure 2. Schematic illustration of the elliptically polarized field of the crystal and polarization characteristics of the photons repeating the geometrical properties of the ellipse in various emission directions

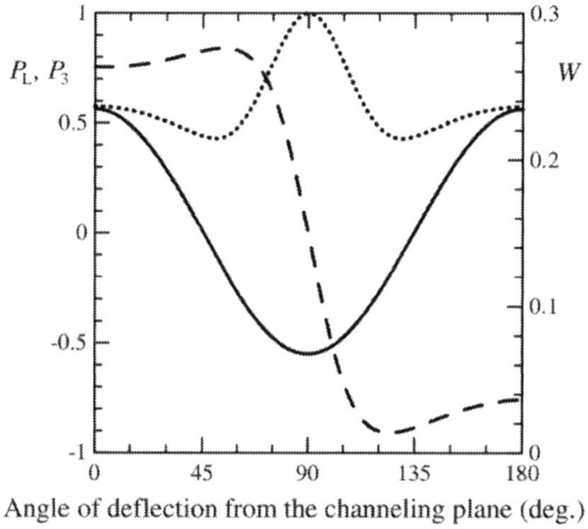

Figure 3. Angular distribution (solid line, in arbitrary units) of the yield and linear polarization degree P_L (dotted line) of photons emitted over the cone with the photon deflection angle 43.5^0 in the laboratory frame for 423 MeV/u Fe^{24+} beam in the 21 μm-thick target(in the ion rest frame these photons are emitted perpendicularly to the beam). Dashed line shows the circular polarization degree P_3 calculated for the $0.9\mu m$-thick target with averaging over one half of the channel.

of the resonating crystal field can produce vector polarization of the trajectory resolved RCE ions which, in turn, can be revealed by detecting their X-ray photons in coincidence with those of the ions whose trajectories lie on one side relative to the central channel plane [14]. Their selection can be performed when the target thickness is of the quarter of the period of the ion transverse oscillations between the walls of the channel (such a procedure was previously used in experiment [17] when measuring the charge state distribution of the RCE ions).

2.3. Stokes parameters of the X-ray radiation from the RCE ions in the laboratory frame
In [18] we presented the total set of the Stokes parameters of the X-rays from resonant excited 423 Mev/u Fe^{24+} ions in the $(2\bar{2}0)$ planar channel of the Si crystal; its 21-μm target

thickness was taken according to the experiment [6]. Here, the energy E_{tr}=6.7 keV of the $1s2p :^1 P_1 \to 1s^2 :^1 S_0$ transition varies in the laboratory frame from 2.75 to 16.82 keV. Note, first of all, a large angle-integrated radiation yield; at the resonance maximum, it is 1.8 photons per incident ion.

Angular distribution $W(\theta_\gamma, \phi_\gamma)$ of photons and their Stokes parameters P_1, P_2 and P_3 are first calculated directly from the submatrix $\rho^{(\gamma)}$ and its statistical tensors ϕ^{γ}_{kq} in the ion rest frame and then are transformed into the lab-frame according to the usual Lorentz transformation rules. Fig.3 presents an example of the results of these calculations. They demonstrate possibilities to vary the frequency of the resulting radiation and select its polarization characteristics in a wide range. An extension of our circular polarization suggestions to such an RCE experiment layout in which the ion beam entering the target is slightly inclined relative to the channel plane is in progress.

3. Conclusion

We have built a continuous line of theoretical description of the RCE process from the basic polarization properties of the resonating in-crystal electric field to the yield, angular distribution and polarization parameters of the characteristic X-ray radiation from resonant coherent excited highly-charged ions under planar channeling in crystals. The calculations show that since the photon yield and their polarization parameters in conditions typical or close to current RCE experiments are expected to be large, the resonance coherent excitation method can, indeed, be considered as a candidate for a tunable source of polarized X-ray radiation in the keV region [6].

Acknowledgments

Author appreciates his customary collaboration on various aspects of the RCE studies with I.Bodrenko, A.Sokolik, A.Stysin and V.Dolinov. The work was supported by RFBR (grants 08-02-08690, 09-02-01266).

References

[1] Datz S, Moak C D, Crawford O H et al 1980 *Nucl. Instr. Meth.* **170** 15
[2] Nakano Y et al 2009 *Phys. Rev. Lett.* **102** 085502
[3] Okorokov V V 1965 *JETP Lett.* **2** 111; *Yad. Phys.* **2** 1009 [*Sov. J. Nucl. Phys.* **2** 719].
[4] Balashov V V, Bodrenko I V 2001 *Moscow University Physics Bulletin* **56** N1 33
[5] Balashov V V, Grum-Grzhimailo A N, Kabachnik N M 2000 *Polarization and Correlation Phenomena in Atomic Collisions* Kluwer Academic (New York)
[6] Azuma T, Takabayashi Y, Kondo C et al 2006 *Phys. Rev. Lett.* **97** 145502
[7] Balashov V V, Sokolik A A, Stysin A V 2008 *JETP* **107** 133
[8] Nakano Y, Masugi S, Muranaka T et al 2007 *J. Phys. Conf. Ser.* **58** 395
[9] Balashov V V, Sokolik A A, Stysin A V 2008 *Nucl. Instr. Meth.* B **267** 905
[10] Nakai Y et al 2008 *Phys. Rev. Lett.* **101** 113201
[11] Balashov V V, Sokolik A A, Stysin A V 2009 *JETP* **108** 1010
[12] Kondo C, Masugi S, Nakano Y et al 2006 *Phys. Rev. Lett.* **97** 135503
[13] Balashov V V, Bodrenko I V 2006 *Nucl. Instr. Meth.* B **245** 52
[14] Balashov V V, Bodrenko I V, Dolinov V K, Sokolik A A, Stysin A V 2009 *J. Phys. Conf. Ser.* **163** 012087
[15] Balashov V V, Sokolik A A, Stysin A V 2009 *Nucl. Instr. Meth.* B **267** 1772
[16] Balashov V V, Bodrenko I V 2006 *Phys. Lett.* A **352** 129; *Nucl. Instr. Meth.* B **245** 52
[17] Komaki K, Azuma T, Ito T et al 1998 *Nucl. Instr. Meth.* B **146** 19
[18] Balashov V V, Dolinov V K, Sokolik A A 2009 *JETP Letters* **89** 399

Journal of Physics: Conference Series **212** (2010) 012029 doi:10.1088/1742-6596/212/1/012029

Coincidence Studies with Antiprotons

M McGovern[1], D. Assafrão[2], J.R. Mohallem[2], Colm T. Whelan[3] and H.R.J. Walters[1]

[1] Department of Applied Mathematics and Theoretical Physics, Queen's University, Belfast BT7 1NN, United Kingdom

[2] Laboratório de Átomos e Moléculas Especiais, Departamento de Física, ICEx, Universidade Federal de Minas Gerais, P.O Box 702, 30123-970 Belo Horizonte, MG, Brazil

[3] Department of Physics, Old Dominion University, Norfolk, VA 23529-0116 , USA

E-mail: mmcgovern06@qub.ac.uk

Abstract. We present a short overview of a new method for calculating fully differential cross sections that is able to describe any aspect of coincidence measurements involving heavy projectiles. The method is based upon impact parameter close coupling with pseudostates. Examples from antiproton impact ionization are shown.

1. Introduction

In [1] we showed how fully differential cross sections for ionization by heavy projectiles could be extracted from an impact parameter coupled pseudostate treatment of the collision. In this paper we give a brief summary of the theory, with some new extensions [2], and illustrate it by some applications to antiproton ionization of atomic hydrogen. Throughout we use atomic units (au) in which $\hbar = m_e = e = 1$.

2. Theory

We consider antiprotons incident with velocity $\boldsymbol{v}_0$ upon a neutral target atom which is stationary in the laboratory. In the relative coordinate system the amplitude for single ionization of the target is

$$f_{ion} = -\frac{1}{2\pi}\langle e^{i\mathbf{k}_f\cdot\mathbf{R}}\psi^-_{\boldsymbol{\kappa}}|V|\Psi^+_0\rangle \tag{1}$$

where $\mathbf{k}_f$ is the final relative momentum of the scattered antiproton, $\boldsymbol{\kappa}$ is the momentum of the ionized electron, $\psi^-_{\boldsymbol{\kappa}}$ is the ionized state of the atom, V is the interaction between the antiproton and the atom, and Ψ^+_0 is the full scattering wave function for the system. Let ψ_n be a set of eigenstates and pseudostates that diagonalise the atomic Hamiltonian H_A:

$$\langle\psi_n|H_A|\psi_m\rangle = \varepsilon_n\delta_{nm} \tag{2}$$

If the ψ_n approximate a complete set then we may write

$$f_{ion} = -\frac{1}{2\pi}\sum_n \langle \psi_{\boldsymbol{\kappa}}^- | \psi_n \rangle \langle e^{i\mathbf{k}_f \cdot \mathbf{R}} \psi_n | V | \Psi_0^+ \rangle \tag{3}$$

Conservation of energy requires that $k_f^2 + 2\mu\varepsilon^+ + \mu\kappa^2 = k_0^2 + 2\mu\varepsilon_0$ where $\mathbf{k}_0 = \mu\boldsymbol{v}_0$ is the relative momentum of the incident antiproton, μ is the reduced mass of the system, ε^+ is the energy of the residual ion, and ε_0 is the energy of the initial atomic state ψ_0. Unless deliberately engineered $k_f^2 + 2\mu\varepsilon_n$ will not be equal to $k_0^2 + 2\mu\varepsilon_0$ and so the amplitude $f_{n0} \equiv \langle e^{i\mathbf{k}_f \cdot \mathbf{R}} \psi_n | V | \Psi_0^+ \rangle$ will normally be off-energy-shell. However, in [1] the pseudostates were chosen so that one state from each angular momentum symmetry had exactly the right energy for conservation, i.e., $\varepsilon_n = \varepsilon^+ + \frac{\kappa^2}{2}$. With this choice it also transpired that $\langle \psi_{\boldsymbol{\kappa}}^- | \psi_n \rangle$ was effectively zero for all states ψ_n with $\varepsilon_n \neq \varepsilon^+ + \frac{\kappa^2}{2}$ so that, in effect, (3) became an expression involving only on-energy-shell amplitudes f_{n0}.

Now, except at very low energies, a wave treatment of the collision, as implied by (1), is not practical. The natural choice then is the impact parameter method where the antiproton is assumed to move along a straight line path with constant velocity $\boldsymbol{v}_0$. One then studies the electronic wave function, Ψ, of the atom as a function of time t. Expanding Ψ in the basis ψ_n according to

$$\Psi = \sum_n a_n(\mathbf{b}, t) e^{-i\varepsilon_n t} \psi_n, \tag{4}$$

substituting into the time dependent Schrödinger equation, and projecting with the ψ_n leads to the coupled equations

$$i\frac{da_n}{dt} = \sum_m e^{i(\varepsilon_n - \varepsilon_m)t} \langle \psi_n | V | \psi_m \rangle a_m \tag{5}$$

These equations must be solved subject to the boundary conditions $a_n(\mathbf{b}, -\infty) = \delta_{n0}$. In [1] it is then shown that, to a very good approximation,

$$f_{n0} = iv_0 \int e^{i\mathbf{q}.\mathbf{b}} [a_n(\mathbf{b}, \infty) - \delta_{n0}] d^2\mathbf{b} \tag{6}$$

where $\mathbf{q} \equiv \mathbf{k}_0 - \mathbf{k}_f$ is the momentum transfer and where f_{n0} is on-energy-shell. Combining (3) and (6) we arrive at the final approximation that is used:

$$f_{ion} = -\frac{iv_0}{2\pi}\sum_n \langle \psi_{\boldsymbol{\kappa}}^- | \psi_n \rangle \int e^{i\mathbf{q}.\mathbf{b}} [a_n(\mathbf{b}, \infty) - \delta_{n0}] d^2\mathbf{b} \tag{7}$$

Exploiting symmetries, (7) can be reduced to a more manageable practical form, see [1].

Given (7) we can now calculate the most fully differential cross section, the triple differential cross section (TDCS). This is given by

$$\frac{d^3\sigma}{dE d\Omega_e d\Omega_p} = \frac{v_f \kappa}{v_0} m_p^2 |f_{ion}|^2 \tag{8}$$

where $\boldsymbol{v}_f = \mathbf{k}_f/\mu$ and m_p is the mass of the projectile. This is the cross section for an electron being ejected with energy in the range E to $E + dE$ and into the solid angle $d\Omega_e$ while the antiproton is scattered into the solid angle $d\Omega_p$. It is very important to specify the frame of reference in which the measurements are to be made. Formula (8) applies to the laboratory frame of reference. In the relative coordinate frame of reference (the natural frame for theortical calculations) the m_p^2 in (8) must be replaced by μ^2 (for Gold ions incident on He this is a factor of 2525 difference!). In all of the results reported here we assume that we are in the laboratory frame. From (8) double differential, single differential, and total cross sections can be calculated by integration.

As described above, the approximation (7) assumes that the pseudostate set ψ_n has been constructed so that one state from each angular momentum symmetry has exactly the energy $\varepsilon^+ + \kappa/2$. This means that for each ejected electron enegy a new set of pseudostates must be constructed and the equations (5) solved. This is highly inconvenient if one neds to average cross sections over experimental energy resolutions or wants to calculate cross sections such as $d\sigma/d\Omega_e$, $d\sigma/dq$ or $d\sigma/dE$ for comparision with experiment. It is therefore important to see if (7) can be easily extended to a range of ejected energies while using a single set of pseudostates. What if we simply take (7) at face value and forget about the on-energy-shell requirement? Recent new work [2] suggests that this is reasonable provided that we do not move too far away from the particular ejected energy for which the pseudostates were constructed. We show some of these new results here.

3. Results

We show a sample of results for antiproton ionization of H($1s$). These have been calculated using the 165 state pseudostate set (H(nl), l =0 to 9, $n = (l + 1)$ to 21) described in [1]. This set has been constructed to describe the ejection of a 5eV electron. As a benchmark, in some of the figures we make comparison with exact first Born calculations (FBA) for which, in the case of atomic Hydrogen, there is an exact analytic expression for the TDCS, see [1].

First let us look at an impact energy of 30 keV. In figure 1 we show a three dimensional picture of the TDCS for an ejected electron energy of 5 eV. Here we see that the FBA gives a cross section with a large binary peak pointing in the direction of $\mathbf{q}$, i.e., in the "forward" region. By contrast, the 165 approximation shows a smaller cross section in which the electron is deflected into the "backwards" region, a result of post collisional interaction between the antiproton and ionized electron, a physical effect not represented in the FBA.

Using our new generalised version of (7) [2] we show in figure 2 the 165 state double differential cross section $d^2\sigma/dEd\Omega_e$ at 30 keV for a range of ejected electron energies up to 5 eV. We see increasing backward repulsion of the electron with decreasing ejected energy. Figure 3(a) shows the single differential cross section $d\sigma/d\Omega_e$ at 30 keV. This cross section is an integral over all ejected electron energies and, like $d^2\sigma/dEd\Omega_e$, an integral over all angles of the scattered antiproton. In $d\sigma/d\Omega_e$ we see a maximum near 70°, a shallow dip near 120° and a subsequent rise to backward ejection. While figure 2 shows a clear preference for backward ejection for the lowest energy electrons, figure 3(a) demonstrates that, overall, the picture is a little bit different, i.e., the higher energy electrons are not so easily stampeded into the backward direction by post collisional interaction. Figure 3(b), which gives $d\sigma/dE$ at 30 keV, shows that there are plenty of electrons with energies higher than those shown in figure 2, i.e., above 5 eV. Also exhibited in figure 3(a) is the FBA which, as one might expect from figure 1, clearly prefers forward ejection.

There has been much discussion in the literature, e.g., [3], about of the role of the nuclear interaction in differential ionization. The interaction V between the antiproton and a hydrogen atom is given by

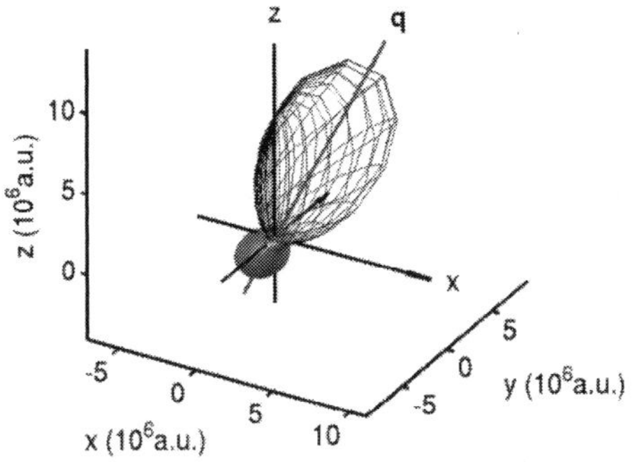

Figure 1. TDCS in three dimensions for an ejected electron of 5 eV and an impact energy of 30 keV with q=0.7 au. Wire cage, FBA; solid surface, full coupled pseudostate calculation. The antiproton is incident along the z-direction.

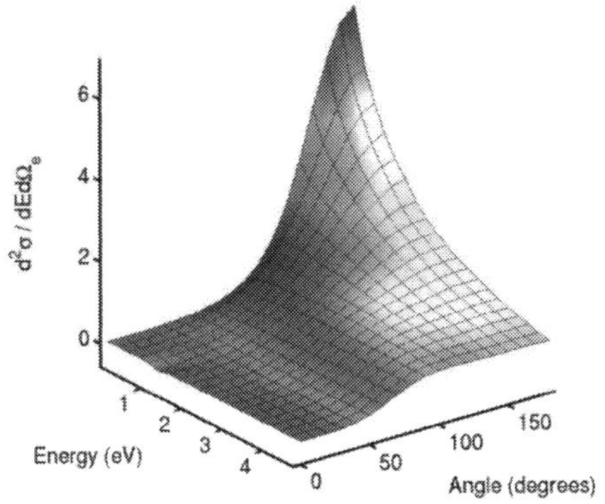

Figure 2. Cross section $d^2\sigma/dEd\Omega_e$ at 30 keV.

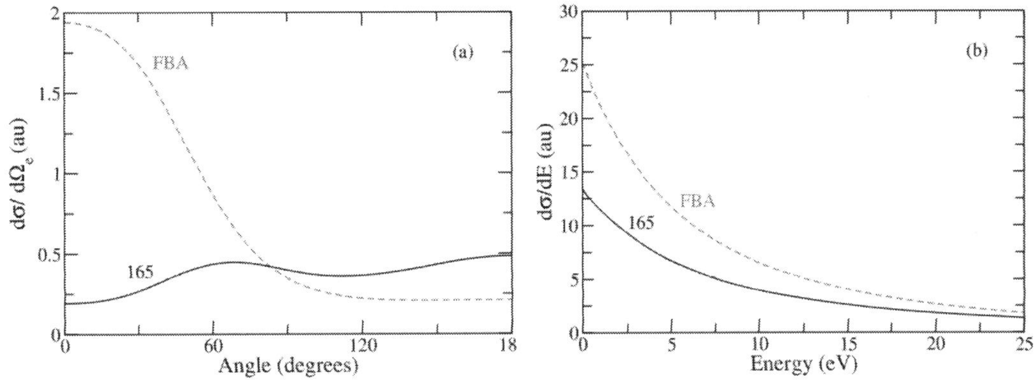

Figure 3. Cross sections (a)$d\sigma/d\Omega_e$ and (b) $d\sigma/dE$ at 30 keV.

$$V = -\frac{1}{R} + \frac{1}{|\mathbf{R} - \mathbf{r}|} \tag{9}$$

where $\mathbf{R}(\mathbf{r})$ is the position vector from the proton to the antiproton (electron). Because of the

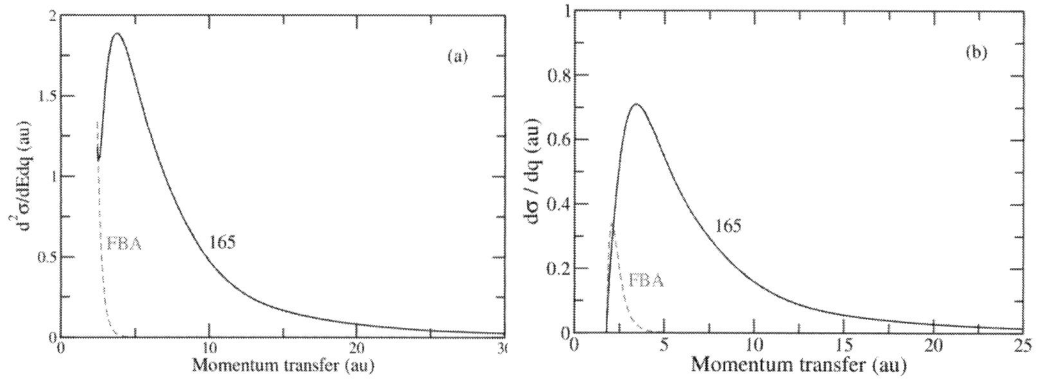

Figure 4. Cross sections (a) $d^2\sigma/dEdq$ for an ejected electron of 5 eV, and (b) $d\sigma/dq$, at 2 keV.

correspondence (6) between the wave treatment and the impact parameter approximation it is clear that the approximation (7) includes a proper representation of the interaction between the antiproton and the atomic nucleus (the $-\frac{1}{R}$ term in (9)). In antiproton scattering the nuclear interaction becomes more and more important as the impact energy is reduced, this is perhaps not surprising since in the low energy limit we approach the onset of significant protonium formation [4]. To illustrate the importance of the nuclear interaction we show in figure 4(a) the cross section $d^2\sigma/dEdq$ for antiproton ionization of H(1s) at an impact energy of 2 keV and for an ejected electron energy of 5 eV. Here we see that the FBA cross section collapses very rapidly with increasing momentum transfer q. From (1) and (9) the first Born amplitude for ionization is

$$f_{ion}^{B1} = -\frac{1}{2\pi}\langle e^{i\mathbf{k}_f\cdot\mathbf{R}}\psi_{\boldsymbol{\kappa}}^{-}(\mathbf{r})| - \frac{1}{R} + \frac{1}{|\mathbf{R}-\mathbf{r}|}|e^{i\mathbf{k}_0\cdot\mathbf{R}}\psi_0(\mathbf{r})\rangle \tag{10}$$

Due to the orthogonality of ψ_0 and $\psi_{\boldsymbol{\kappa}}^{-}$ the nuclear interaction $-\frac{1}{R}$ disappears from the approximation which can therefore only represent scattering of the antiproton by the electron. The light electron is unable to provide large momentum transfers to the antiproton and so the FBA declines rapidly with increasing q. Initially the 165 coupled pseudostate approximation falls like the FBA but then, with increasing q, quickly reaches a sharp mimimum and then rises to a pronounced maximum and continues to show a sustained cross section up to large values of q. In the initial fall of the 165 state cross section we see the longer range interactions which are included in the FBA, but in the minimum and subsequent maximum we see the nuclear interaction start to exert its influence.

Figure 4(b) shows the single differential cross section $d\sigma/dq$. This is obtained by integrating $d^2\sigma/dEdq$ over all allowed ejection energies for a given allowed q, i.e.,

$$\frac{d\sigma}{dq} = \int_0^{(qv_0+\varepsilon_0-\varepsilon^+)} \frac{d^2\sigma}{dEdq}dE \tag{11}$$

$$q \geq q_{min} = \frac{(\varepsilon^+ - \varepsilon_0)}{v_0} + O\left(\frac{1}{\mu}\right) \tag{12}$$

Note that $d\sigma/dq$ is zero at the threshold $q = q_{min}$, figure 4(b), unlike $d^2\sigma/dEdq$ which is non-zero at the minimum q allowed for the given ejected electron energy, figure 4(a). Apart from

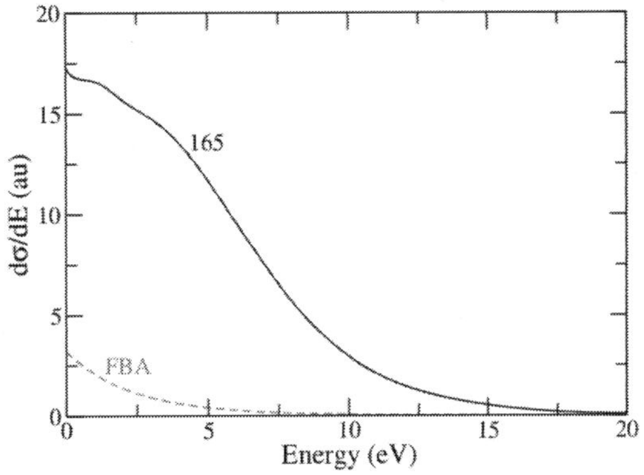

Figure 5. Cross section $d\sigma/dE$ at 2 keV.

rising from zero at $q = q_{min}$ the FBA and 165 cross sections of figure 4(b) are similar to those of figure 4(a). This similarity is perhaps not a surprise since the nuclear effect is probably not that different for each ejected electron energy.

Figure 5 shows $d\sigma/dE$ at 2 keV. As at 30 keV, figure 3(b), and as in the FBA, the 165 pseudostate cross section is greatest at zero ejected energy. However, the shape of the cross section is different. being mainly convex upwards at the lower energies while the other cross sections are concave downwards which seems to be a feature of the FBA and therefore of the higher impact energy regime.

4. Conclusions

In [1] and [2] a powerful method based on impact parameter pseudostate close coupling has been introduced. This method can give fully differential cross sections able to describe any aspect of coincidence measurements using heavy projectiles. Here we have shown but a few simple examples from antiproton impact ionization of atomic hydrogen. Results for ionization by carbon and gold ions have also been obtained but will be published elsewhere.

Acknowledgments

One of us (MMcG) acknowledges support from the European Social Fund and Queen's University Belfast for a postgraduate studentship. DA and JRM acknowledge financial support from the Brazilian agencies CNPq, Capes and Fapemig.

References

[1] McGovern M, Assafrão D, Mohallem J R, Whelan C T and Walters H R J 2009 *Phys. Rev. A* **79** 042707
[2] McGovern M, Assafrão D, Mohallem J R, Whelan C T and Walters H R J, in preparation
[3] Schulz M, Dürr, Najjari B, Moshammer R and Ullrich J 2007 Phys. Rev. A **76** 032712
[4] Schultz D R, Kristić, Reinhold C O and Wells J C 1996 Phys. Rev. Lett. **76** 2882

Journal of Physics: Conference Series **212** (2010) 012030 doi:10.1088/1742-6596/212/1/012030

Projectile scattering and electron-electron interaction in ion-atom collisions

Myroslav Zapukhlyak[1], **Nils Henkel**[1] **and Tom Kirchner**[2]

[1] Max-Planck-Institut für Physik komplexer Systeme, D-01187 Dresden, Germany

[2] Department of Physics and Astronomy, York University, Toronto, Ontario, Canada M3J 1P3

E-mail: zmi@pks.mpg.de, henkel@pks.mpg.de, tomk@yorku.ca

Abstract. One- and two-electron processes involving electron transfer and excitation are considered for a few basic ion-atom collision systems at intermediate impact energies. We have used the two-center basis generator method in the framework of the independent electron model for orbital propagation and the eikonal approximation for the calculation of scattering-angle differential cross sections. Our results shed some light on the roles of the nucleus-nucleus and the electron-electron interactions in these collisions.

1. Introduction

Recent studies on ion-atom collisions have demonstrated that the role of the nucleus-nucleus interaction is less well understood than one might have thought. The issue appeared on the agenda in 2003, when unexpected discrepancies between experiment and theory were found in kinematically complete data for ionization processes [1]. Quite a few papers offering illuminating analysis and explanations have appeared since then, but none of them was able to clarify the situation completely and solve the problem once and for all. As it seems to date, the nucleus-nucleus interaction poses one of the largest remaining problems in atomic collision theory.

Supportive of this conclusion are investigations of processes involving electron transfer, which is an (almost) unique feature of ion-atom collisions. Even though such processes have been studied for some time for basic collision systems such as p-He and He^{2+}-He the state of affairs has become clearer only recently when well-resolved and detailed experimental and theoretical results became available.

In this contribution, we present calculations for one- and two-electron processes involving electron transfer and target excitation [2, 3]. In particular, we address the question whether it is possible to disentangle effects of the heavy-particle motion from those of electronic correlations, which constitute the other unsolved big issue of atomic collision physics.

Our calculations are based on the independent electron model, the two-center basis generator method for orbital propagation, and the eikonal approximation to extract angular-differential cross sections (section 2). Certainly, this framework does not suffice to describe the physics in full, but as shown in section 3 it sheds some light on it and helps answer some of the pending questions.

2. Theory

2.1. Independent electron model (IEM)

The problem is treated in the semiclassical, nonrelativistic approximation. The nuclei are assumed to be moving on classical trajectories with the relative vector $\mathbf{R}(t)$. For the calculation of the Hamiltonian we assume this trajectory to be straight with impact parameter b, and therefore $R(t) = \sqrt{b^2 + v^2t^2}$, where v is the relative velocity. The total Hamiltonian is $\hat{H} = \hat{H}_e + \hat{V}_{nn}$ with the nuclear repulsion $V_{nn}(\mathbf{R}, t) = \frac{Z_p Z_t}{R(t)}$ (atomic units characterized by $\hbar = e^2 = m_e = 1$ are used unless stated otherwise). Within the IEM, $\hat{H}_e$ is approximated as a sum of single-particle Hamiltonians $\hat{h}_i$:

$$\hat{H}_e = \sum_{i=1}^{N} \hat{h}_i. \tag{1}$$

The $\hat{h}_i$ take the electron-electron interaction into account via an effective potential $\hat{V}_{eff}$. We use the helium ground-state potential obtained from the optimized potential method [4], i.e. response effects are neglected. Finally, the single electron Hamiltonian reads in spatial representation

$$h(\mathbf{r}_t, t) = -\frac{1}{2}\Delta + V_{eff}(r_t) - \frac{Z_p}{|\mathbf{R}(t) - \mathbf{r}_t|}, \tag{2}$$

where $\mathbf{r}_t$ is the relative vector between target nucleus and electron.
In the IEM the wave function can be approximated as a Slater determinant consisting of one-particle states $|\psi_i\rangle$. The resulting time-dependent Schrödinger equations (TDSEs)

$$i\partial_t|\psi_i(t)\rangle = \hat{h}|\psi_i(t)\rangle \tag{3}$$

are solved by the two-center basis generator method (TC-BGM), a problem-adapted, finite basis representation [5, 6]. The TC-BGM basis is constructed from

- target states (all $KLMN$ shell eigenstates of the unperturbed target atom) and from
- projectile states (analogously)

taking Galilean invariance into account by suitable phase factors. The resulting states $|\varphi_u^0\rangle$ are well-suited to describe electrons bound to the target or the projectile. Additionally, we need states that can describe ionization. These so-called pseudo scattering states $|\chi_u^\mu\rangle$ are constructed by μ-fold application of the (regularized) projectile-electron potential onto the target states.

2.2. Calculation of observables

The single electron transition amplitude a_{if} for the transition of the ith electron into the final state $|\varphi_f^0\rangle$ is

$$a_{if} = \lim_{t\to\infty} \langle\varphi_f^0|\psi_i(t)\rangle. \tag{4}$$

Applying the IEM and taking into account that for He targets both electrons start from the same (1s-)orbital (which implies $a_{1f} = a_{2f}$), the 2-electron amplitudes $a^{ff'}$ for transfer of one electron each into $|\varphi_f^0\rangle$ and $|\varphi_{f'}^0\rangle$ are

$$\begin{aligned} a^{ff'} &= \frac{1}{\sqrt{2}}\left(a_{1f}a_{2f'} + a_{1f'}a_{2f}\right) = \sqrt{2}a_{1f}a_{1f'} \text{ for } f \neq f', \\ a^{ff} &= a_{1f}^2, \end{aligned} \tag{5}$$

where the factor $\sqrt{2}$ takes the indistinguishability of the electrons into account.
While the electronic Hamiltonian is calculated under the assumption of a straight-line trajectory

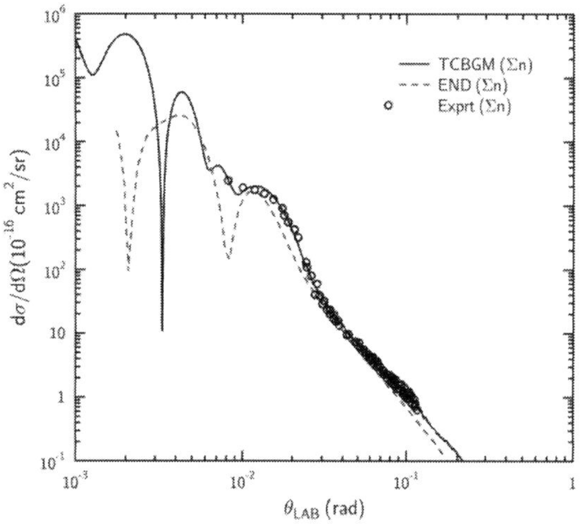

Figure 1. Differential capture cross section as a function of scattering angle in p-H collisions at $E_P = 1$ keV. Theory: present TC-BGM calculation, END calcuation [12]. Experiment: [11]. In all cases capture into all states ($\sum n$) is considered.

for the projectile, projectile scattering can still be described by the eikonal approximation [7],[8],[9]. The eikonal scattering amplitude $g^{ff'}(\theta)$ is

$$g^{ff'}(\theta) = ik \int_0^\infty \mathrm{d}b\; bJ_{\Delta M}\left(2kb\sin\left(\tfrac{\theta}{2}\right)\right) A^{ff'}(b), \tag{6}$$

where k is the initial projectile momentum, ΔM is the difference of initial and final state magnetic quantum numbers, and $J_{\Delta M}$ is a Bessel function.

$$A^{ff'}(b) = a^{ff'}(b) \exp\left(-\tfrac{2i}{v} \int_0^\infty \mathrm{d}z\; V_{nn}(\sqrt{b^2+z^2})\right) \tag{7}$$

is the electronic transition amplitude $a^{ff'}(b)$ multiplied by a phase factor that takes the nucleus-nucleus interaction into account.
Finally, the differential cross section (DCS) in the center-of-mass-system is

$$\left(\frac{\mathrm{d}\sigma^{ff'}}{\mathrm{d}\Omega}\right)(\theta) = |g^{ff'}(\theta)|^2. \tag{8}$$

A technical difficulty consists in the calculation of the integral (6) since the phase factor in (7) becomes singular at $b \to 0$, which makes straightforward numerical integration difficult and time-consuming. Two other methods have been tried to overcome this, namely truncated series expansion of the integral and Hankel-transform-Laguerre-expansion [10]. As it turned out, all these methods including direct numerical quadrature yield the same results.

3. Results and Discussion

3.1. p-H system

We have selected the p-H collision system as a test system in our work. It is obvious that there is no electron correlation in this three-body system. Nevertheless, it can help to assess the validity of the straight-line trajectory approximation for the heavy-particle motion. Limitations should be most visible at low projectile energies. However, we obtain good agreement at $E_P = 1$ keV with the experiment [11] and an electron nuclear dynamics (END) calculation [12] except for very small scattering angles (figure 1). In the END calculation a classical (Coulomb) trajectory for the nuclear motion was used. Moreover, we were not able to identify interference effects associated with the indistinguishability of the nuclei, because the backward scattering probability is just

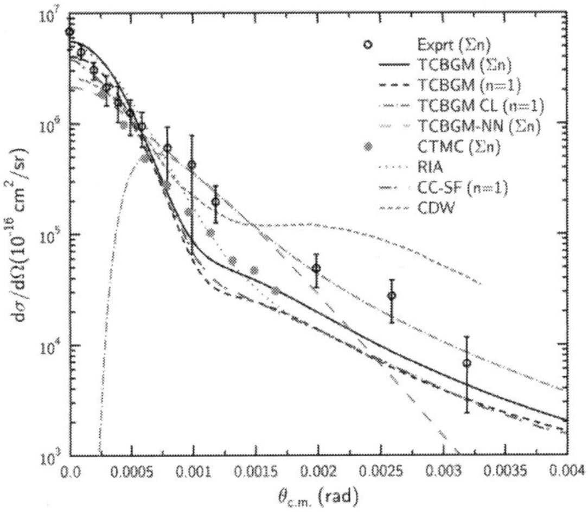

Figure 2. Differential capture cross section as a function of scattering angle in p-H collisions at $E_P = 25$ keV. Theory: present TC-BGM calculation for transfer into n=1 and $\sum n$, present TC-BGM calculation with classical scattering (TCBGM CL), present TC-BGM calculation without phase factor for the nucleus-nucleus interaction (TCBGM-NN), classical trajectory Monte-Carlo (CTMC) [20], reformulated impulse approximation (RIA), and continuum distorted wave (CDW) calculations [13], close coupling calculation with Sturmian functions (CC-SF) [16, 17, 18]. Experiment: capture into all states [21].

too small. Our calculation with explicit inclusion of these effects did not show any noticeable changes in the DCS.

There are many calculations of the DCS for charge transfer in p-H collisions in the 25–125 keV energy region [13, 14, 15, 16, 17, 18, 19, 20], and they are more or less all successful in describing the experimental DCS reported in [21]. Our own calculation is no exception (figure 2). To illustrate the importance of the nucleus-nucleus interaction we have also included a DCS curve in figure 2 that is obtained by neglecting the corresponding phase factor, i.e. by assuming $A^{ff'}(b) = a^{ff'}(b)$ in equation (7). It clearly has the wrong shape.

There is another fact we would like to point out here, which was not touched upon in earlier works. Interestingly, if we combine our TC-BGM calculation with the DCS for classical elastic (Coulomb) scattering we reproduce the experimental data in the angular region $\theta > 1/k$, in which this approximation is valid [22]. In contrast, all other, supposedly more refined theories yield smaller cross sections at these relatively large scattering angles. This might be just fortuitous, but new experiments are needed to clarify this case.

3.2. p-He system

For He target atoms the question about the role of the electron-electron interaction comes into play. It has been debated for many years, and it seems that one generally accepted conclusion is that two-electron processes, such as double capture are largely determined by electron correlation effects. The results presented in this and the following subsections indicate that this is not necessarily true.

Recently, the DCS for transfer excitation (TE), in which one electron is transferred to the projectile, while the other one is promoted to an excited target state was measured at intermediate impact energies [23]. When combined with DCSs for single transfer (ST), double excitation (DE), and single excitation (SE) it turned out that the double ratio

$$R = \frac{TE}{ST} / \frac{DE}{SE} \tag{9}$$

exhibited a peak structure around the scattering angle θ_{LAB}=0.5 mrad. A simple analysis shows that if one assumes all DCSs involved in R to be products of one classical DCS for elastic scattering and electronic transition probabilities calculated within the IEM one obtains a constant double ratio R=2. When using different potentials for the different processes the

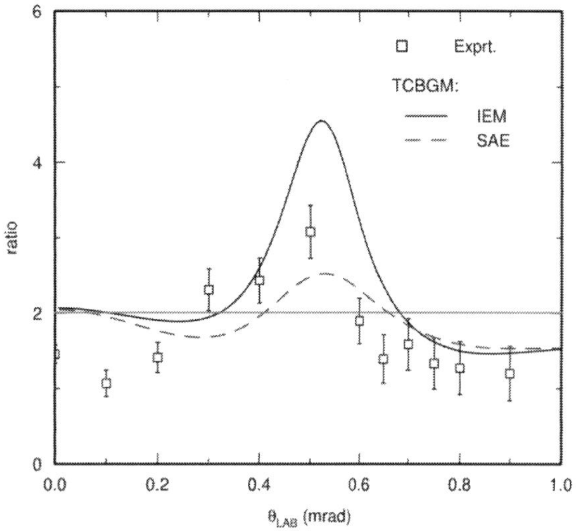

Figure 3. Ratio of transfer-excitation to single transfer versus double to single excitation cross sections as a function of laboratory scattering angle for 50 keV p-He collisions. Theory: TC-BGM calculation for all amplitudes within IEM, calculation within single-active-electron model (SAE) for single transfer and single excitation amplitudes. The constant ratio $R = 2$ corresponds to the simple model described in the text. Experiment: [23].

calculated double ratio is no longer constant, but it does not show a peak structure either [23]. At this point it is unclear whether the measured structure is associated with electron-electron interactions beyond the IEM or with the nucleus-nucleus interaction beyond classical scattering or with both.

In order to clarify this issue we have replaced the classical scattering ansatz by the eikonal approximation, but have adhered to the IEM. The results for R are shown in figure 3. Clearly, peak structures in qualitative agreement with experiment are present. We have found that the height of the peak is quite sensitive to the details of how the two-electron transition amplitudes are composed from one-electron amplitudes. The full (black) curve in figure 3 is obtained from the standard IEM according to equation (5). If one uses a slightly modified analysis in which the one-electron processes ST and SE are evaluated in terms of the single-active electron model [2] one obtains the dashed (blue) curve. This might indicate that the peak height is sensitive to electron-electron interaction effects, which are treated differently in both analyses. The fact that it is there, however, is due to nucleus-nucleus interaction effects, or more precisely to quantum mechanical heavy-particle-electron couplings, which are taken into account in the eikonal approximation.

3.3. He^{2+}-He system

Finally, we briefly consider the He^{2+}-He collision system, in which two-electron transfer is an important reaction channel. Figure 4 shows the DCS for this process at the impact energy E_P=375 keV/amu. It was measured some time ago and compared with continuum distorted-wave (CDW) calculations, in which correlation effects in the initial and final states were included in terms of a configuration interaction (CI) ansatz [24, 25]. Furthermore, the calculations were folded with the experimental resolution function. In addition to those previous results the figure includes our (unfolded) DCSs for double capture into the ground state ($n = 1$) and into all shells ($\sum n$). The latter calculation is in very good agreement with the measurements at scattering angles $\theta_{LAB} \geq 0.2$ mrad, but lies above the data at smaller angles. In this region, the CDW results show that the convolution with the experimental resolution function reduces the DCS. Assuming that the convolution would have the same effect on our results we would most likely obtain excellent agreement with the experimental data in the entire angular range. In other words, there would be no indication for electron correlation effects for this two-electron cross section. Unfortunately, the experimental resolution function was not available to us so that we were not able to prove this conjecture.

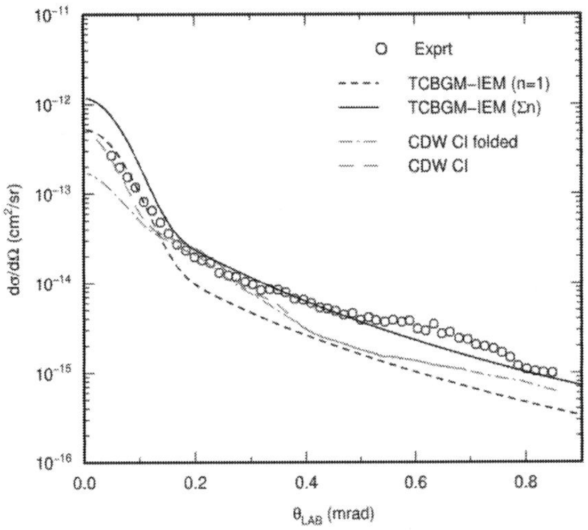

Figure 4. Differential two-electron transfer cross section as a function of laboratory scattering angle for E_P = 375 keV/amu $^4He^{2+}$ – ^4He collisions. Theory: present TC-BGM calculation within IEM for two-electron transfer into the ground state (n=1) and into all ($\sum n$) shells of the projectile; CDW CI calculation [24, 25]; CDW CI folded, same CDW calculation folded with the experimental resolution function [24, 25]. Experiment: two-electron transfer into all shells [25].

4. Conclusions

Fully differential cross sections for one- and two-electron processes in ion-atom collisions exhibit interesting structures, which are not always well understood. Often they have been blamed on electron-electron interaction effects, but not in all cases with good reason. As it turns out, calculations on the level of the independent electron model are able to describe several experimental data sets at intermediate impact energies, if they are based on a reliable propagation method and the eikonal approximation for the extraction of differential cross sections. Our studies provide ample evidence for the importance of the nucleus-nucleus interaction, but not so much for electron correlation effects.

References

[1] Schulz M, Moshammer R, Fischer D, Kollmus H, Madison D H, Jones S and Ullrich J, 2003 *Nature* (London) **422**, 48.
[2] Zapukhlyak M, Kirchner T, Hasan A, Tooke B and Schulz M, 2008 *Phys. Rev.* A **77**, 012720.
[3] Zapukhlyak M and Kirchner T, 2009 submitted to *Phys. Rev.* A
[4] Engel E and Vosko S H, 1993 *J. Phys.* B **29**, 4423
[5] Kroneisen O J, Lüdde H J, Kirchner T and Dreizler R M, 1999 *J. Phys.* A **32**, 2141.
[6] Zapukhlyak M , Kirchner T, Lüdde H J, Knoop S, Morgenstern R and Hoekstra R, 2005 *J. Phys.* B **38**, 2353.
[7] McCarroll R and Salin A, 1968 *J. Phys.* B **1**, 163.
[8] Wilets L and Wallace S J, 1968 *Phys. Rev.* **169**, 84.
[9] Glauber R J, 1959 *Lectures in Theoretical Physics* Vol 1, pp 315–414
[10] Miraglia J E, Piacentini R D and Rivarola R D, 1980. *Comp. Phys. Commun.* **19**, 299–303
[11] Houver J C, Fayeton J, and Barat M, 1993 *J. Phys.* B **7**, 1358
[12] Killian B J, Cabrera-Trujillo R, Deumens E and Öhrn Y, 2004 *J. Phys.* B **37**, 4733
[13] Belkić D, 2004 *Principles of Quantum Scattering Theory* (Bristol: Institute of Physics Publishing)
[14] Lin C D and Soong S C, 1978 *Phys. Rev.* A **17**, 1646
[15] Lin C D and Soong S C, 1978 *Phys. Rev.* A **18**, 499
[16] Shakeshaft R, 1976 *Phys. Rev.* A **14**, 1626
[17] Shakeshaft R, 1978 *Phys. Rev.* A **18**, 307
[18] Shakeshaft R, 1978 *Phys. Rev.* A **18**, 1930
[19] Alt E O, Kadyrov A S and Mukhamedzhanov A M, 1999 *Phys. Rev.* A **60**, 314
[20] Olson R E, 1983 *Phys. Rev.* A **27**, 1871
[21] Martin P J, Blankenship D M, Kvale T J, Redd E, Peacher J L and Park J T, 1981 *Phys. Rev.* A **23**, 3357
[22] Greenland T, 1981 *J. Phys.* B **14**, 3707
[23] Hasan A, Tooke B, Zapukhlyak M, Kirchner T and Schulz M, 2006 *Phys. Rev.* A **74**, 032703
[24] Deco G and Grün N, 1991 *Z. Phys.* D **18**, 339
[25] Schuch R, Justiniano E, Vogt G, Deco G and Grün N, 1991 *J. Phys.* B **24**, L133

International Symposium on (e,2e) and the 15th ISPCEAC IOP Publishing
Journal of Physics: Conference Series 212 (2010) 012031 doi:10.1088/1742-6596/212/1/012031

Four-Body Charge Transfer Processes in Heavy Particle Collisions

A.L. Harris, J.L. Peacher, M. Schulz, and D.H. Madison

Physics Department, Missouri University of Science and Technology, 1315 N. Pine St., Rolla, MO, USA 65401

E-mail: alh4b9@mst.edu

Abstract. Fully differential cross sections (FDCS) for proton + helium single capture and transfer-excitation collisions are presented using the Four-Body Transfer-Excitation (4BTE) model. This is a first order perturbative model that allows for any two-particle interaction to be studied. For single capture, the effect of the projectile-nuclear term in the perturbation is examined. It is shown that inclusion of this term results in an unphysical minimum in the FDCS, but is required to correctly predict the magnitude of the experimental results. For transfer-excitation, the role of electron correlation in the target helium atom is studied, and shown to be unimportant in the calculation of the FDCS.

1. Introduction

The few-body problem is one of the most fundamental unsolved problems in physics, and arises from the fact that the Schrödinger equation is not analytically soluble for more than two mutually interacting particles. As a result, theory must resort to approximations, the validity of which is determined by comparison with experiment.

There has been much work done on the three-body problem, and in many cases, theory and experiment agree very well [1, 2, 3, 4]. Recent advancements in experimental techniques and computing capabilities now allow for the study of more complicated collision systems, such as four-body collisions. One of the simplest four-body problems is a charged particle collision with a helium atom, in which both atomic electrons change state. This type of collision can result in many different outcomes, such as excitation-ionization, double excitation, and double ionization for either positively or negatively charged projectiles; as well as single charge transfer, double charge transfer, transfer-excitation, and transfer-ionization for positively charged projectiles. This paper will focus on single charge transfer and transfer-excitation for proton-helium collisions.

The process of single charge transfer (or single capture, SC) has been studied a great deal, beginning early in the 20th century. For the case of proton collisions, an incident proton collides with a target atom, captures a single electron, and leaves the collision as a neutral hydrogen atom. The first theoretical model for charge transfer was developed by Thomas [5] in 1927 for alpha particle collisions with hydrogen, and was strictly a classical calculation consisting of a two-step process. Soon after Thomas' model, Oppenheimer [6] and Brinkman and Kramers [7] performed the first quantum mechanical calculations for the SC process.

Most models treat the SC process as a three-body process using the independent particle model, regardless of the target being used. However, for low target nuclear charge, electron correlation can be important. For helium atoms, there have only been a couple of calculations that include the interaction of the passive electron through a correlated target wave function [8, 9]. In the case of four-body problems, this correlation is expected to be much more important. One such four-body problem is transfer-excitation (TE). In this case, not only is one electron captured by the projectile, but the remaining atomic electron is promoted to an excited state, making it a full four-body process.

Many of these calculations have been performed for total cross sections, but differential cross sections provide a much more stringent test of theory, and will be the focus of the work presented here. In this paper, theoretical differential cross sections for p + He single charge transfer and transfer-excitation will be presented using the Four-Body Transfer-Excitation model (4BTE). Atomic units are used unless otherwise noted.

2. 4BTE Theory

The FDCS for single charge transfer and transfer-excitation is differential only in projectile scattering angle, and is given by

$$\frac{d\sigma}{d\Omega} = (2\pi)^4 \mu_{pa}\mu_{pi}\frac{k_f}{k_i}\left|T_{fi}\right|^2, \tag{1}$$

where μ_{pa} is the reduced mass of the projectile and target atom, μ_{pi} is the reduced mass of the hydrogen atom and the residual He^+ ion, k_i is the momentum of the incident projectile, and k_f is the momentum of the scattered hydrogen atom.

In the 4BTE model [10], the transition matrix, T_{fi}, is given by

$$T_{fi} = \left\langle \chi_p^f \phi_H \psi_{He^+} \middle| V_i \middle| \chi_p^i \Phi_{He} \right\rangle, \tag{2}$$

where V_i is the initial state perturbation given by

$$V_i = \frac{Z_p Z_{nuc}}{r_1} + \frac{Z_p Z_e}{r_{12}} + \frac{Z_p Z_e}{r_{13}}. \tag{3}$$

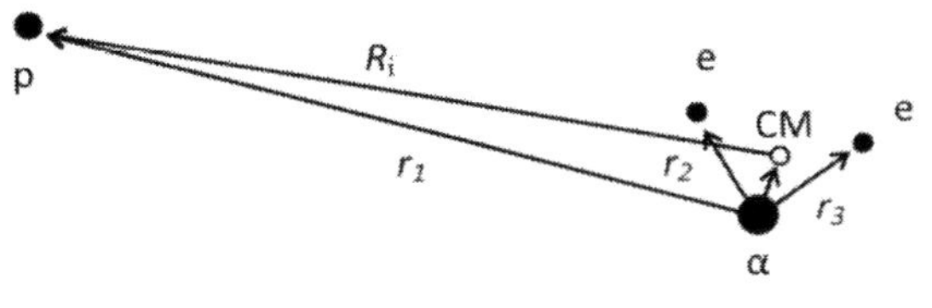

Figure 1. Jacobi coordinate system for the initial state projectile-helium atom system.

Here, Z_p, Z_e, Z_{nuc} are the charges of the projectile, electron, and target nucleus respectively; and $\vec{r}_1, \vec{r}_2, \vec{r}_3$ are the coordinates of the projectile and two atomic electrons with respect to the target nucleus. Also, $\chi_p^i \left(\chi_p^f\right)$ is the wave function for the incident (scattered) projectile, Φ_{He} is the wave function for the groundstate helium atom, $\psi_{\mathrm{He}+}$ is the wave function for the final state He^+ ion, and ϕ_H is the wave function for the captured electron. Both ϕ_H and ψ_{He^+} are simply hydrogenic wave functions, and thus known exactly. The ground state helium atom wave function is given by either a 20-term Hylleraas wave function [11] that includes both radial and angular correlations or a Hartree-Fock wave function [12] that has no correlation.

The calculations are performed in the center of mass frame, using the Jacobi coordinates shown in figures 1 and 2. Here, $\vec{R_i}$ is the relative vector between the projectile and the center of mass of the helium atom, and $\vec{R_f}$ is the relative vector between the center of mass of the hydrogen atom and the center of mass of the He^+ ion. They are given by

$$\vec{R_i} = \vec{r_1} - \frac{m_e}{m_a + 2m_e}(\vec{r_2} + \vec{r_3}) \tag{4}$$

and

$$\vec{R_f} = \frac{m_e\vec{r_2} + m_p\vec{r_1}}{m_p + m_e} - \frac{m_e}{m_e + m_a}\vec{r_3}, \tag{5}$$

where m_e, m_a, and m_p are the masses of the electron, alpha particle, and projectile respectively.

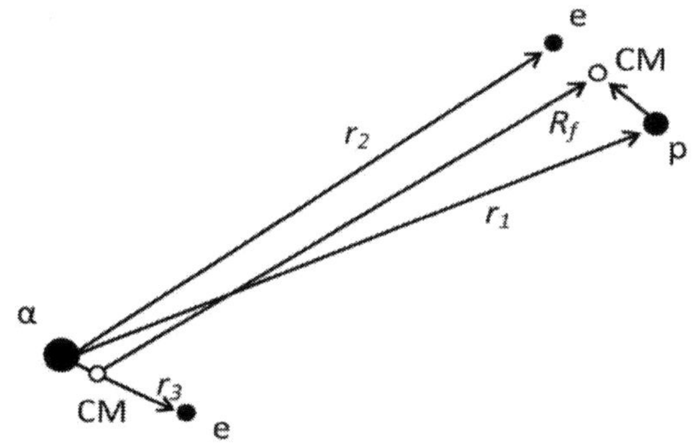

Figure 2. Jacobi coordinate system for the final state hydrogen-helium ion system.

The incident projectile wave function is represented by a plane wave given by

$$\chi_p^i = \frac{e^{i\vec{k_i}\cdot\vec{R_i}}}{(2\pi)^{3/2}}, \tag{6}$$

and the scattered projectile wave function is a Coulomb wave given by

$$\chi_p^f = \frac{e^{i\vec{k_f}\cdot\vec{R_f}}}{(2\pi)^{3/2}} e^{-\pi\gamma/2}\Gamma(1 - i\gamma)_1F_1\left(i\gamma, 1, -ik_f(r_1 + z_1)\right). \tag{7}$$

Here, $\gamma = 1/v_H$, $\Gamma(1 - i\gamma)$ is the gamma function, ${}_1F_1\left(i\gamma, 1, -ik_f(r_1 + z_1)\right)$ is a confluent hypergeometric function, and v_H is the speed of the hydrogen atom.

3. Results

3.1. Single Capture

There are many sets of experimental results for SC, and the calculations shown in figure 3 were performed for only a small subset of these. These results exhibit some well-known features and trends. The unphysical minimum seen in the calculation with the full perturbation is a result of cancellation of the terms in the perturbation [13]. Note that this minimum becomes deeper and shifts to smaller angles as the projectile energy increases, as was previously observed by Band [14] and Sil *et al.* [15]. The removal of the projectile-nuclear term in the perturbation results in the removal of this minimum, and an increase in the overall magnitude of the cross section, something seen by Belkić and Salin [13].

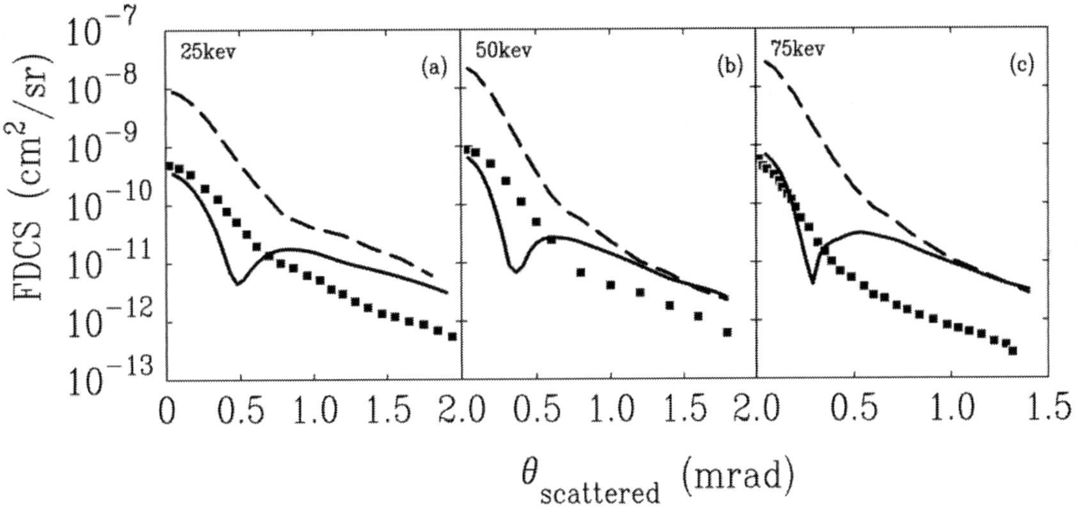

Figure 3. FDCS as a function of projectile scattering angle for p + He SC. The squares are the experimental results of Schulz, *et al.* [16] for the incident projectile energies shown in the figure. Both theoretical curves are from the 4BTE model with a plane wave for the incident projectile, Hylleraas wave function for the helium atom, and Coulomb wave for the scattered projectile. The solid line was calculated with all three terms in the perturbation, and the dashed line was calculated without the projectile-nuclear term in the perturbation.

3.2. Transfer-Excitation

Currently, only two sets of experimental data for fully differential cross sections for TE collisions are available [17, 18]. Experimentally, it is known that the outgoing hydrogen atom is in the ground state, and the residual helium ion is in an excited state. However, the exact excited state is not determined experimentally. Therefore, the cross sections must be summed over all possible excited states. Calculations have shown that contributions from excited states above $n = 4$ are negligible, as are contributions from higher angular momentum states. Because of this, the present results include only s and p excited states for $2 \leq n \leq 4$.

In figure 4, the effect of initial state correlation is shown [10]. One would be inclined to think that correlation would play an important role in the first order model of a four body process because the only interactions included in the perturbation are between the projectile and each individual electron, as well as the projectile-nuclear interaction. Thus, in order for two electrons to change state, some correlation should be required. However, figure 4 shows this expectation to be incorrect. Two calculations are shown in figure 4; one using a fully correlated Hylleraas wave function and the other using an uncorrelated Hartree-Fock wave function [12]. There is very little difference between the results of these two calculations, indicating that correlation is not important in this process.

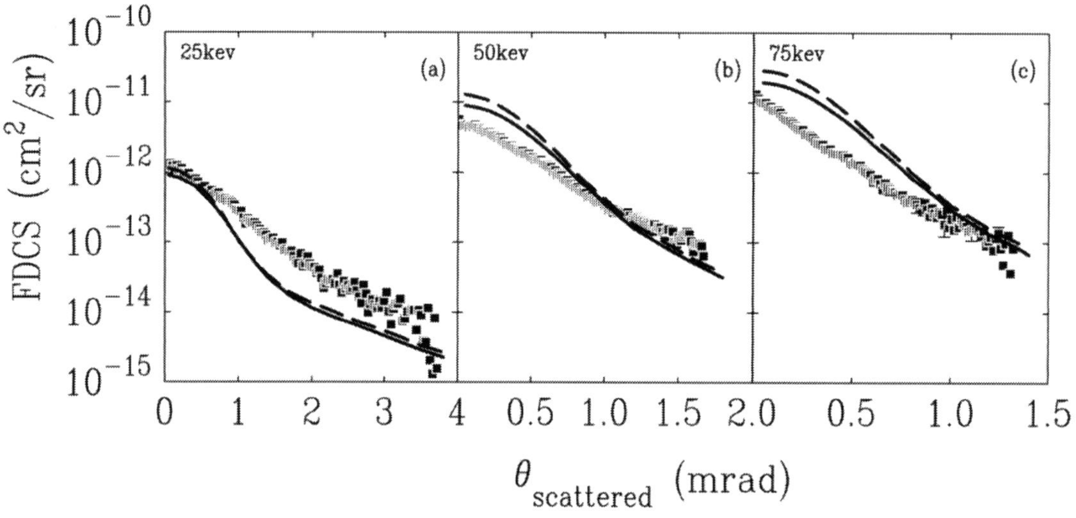

Figure 4. FDCS as a function of scattering angle for p + He TE showing the effect of electron correlation in the target atom wave function. The squares are the experimental results of Hasan, *et al.* [17] for the incident projectile energies shown in the figure. The solid line is the 4BTE model with a plane wave for the incident projectile, Hylleraas wave function for the helium atom, and Coulomb wave for the scattered projectile. The dashed line is the 4BTE model with a plane wave for the incident projectile, Hartree-Fock wave function for the helium atom, and Coulomb wave for the scattered projectile.

4. Conclusion

Results have been presented here for SC and TE in p + He collisions. The SC results exhibit the expected behavior of a first order perturbative model. A pronounced minimum in the FDCS is observed when all three terms in the perturbation potential are included in the calculation, and this minimum is removed when the projectile-nuclear term is excluded from the perturbation. However, using the full perturbation potential gives results closer to the magnitude of the experiment for small scattering angles.

The role of initial state electron correlation was explored for the TE process. It was expected that correlation in the ground state helium atom would play an important role in TE. However, results with and without correlation were nearly identical, showing that electron correlation has a negligible effect in the TE process.

Acknowledgments

This research was supported by the National Science Foundation (Grant No. PHY-0757749) and TeraGrid resources (Grant No. TG-MCA07S029) provided by the Texas Advanced Computing Center.

References

[1] Kheifets A S and Bray I 1998 *J. Phys. B: At. Mol. Opt. Phys.* **31** L447
[2] Malegat L, Selles P and Kazansky A K 2000 *Phys. Rev. Lett.* **85** 4450
[3] Colgan J, Pindzola M S and Robicheaux F 2001 *J. Phys. B: At. Mol. Opt. Phys.* **34** L457
[4] McCurdy C W, Horner D A, Resigno T N and Martín F 2004 *Phys. Rev.* A **69** 032707
[5] Thomas L H 1927 *Proc. R. Soc. Lond.* A **114** 561
[6] Oppenheimer J R 1928 *Phys. Rev.* **31** 349
[7] Brinkman H C and Kramers H A 1930 *Proc. Acad. Sci. Amsterdam* **33** 973

[8] Mapleton R A 1963 *Phys. Rev.* **130** 1839
[9] Crothers D S F and McCarroll R 1987 *J. Phys. B: At. Mol. Opt. Phys.* **20** 2835
[10] Harris A L, Peacher J L, Madison D H, and Colgan J 2009 *Phys. Rev.* A **80** 062707
[11] Hart J F and Herzberg G 1957*Phys. Rev.* **106** 79
[12] Byron F W Jr and Joachain C J 1966 *Phys. Rev.* **146** 1
[13] Belkić Dž. and Salin A 1978 *J. Phys. B: At. Mol. Opt. Phys.* **11** 3905
[14] Band Y B 1973 *Phys. Rev.* A **8** 243
[15] Sil N C, Chaudhuri J and Ghosh A S 1975 *Phys. Rev.* A **12** 785
[16] Schulz M, Vajnai T and Brand J A 2007 *Phys. Rev.* A **75** 022717
[17] Hasan A, Tooke B, Zapukhlyak M, Kirchner T and Schulz M 2006 *Phys. Rev.* A **74** 032703
[18] M.S. Schöffler, Ph.D. thesis, University of Frankfurt am Main, 2006 (unpublished).

Journal of Physics: Conference Series **212** (2010) 012032 doi:10.1088/1742-6596/212/1/012032

Polarization studies on the dielectronic recombination hypersatellite x–ray lines

A Surzhykov[1,2], **L Sharma**[1,2], **Th Stöhlker**[1,2] **and S Fritzsche**[2,3]

[1] Physikalisches Institut, Universität Heidelberg, D–69120 Heidelberg, Germany
[2] GSI Helmholtzzentrum für Schwerionenforschung GmbH, D–64291 Darmstadt, Germany
[3] Frankfurt Institute for Advanced Studies (FIAS), D–60438 Frankfurt am Main, Germany

E-mail: surz@physi.uni-heidelberg.de

Abstract. The dielectronic recombination of heavy ions has been studied theoretically within the framework of the density matrix theory and the multiconfiguration Dirac–Fock approach. Emphasis was placed especially on the (linear) polarization of the characteristic x–ray emission and how it is affected by the non–dipole terms in the expansion of the electron–photon interaction operators. In particular, it is found that the interference between the leading electric–dipole (E1) and higher—electric and magnetic—multipole channels may significantly enhance the degree of linear polarization of the Kα hypersatellite lines. Detailed calculations for such a multipole–mixing effect are performed for the decay of the doubly–excited $L_{3/2}L_{3/2}$ resonances following the dielectronic recombination of (initially) hydrogen–like uranium U^{91+} ions.

1. Introduction

In the dielectronic recombination (DR), a free electron is captured by an ion under the simultaneous excitation of an (initially bound) electron, thus, producing a multiply excited ion state that may stabilize itself under photon emission. For many years, this resonant capture of free (or quasi–free) electrons has attracted much attention both, by experiment and theory, since it provides a unique tool for studying relativistic, quantum electrodynamical (QED) and many–body effects on the structure and dynamics of few–electron, heavy ions [1, 2, 3]. While, in the past, these studies have mainly dealt with the total DR rates, much of today's interest is focused on the *angular–resolved* properties of the characteristic x–ray emission following the resonant electron capture. For the $K-LL$ dielectronic recombination of initially hydrogen–like uranium U^{91+} ions, for example, the angular distribution of the (hyper–)satellite K$\alpha_{1,2}$ lines have been recently measured at the GSI facility in Darmstadt [4, 5]. When compared with theoretical predictions based on Dirac's theory [6, 7], these measurements reveal important information on the magnetic corrections to the electron–electron interaction as well as the mixing of different multipoles in the expansion of the radiation field.

Beside the angular–resolved measurements, polarization studies on the DR characteristic radiation provide an alternative and very promising route for a systematic analysis of the relativistic and retardation effects in the electron–electron and electron–photon interactions and in the presence of strong electromagnetic fields. During the last few years such investigations became feasible owing to the recent advancements in the design of position–sensitive solid–state detectors [8]. Therefore, the application of these x–ray detectors for analyzing the *linear polarization* of the characteristic emission from heavy ions is presently discussed, and this will lead very likely to a new kind of electron-ion collision experiments at the GSI facility in Darmstadt in forthcoming years.

In this contribution, we apply the density matrix theory together with the multiconfiguration Dirac–Fock approach in order to predict the polarization properties of the DR hypersatellite lines. Although the basic theory, which is presented in Sections 2 and 3, can be utilized for the (polarization) analysis of *any* bound–bound transition, and independent of the particular shell structure of the ions in their initial and final state, here we restrict ourselves to the decay of the $L_{3/2}L_{3/2}$ resonances in helium–like U^{90+} uranium ions. For these ions, the two $2p_{3/2}2p_{3/2} : J = 0, 2$ levels belonging to the $L_{3/2}L_{3/2}$ group are well isolated from any other doubly–excited LL resonances, allowing thus their *exclusive* population in ion–electron collision experiments. Moreover, only the $J = 2$ level can be aligned and may, hence, give rise to the emission of polarized Kα_1 ($2p_{3/2}2p_{3/2} : J = 2 \to 1s_{1/2}2p_{3/2} : J = 1, 2$) photons. In Section 4, we describe how the (degree of) linear polarization can be computed for the Kα_1 radiation and how it is modified by the interference of the dominant electric–dipole (E1) with other, higher–multiple channels in the coupling of the radiation field. Finally, a short summary of our results is given in Section 5.

2. Theoretical background

Not much has to be said about the basic formalism for studying the dielectronic recombination of few–electron relativistic ions. During the last two decades this formalism has been widely applied to investigate not only the total DR rates but also the angular distribution and polarization of the satellite radiation [6, 7, 9, 10]. For the analysis of these—angle– and polarization–dependent—properties, one usually treats the dielectronic recombination as a *two–step* process: a resonant electron capture that is followed by the radiative stabilization of the ions. Within this two–step approach, it is then convenient to make use of the density matrix formalism [11, 12] in order to express the magnetic sublevel population of the intermediate (doubly–excited) ionic states in terms of the *alignment parameters* $\mathcal{A}_{kq}$. Using the standard techniques of angular momentum algebra, the evaluation of the alignment parameters can be traced back to the DR reduced matrix elements $\langle \alpha_d J_d \,||V||\, \alpha_0 J_0, lj : J_d \rangle$ that describe the formation of the resonance $|\alpha_d J_d\rangle$ due to the interelectronic interaction between the bound electrons in the initial state $|\alpha_0 J_0\rangle$ and the free electron (see Refs. [7, 10] for further details):

$$\begin{aligned}\mathcal{A}_{kq}(\alpha_d J_d) \;=\;& \delta_{q0}\, \frac{1}{(2J_d+1)N} \sum_{ll'jj'} (-1)^{J_d+J_0-1/2}\, [l, l', j, j']^{1/2} \, \langle l0\, l'0 \,|\, k0 \rangle \left\{ \begin{array}{ccc} j & l & 1/2 \\ l' & j' & k \end{array} \right\} \\ & \times \left\{ \begin{array}{ccc} j & J_d & J_0 \\ J_d & j' & k \end{array} \right\} \langle \alpha_d J_d \,||V||\, \alpha_0 J_0, lj : J_d \rangle \, \langle \alpha_d J_d \,||V||\, \alpha_0 J_0, l'j' : J_d \rangle^{*} \, . \qquad (1)\end{aligned}$$

Here, we have supposed that neither the ion nor the electron are polarized, and that the quantization (z–) axis is chosen along the incoming electron momentum *as seen* within the rest frame of the ion. In Eq. (1), moreover, $N = \sum_{lj} |\, \langle \alpha_d J_d \,||V||\, \alpha_0 J_0, lj : J_d \rangle \,|^2$, $\langle l0\, l'0 \,|\, k0 \rangle$ is the Clebsch-Gordan coefficient, $[a, b, ...] \equiv (2a+1)(2b+1)...$, and we use the standard notation for the Wigner 6j symbols.

The alignment parameters (1) are known to obey several properties. In particular, the parameter $\mathcal{A}_{k0}(\alpha_d J_d)$ is *nonzero* only if k is even and satisfies the condition $k \le 2J_d$. Therefore, the magnetic sublevel population of the doubly–excited states with $J_d = 1$ is described by a single parameter $\mathcal{A}_{20}(\alpha_d J_d = 1)$, while two parameters $\mathcal{A}_{20}(\alpha_d J_d = 2)$ and $\mathcal{A}_{40}(\alpha_d J_d = 2)$ are required for all resonances with $J_d = 2$.

Owing to the alignment of the excited ion as produced in course of the resonant electron capture, the subsequent photon emission is in general anisotropic and linearly polarized. Indeed, both the angular and polarization properties of the (hyper–)satellite photons are closely related to the sublevel population of the excited ionic state $|\alpha_d J_d\rangle$ and, hence, to the parameters $\mathcal{A}_{k0}$. For a given $|\alpha_d J_d\rangle \to |\alpha_f J_f\rangle$ transition, for example, the degree of linear polarization of the emitted x–ray photons can be expressed as:

$$P_L(\theta) = \frac{\sum\limits_{k=2,4,...} \mathcal{A}_{k0}\left(D^k_{0-2}(0,\theta,0) + D^k_{0,2}(0,\theta,0)\right) g_k(\alpha_d J_d, \alpha_f J_f)}{1 + \sum\limits_{k=2,4,...} \mathcal{A}_{k0}\, f_k(\alpha_d J_d, \alpha_f J_f)\, P_k(\cos\theta)} \tag{2}$$

where $P_k(\cos\theta)$ denotes a Legendre polynomial of order k, $D^k_{0,q}(0,\theta,0)$ a Wigner's D–function, and where θ refers to the angle between the propagation direction of the emitted photons and the quantization axis. In addition to the alignment parameters $\mathcal{A}_{k0}$, there appear the so–called *structure functions* $f_k(\alpha_d J_d, \alpha_f J_f)$ and $g_k(\alpha_d J_d, \alpha_f J_f)$ in Eq. (2) which purely depend on the given bound-state transition of the ions. In general, these structure functions contain the matrix elements of different multipole components as they arise in the expansion of the electron–photon interaction operator [13]; they are given by:

$$\begin{aligned} f_k(\alpha_d J_d, \alpha_f J_f) &= \frac{\sqrt{2J_d+1}}{2} \sum_{LpL'p'} i^{L'+p'-L-p} (-1)^{J_f+J_d+k+1} [L, L']^{1/2} \langle L1L'-1 \,|\, k0\rangle \\ &\times \left(1 + (-1)^{L+p+L'+p'-k}\right) \begin{Bmatrix} L & L' & k \\ J_d & J_d & J_f \end{Bmatrix} \langle \alpha_d J_d \,||H_\gamma(pL)||\, \alpha_f J_f\rangle^* \\ &\times \langle \alpha_d J_d \,||H_\gamma(p'L')||\, \alpha_f J_f\rangle \left[\sum_{Lp} |\langle \alpha_d J_d \,||H_\gamma(pL)||\, \alpha_f J_f\rangle|^2\right]^{-1}, \end{aligned} \tag{3}$$

and

$$\begin{aligned} g_k(\alpha_d J_d, \alpha_f J_f) &= \frac{\sqrt{2J_d+1}}{2} \sum_{LpL'p'} i^{L'+p'-L-p} (-1)^{J_f+J_d+k+1+p'} [L, L']^{1/2} \langle L1L'1 \,|\, k2\rangle \\ &\times \begin{Bmatrix} L & L' & k \\ J_d & J_d & J_f \end{Bmatrix} \langle \alpha_d J_d \,||H_\gamma(pL)||\, \alpha_f J_f\rangle^* \langle \alpha_d J_d \,||H_\gamma(p'L')||\, \alpha_f J_f\rangle \\ &\times \left[\sum_{Lp} |\langle \alpha_d J_d \,||H_\gamma(pL)||\, \alpha_f J_f\rangle|^2\right]^{-1}, \end{aligned} \tag{4}$$

if all the *allowed* multipoles are taken into account, and where $\langle \alpha_d J_d \,||H_\gamma(pL)||\, \alpha_f J_f\rangle$ denotes the reduced matrix element for the particular magnetic ($p = 0$) or electric ($p = 1$) transition of order L [14].

Equations (2)–(4) represent the most general form of the linear polarization of the characteristic radiation following the resonant electron capture. They include especially the summation over all the multipoles (pL) of the radiation field that are allowed owing to the parity and angular momentum selection rules. If several multipoles are allowed for a given transition, one of them often dominates the radiative decay. In many cases, therefore, the summation over the multipole components of the radiation field can be restricted to just a single term which results in a significant simplification of Eqs. (2)–(4). For electric dipole (E1) radiation, for example, the degree of linear polarization is given by the well–known formula [12]:

$$P_L^{E1}(\theta) = \sqrt{\frac{3}{2}}\, \frac{\mathcal{A}_{20}\, g_2^{E1}(\alpha_d J_d, \alpha_f J_f)\, \sin^2\theta}{1 + \mathcal{A}_{20}\, f_2^{E1}(\alpha_d J_d, \alpha_f J_f)\, P_2(\cos\theta)}, \tag{5}$$

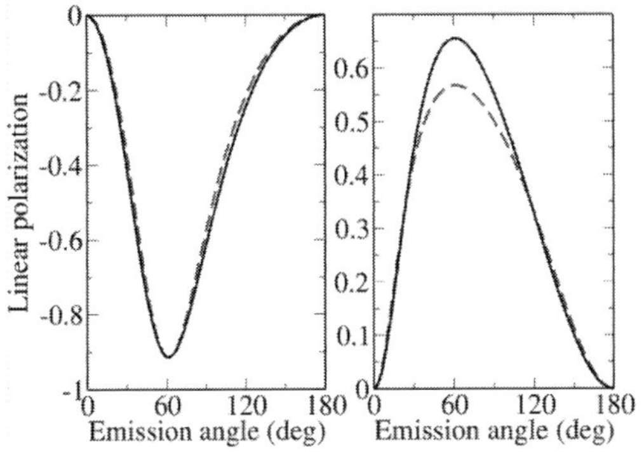

Figure 1. Degree of linear polarization of the $2p_{3/2}2p_{3/2} : J = 2 \rightarrow 1s_{1/2}2p_{3/2} : J = 1$ (left panel) and $2p_{3/2}2p_{3/2} : J = 2 \rightarrow 1s_{1/2}2p_{3/2} : J = 2$ (right panel) transitions in the helium-like uranium ions following the K–$L_{3/2}L_{3/2}$ DR. Calculations are performed within the electric–dipole approximation (dashed line) and by taking into account all of the allowed multipoles (solid line). Results are presented in the laboratory frame (i.e. rest frame of electron target).

where the structure functions

$$f_2^{E1}(\alpha_d J_d, \alpha_f J_f) = -\sqrt{\frac{2}{3}}\, g_2^{E1}(\alpha_d J_d, \alpha_f J_f) = (-1)^{1+J_d+J_f} \sqrt{\frac{3(2J_d+1)}{2}} \left\{ \begin{array}{ccc} 1 & 1 & 2 \\ J_d & J_d & J_f \end{array} \right\}, \quad (6)$$

coincide with the anisotropy parameter α_2^{ph} as introduced earlier in Ref. [12]: $f_2(\alpha_d J_d, \alpha_f J_f; \mathrm{E1}) \equiv \alpha_2^{ph}$. As seen from this equation, the structure function for the electric–dipole case *does not* depend on the bound–bound reduced transition amplitudes and, hence, on the nuclear charge Z of the ion.

3. Computations

As seen from the formalism above, the computation of the polarization properties of the DR hypersatellite radiation can be traced back to the reduced matrix elements $\langle \alpha_d J_d || V || \alpha_0 J_0, lj : J_d \rangle$ and $\langle \alpha_d J_d || H_\gamma(pL) || \alpha_f J_f \rangle$ for the resonant electron capture and the multipole radiative transitions, respectively. Matrix elements of this or similar type occur frequently in atomic structure studies and have therefore been considered in various approaches. In the calculations below, we made use of the multiconfiguration Dirac–Fock (MCDF) method which has been found a versatile tool to account for the relativistic and many–body effects in high–Z ions and which is implemented in the RATIP computer program [15, 16]. This program facilitates now the computation of the DR cross sections, alignment parameters and polarization characteristics within a distorted–wave approximation and includes, moreover, a complete multipole decomposition of the *electron–photon* interaction as well as the full operator $V = V^{\mathrm{Coulomb}} + V^{\mathrm{Breit}}$ for the *electron–electron* interaction.

4. Results and discussion

After this short discussion of the basic formulas for the resonant electron capture and the subsequent radiative decay, we are ready now to analyze the effects of multipole–mixing on the linear polarization of the dielectronic recombination hypersatelite lines for ions from the helium isoelectronic sequence. In the present work, we shall restrict ourselves to the formation of the $2p_{3/2}2p_{3/2} : J = 0, 2$ resonances following the $K - L_{3/2}L_{3/2}$ DR of initially hydrogen–like uranium U^{91+}. From these two intermediate levels, only the resonance with $J = 2$ may become aligned and, hence, may give rise to a linearly polarized photon emission. As seen from Eqs. (2)–(4), the degree of this linear polarization is determined by the two parameters $\mathcal{A}_2$ and $\mathcal{A}_4$, which characterize the magnetic sublevel population of the $2p_{3/2}2p_{3/2} : J = 2$ state and which take the values 1.116 and 1.249, respectively, as well as by the structure functions f_k and g_k that account for the interference between different allowed decay channels. The explicit form of these functions

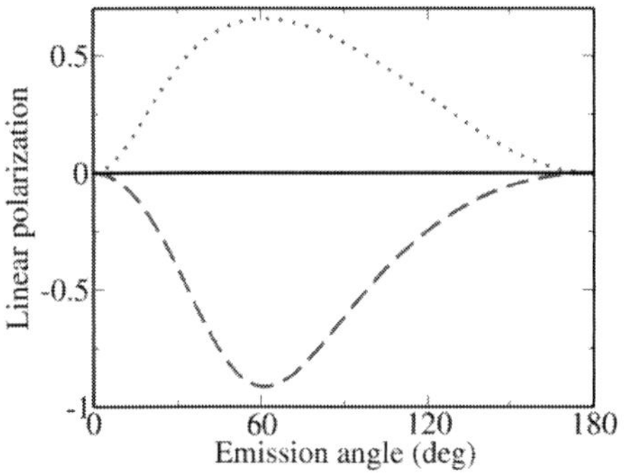

Figure 2. Degree of linear polarization of the $2p_{3/2}2p_{3/2}: J = 2 \rightarrow 1s_{1/2}2p_{3/2}: J = 1$ (dashed line) and $2p_{3/2}2p_{3/2}: J = 2 \rightarrow 1s_{1/2}2p_{3/2}: J = 2$ (dotted line) fine–structure transitions together with their average (solid line) if the two line components are not resolved explicitly. Results are shown for the DR into hydrogen–like U^{91+} ions and only for the full account multipole mixing effects.

and, hence, the strength of the multipole–mixing effects depend of course on the particular bound–bound transition. In order to study this dependence, Fig. 1 displays the (degree of the) linear polarization for the two $K\alpha_1$ hypersatellite lines $2p_{3/2}2p_{3/2}: J = 2 \rightarrow 1s_{1/2}2p_{3/2}: J = 1$ (left panel) and $2p_{3/2}2p_{3/2}: J = 2 \rightarrow 1s_{1/2}2p_{3/2}: J = 2$ (right panel) for helium–like U^{90+} ions following the $K - L_{3/2}L_{3/2}$ DR. Two different approximations are shown in this figure: while the dashed line refers to the electric–dipole approximation (i.e., when the contributions of all the higher multipoles are neglected for a particular line), the solid line represents the degree of linear polarization P_L as defined by Eq. (2).

Moreover, since the DR of high–Z ions is usually measured in collision with electrons and low–Z atoms [5], the linear polarization in Fig. 1 is presented for angles as measured in the laboratory system (i.e. the rest frame of the target). As seen from this figure, the multipole–mixing affect the polarization properties of two hypersatellite lines rather differently. While, for the $J = 2 \rightarrow J = 1$ transition, the non–dipole effects give rise to only a minor change in the shape of the linear polarization $P_L(\theta)$, the degree of polarization is remarkably enhanced for the $J = 2 \rightarrow J = 2$ line and, especially, for $\theta = 62^\circ$ that corresponds to an angle of 90° in the frame of the projectile. At this particular angle, we expect an enhancement of the linear polarization from 0.56 in the E1 approximation to 0.65 if all the (higher) multipoles are taken into account.

Unfortunately, a direct measurement of the multipole–mixing effects is not that simple in practice for (high–Z) ions from the helium-like sequence. For U^{90+} ions, for example, the two fine–structure components $2p_{3/2}2p_{3/2}: J = 2 \rightarrow 1s_{1/2}2p_{3/2}: J = 1,2$ of the $K\alpha_1$ radiation cannot (so easily) be resolved by the currently available x–ray detectors as these lines with transition energies 101.9 and 102.0 keV are separated from each other by only 70 eV about. In present–day experiments, therefore, only a superposition of the two fine–structure lines with their corresponding branching fractions can be observed. However, since the $J = 2 \rightarrow J = 1$ and $J = 2 \rightarrow J = 2$ components are polarized in perpendicular directions to each other, such a superposition will lead to a complete depolarization of the resulting $K\alpha_1$ line as seen from Fig. 2. But although these multipole–mixing effects might be difficult to resolve with current detector technology, the rapid progress in the field of position and energy–sensitive x–ray detectors may facilitate such measurements already in the near future, for instance at the future Facility for Antiproton and Ion Research [17].

5. Summary and outlook

In summary, the linear polarization of the $K\alpha_1$ hypersatellite radiation has been explored for the helium–like U^{90+} ions as produced in course of the $K - L_{3/2}L_{3/2}$ dielectronic recombination. In this study, emphasis was placed on the effects which arise from the higher non–dipole terms in the expansion of the electron–photon interaction. It is shown that the interference between the multipole components can enhance the (degree of the) linear polarization of the $2p_{3/2}2p_{3/2}: J = 2 \rightarrow 1s_{1/2}2p_{3/2}: J = 2$ fine–structure transition, and that the effect can be as large as 16 % if the photons are emitted perpendicular to the beam direction. However, this rather strong effect may remain hidden in the x–ray spectrum if the individual fine–structure components of the $K\alpha_1$ hypersatellite line cannot be resolved experimentally. — We therefore hope that this theoretical analysis will stimulate the further development of high–resolution polarization detectors for hard x–rays and will thus allow a much more detailed analysis of the relativistic and non–dipole effects on the resonant electron capture of high–Z, few–electron ions and their subsequent radiative stabilization.

Acknowledgements

A.A. and L.S. acknowledge support from the Helmholtz Gemeinschaft and GSI (Nachwuchs-gruppe VH–NG–421).

References

[1] Beiersdorfer P, Phillips T W, Wong K L, Marrs R E and Vogel D A 1992 *Phys Rev* A **46** 3812
[2] Brandau C et al. 2008 *Phys Rev Lett* **100** 073208
[3] Nakamura N et al. 2008 *Phys Rev Lett* **100** 073203
[4] Although the notation of the $K\alpha$ (hyper–) satellite photons is a bit confusing, it occurs frequently in the literature and has been adopted here also for the helium–like ions; historically, the $K\alpha$, satellites refer to transitions with additional vacancies in the L shell, and the hypersatellites to those with an initially empty K shell.
[5] Ma X, Mokler P H, Bosch F, Gumberidze A, Kozhuharov C, Liesen D, Sierpowski D, Stachura Z, Stöhlker Th and Warczak A 2003 *Phys Rev* A **68**, 042712
[6] Zakowicz S, Harman Z, Grün N and Scheid W 2003 *Phys Rev* A **68** 042711
[7] Fritzsche S, Kabachnik N M and Surzhykov A 2008 *Phys Rev* A **78** 032703
[8] Tashenov S et al. 2006 *Phys Rev Lett* **97** 223202
[9] Chen M H and Scofield J H 1995 *Phys Rev* A **52** 2057
[10] Fritzsche S, Kabachnik N M, Surzhykov A and Stöhlker Th 2009 *Nucl Instr Meth* B **267** 257
[11] Balashov V V, Bodrenko I V, Dolinov V K and Strakhova S I 1994 *Opt Spectrosc* **77** 801
[12] Balashov V V, Grum–Grzhimailo A N and Kabachnik N M 2008 *Polarization and Correlation Phenomena in Atomic Collisions* (Kluwer Academic Plenum Publishers, New York)
[13] Surzhykov A et al. 2002 *Phys Rev Lett* **88** 153001
[14] Fritzsche S, Surzhykov A and Stöhlker Th 2005 *Phys Rev* A **72** 012704
[15] Fritzsche S 2001 *J Electron Spectrosc Relat Phenom* **114–16** 1155
[16] Fritzsche S et al. 2003 *Nucl Instr Meth* B **205** 93
[17] Gumberidze A et al. 2008 *Nucl Instr Meth* B **267** 248

Journal of Physics: Conference Series **212** (2010) 012033 doi:10.1088/1742-6596/212/1/012033

Which-pass information in the double-slit experiment of diatomic molecules

Kiyoshi Ueda

Institute of Multidisciplinary Research for Advanced Materials, Tohoku University, Sendai 980-8577, Japan

E-mail: `ueda@tagen.tohoku.ac.jp`

Abstract. Analogy between the double-slit experiment in quantum optics and the photoelectron emission from diatoms is discussed in the light of availabilities of which-pass information and quantum eraser. The availability of which-pass information and the degree of the predictability may be determined by the intrinsic molecular processes involved, whereas the erasing the which-pass information relies on the observation.

1. Introduction

Interference has been continuously capturing imagination of physicists since Young's double-slit experiment. The present work is concerned with the double-slit interference by the matter wave, i.e., the interference of the coherent photoelectron emission from the diatoms. As the introduction to the double-slit experiment, let us first briefly review two experiments by the coherent light, in order to introduce the concepts of a "which-pass marker" and a "quantum eraser".

Eichmann *et al.* investigated interference effects in the light scattering from the two atoms [1]. In this experiment, two $^{198}Hg^{+}$ ions are localized in a linear Paul trap at an appropriate separation, to form a double-slit. The two ions are then irradiated by a linearly polarized, traveling wave laser beam tuned below the resonance frequency of the $^{198}Hg^{+}$ $6s\,^{2}S_{1/2}-6p\,^{2}P_{1/2}$ transition at 194 nm. They found that interference fringes appear (disappear) in a distant detection region when the polarization direction of the scattered photon is parallel (perpendicular) to the exciting laser polarization direction. They explained their observation based on the "witch pass" considerations. In case of the perpendicular polarization, the magnetic quantum number of the scattering atom is different from the other one. Thus, the different magnetic quantum number can be regarded as a "which-pass marker". The interference fringes disappear whenever a which-pass marker is present. Measuring the which-pass marker is not necessary to destroy interference.

The second example is the case where two photons are entangled, as Walborn *et al.* studied [2]. When one photon of an entangled pair is incident on a double-slit of appropriate dimensions, interference fringes appear in a distant detection region. When quarter-wave plates, oriented so that their fast axes are orthogonal, are placed in front of each slit, the fringes disappear because a which-pass marker is present. If the polarization of the other entangled photon is measured in coincidence with the first photon, the fringes can be recovered because the which-way marker can be erased by coincident detection of the entangled photon at a certain polarization direction. Such an eraser of the which-way marker is called a "quantum eraser".

Electron emission from a gerade or ungerade orbital in a homonuclear diatom causes interference that is analogous to double-slit interference. Here, the photoelectron wave is described as a superposition of two phase-coherent waves emitted from the two atoms (slits). Cohen and Fano were the first who predicted that the photoionization cross sections of homonuclear diatomic molecules oscillate as a function of the photoelectron momentum, due to double-slit interference [3]. Cohen-Fano (CF) interference has been first observed in electron emission from H_2 molecules in ion-impact experiments [4, 5]. For photoionization experiments of H_2 molecules, Frankfurt group clearly demonstrated double-slit interference of the coherent photoelectron emission from the two sites [6, 7], fixing the molecule in space using reaction microscope [8, 9].

Coherent photoelectron emission from the $1\sigma_g$ or $1\sigma_u$ core orbital of N_2 is another example that exhibits double-slit interference [10, 11] and indeed Cohen-Fano oscillation has been observed by the author's group in the ratio of the $1\sigma_g$ and $1\sigma_u$ photoionization cross sections of N_2 [12, 13, 14, 15].

2. Core-level photoemission of N_2

2.1. Cohen-Fano interference

The symmetry-adapted core orbitals in nitrogen molecules are $1\sigma_g$ and $1\sigma_u$, which can be described as a linear combination of the atomic N $1s$ core orbitals, $1s_1$ and $1s_2$, localized at two N atoms N(1) and N(2), respectively:

$$1\sigma_{g,u} = \frac{1s_1 \pm 1s_2}{\sqrt{2}} . \tag{1}$$

Ab initio calculations [16, 17] predicts the energy splitting ($\sim$ 100 meV) for the $1\sigma_g$ and $1\sigma_u$ core-hole states and high-resolution X-ray photoelectron spectroscopy using synchrotron radiation as a light source indeed confirmed not only this g-u splitting [18] but also the different equilibrium bond lengths for the $1\sigma_g$ and $1\sigma_u$ core-hole states [17]. The g-u splitting in the core-hole states was confirmed also in high-resolution Auger electron spectra [19, 20]. If one observes photoelectron emission from the $1\sigma_g$ or $1\sigma_u$ orbital, the photoelectron emission can be considered as a coherent superposition of the two waves emitted from the two atoms. As a result, the quantum interference, similar to the Young's double-slit experiment, may appear. Indeed we have observed CF oscillation in the ratio of the core-level photoemission cross sections for $1\sigma_g$ and $1\sigma_u$ single hole creation of N_2 [12, 13, 14, 15].

Figure 1 depicts the ratios for the partial cross sections of the $v = 0 \rightarrow v' = 0$ and $v = 0 \rightarrow v' = 1$ vibrational transitions, as calculated and as measured, as a function of the photoelectron momentum $k = \sqrt{2E}$, where E is the kinetic energy of the photoelectron in atomic units. The calculations were carried out using the random-phase approximation (RPA) with the use of the relaxed-core Hartree-Fock (RCHF) wave functions as a zero-order approximation, employing the fractional-charge RCHF method [21]. The measurements [12, 13, 14, 15] were carried out on beam line 27SU [22, 23, 24] at SPring-8, Japan, using a high-resolution electron spectroscopy end station [25], whose heart is 20-cm radius hemispherical electron analyzer SES2002 (GammaData-Scienta). The agreement between the measurements and calculations is reasonable. The peak structure appears at $k \simeq 1$ a.u. It corresponds to the σ^* shape resonance as discussed in detail in [26]. In the region of $1.5 < k < 6.5$ a.u., the ratios exhibit the oscillatory structure, which is due to CF interference.

In the case of photoemission from a nitrogen molecular core orbital, which is a superposition of $1s$ orbitals localized in two individual N atoms, $N^{(1)}$ and $N^{(2)}$, the two N atoms emit coherently phase-shifted electronic waves $\propto \exp(i\mathbf{k} \cdot \mathbf{R}_1)$ and $\propto \exp(i\mathbf{k} \cdot \mathbf{R}_2)$, where $\mathbf{R}_1$ and $\mathbf{R}_2$ are the

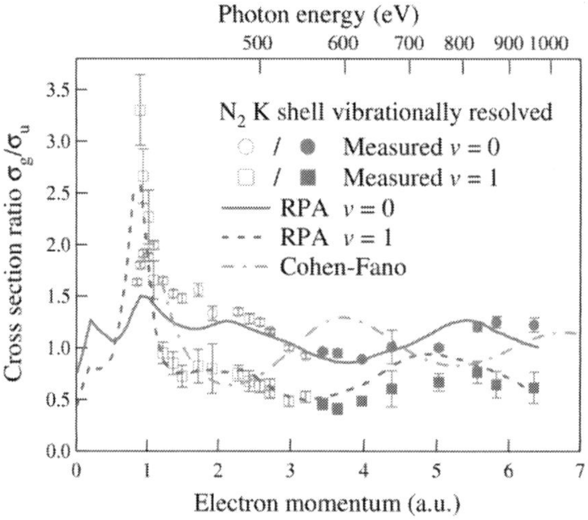

Figure 1. Comparison of experimental cross section ratios (circles, for $v' = 0$; squares, for $v' = 1$; open symbols, from [21, 26]; closed symbols, from [12, 13]) with *ab initio* calculations (solid red curve, for $v' = 0$; dashed blue curve, for $v' = 1$). A prediction based on the original Cohen-Fano formula given by Eq. (3) with $R = 2.068$ a.u. is shown by a dash-dotted green curve.

position vectors of $N^{(1)}$ and $N^{(2)}$. When the electron with the momentum $\mathbf{k}$ is ejected from the gerade or ungerade core orbital given by Eq. (1), the interference of waves emitted coherently from the two atoms leads to interference in the cross section of K-shell photoinization

$$\sigma_{g,u}(\omega) \propto \frac{1}{2}\left|e^{i\mathbf{k}\cdot\mathbf{R}_1} \pm e^{i\mathbf{k}\cdot\mathbf{R}_2}\right|^2 = 1 \pm \cos(\mathbf{k}\cdot\mathbf{R}) \,, \tag{2}$$

with $\mathbf{R} = \mathbf{R}_1 - \mathbf{R}_2$. Molecules are randomly oriented in gas phase. The orientational averaging of the cross section of Eq. (2) results in the CF interference pattern [3]:

$$\sigma_{g,u}(\omega) = \sigma_0(\omega)\left[1 \pm \chi_{\rm CF}(k)\right] \,, \tag{3}$$

where

$$\chi_{\rm CF}(k) = \frac{\sin kR}{kR} \tag{4}$$

for $kR \gg 1$. Here, $\sigma_0(\omega) \propto |d(\omega)|^2(\mathbf{e}\cdot\mathbf{k})^2$ is the K-shell photoionization differential cross section of a single atom by monochromatic X-rays with the frequency ω and polarization vector $\mathbf{e}$, and $d(\omega)$ is the transition dipole moment of atomic photoionization. The prediction by the CF formula is also given in Fig. 1 by the dashed line. Surprisingly, the oscillatory structure is completely out of phase.

2.2. Role of photoelectron scattering and recoil

We note that CF formula assumes that the wave function of the photoelectron is a plane wave outside of the atom. Thus the shift of the ab initio and experimental oscillatory structure from the CF prediction may be attributed to photoelectron scattering by the neighboring atom [13, 14]. The expression for the interference term is

$$\chi_{g,u}(k) = \frac{\sin\left[kR + 2\delta_1(k)\right]}{kR} \,, \tag{5}$$

where δ_1 is the phase of the atomic scattering of the p electron wave. It is worth noting that the photoelectron scattering does not reduce the amplitude of the oscillatory structure, i.e., the visibility of the fringes, indicating the scattering does not provide any "which-pass" information, or core-hole localization.

Not only the photoelectron diffraction but also the photoelectron recoil cause the shift of the CF oscillation [15]. The recoil phase shift depends strongly on the vibrational quantum number v of the core hole state. The interference term $\chi_{g,u}(k,v)$ experiences a phase shift due to the recoil which is twice the phase of the one-center amplitude:

$$2\psi_{g,u}(v,k) = \arctan(k\Delta R_v) \,, \tag{6}$$

where

$$\Delta R_v = \frac{\langle 0|q|v\rangle_{g,u}}{\langle 0|v\rangle_{g,u}} \tag{7}$$

is the effective *displacement* caused by the photoelectron recoil. The final expression for the interference term is

$$\chi_{g,u}(k,v) = \frac{\eta_v}{kR}\sin\left[kR + 2\psi_{g,u}(v,k) + 2\delta_1(k)\right] \,, \tag{8}$$

where $\eta_v = \sqrt{1+(k\Delta R_v)^2}$. We note that the photoelectron recoil also does not reduce the amplitude of the oscillatory structure, or the visibility of the fringes, indicating the recoil does not provide any "which-pass" information, or core-hole localization.

In the conditions described above, one cannot say which atom emitted an electron and thus has a core hole, in the same analogy as Young's double-slit experiment.

2.3. Role of the Auger emission

Recently, we demonstrated, experimentally and theoretically, that a high-resolution Auger electron spectrum, in which the $1\sigma_g$ and $1\sigma_u$ core-hole states are partially resolved, is accurately described as an incoherent sum of the two separate contributions from the decays of the $1\sigma_g$ and $1\sigma_u$ core-hole states [20]. Namely, the intensities of the N $1s\sigma_{g,u}^{-1}$ Auger transitions to a final state $|f\rangle$ can be described by $I^f_{g,u} = |D_{g,u}|^2|Q^f_{g,u}|^2$ with D being the dipole matrix element for the photoionization and Q the Coulomb matrix element for the decay. This finding suggests that the Auger electron alone does not provide any information about the which-pass information, or core-hole localization.

The energy separation between the $1\sigma_g$ and $1\sigma_u$ core-hole states of N2 is 100 meV, which is in fact smaller than the lifetime width of 120 meV. Thus, the separations in the photoelectron and Auger electron spectra are possible only via least-squares curve fitting. The Cohen-Fano oscillation disappears when $1\sigma_g$ and $1\sigma_u$ core-hole states are not resolved. Without resolving the $1\sigma_g$ and $1\sigma_u$ core-hole states, Schöffler *et al.* measured molecular-frame N $1s$ photoelectron angular distributions (MFPAD) of N_2, resolving also the direction of the Auger electron emission [27]. They demonstrated that, depending on the direction of the Auger electron emission, the MFPAD exhibits the emission from the non-localized $1\sigma_g$ or $1\sigma_u$ orbital as well as from the atomic orbital localized to the right or left atom.

In the atomic sequential doublephotoionization, it is well known that the angular correlations between the photoelectron and the Auger electron are well described by a general expression in which the sequential double-photoionization is treated as a single scattering process [28, 29, 30, 31, 32]:

$$M_f(\epsilon,\epsilon_A) = \langle f\epsilon,\epsilon_A\,|\,D\,|\,0\rangle + \sum_n \int \frac{\langle f\epsilon,\epsilon_A\,|\,Q\,|\,n\tau\rangle\langle n\tau\,|\,D\,|\,0\rangle d\tau}{\hbar\omega - E_\tau + i\Gamma_n/2} \,. \tag{9}$$

Here the summation is over the intermediate states $|n\tau\rangle$, which are characterized by their energy $E_{n\tau}$ and life-time width Γ_n. The total intermediate-state energy is $E_\tau = E_0 + I_n + \tau$, where E_0 is the energy of the ground state $|0\rangle$. These states, excited by the dipole photon with the energy $\hbar\omega$ from the ground state $|0\rangle$, further decay to the Auger final state $|f\rangle$ via the Coulomb interaction Q. The total final-state energy is $E_f = E_0 + I_f + \epsilon + \epsilon_A$, with ϵ and ϵ_A being photoelectron and Auger electron kinetic energies. This general formula is often called Kramers-Heizenberg formula. From this general formula it is straightforward to obtain the expression for the current N_2 1s photoemission:

$$M_f(\epsilon, \epsilon_A, \tau) \propto \langle f\epsilon, \epsilon_A \,|\, V \,|\, g\tau\rangle\langle g\tau \,|\, D \,|\, 0\rangle + \langle f\epsilon, \epsilon_A \,|\, V \,|\, u\tau\rangle\langle u\tau \,|\, D \,|\, 0\rangle \,. \tag{10}$$

Here we neglected direct double photoionization term and the energy splitting of N $1s\sigma_{g,u}$. It is clear that the photoelectron and the Auger electron are entangled to form a Bell state.

The above situation reminds us the quantum eraser experiment with the two entangled photons described in the introduction. In the quantum eraser experiment, quarter-wave plates oriented so that their fast axes are orthogonal are placed in front of each slit and one photon of an entangled pair is incident on the double-slit. The polarization of the other entangled photon is measured in coincidence. Then the double-slit interference fringes appear and disappear depending on the direction of the polarization of the entangled photon detected in coincidence. Analogously, in the experiment by Schöffler *et al.*, the photoelectron angular distributions in the molecular frame mimic the localized core-hole or non-localized core-hole, depending on the direction of the Auger emission detected in coincidence. Thus, the experiment by Schöffler *et al.* is in a sense an analog of the quantum eraser experiment.

3. Concluding remark

We have discussed analogy of the double slit in quantum optics with coherent electron emission from diatoms, using the N_2 molecule as a specific example. We have observed a separated g or u core-hole level both in the core-level photoemission and Auger emission. We have observed Cohen-Fano interference oscillation in the g-u split partial cross sections or randomly oriented molecules. This oscillation is equivalent to Young's double slit interference "fringes", indicating that the core-level photoemission is indeed a coherent two-center emission. Intramolecular elastic scattering and photoelectron recoil shift the phase of the interference fringes but hardly affect the fringe visibility, illustrating that these effects would not provide the which-pass information. If we do not resolve g and u states, double slit interference fringes disappear. Even if one does not resolve g and u symmetry of the intermediate core-hole state but detects the entangled Auger electrons with well defined g or u final state, one can restore or destroy the double slit interference, as demonstrated by Schöffler *et al.*. Here the observation plays a vital role. On the other hand, in certain cases, atomic ion left after the photoemission may provide a partial which-pass information, as demonstrated for the Ne_2 case [33, 34]. The availability of which-pass information and the degree of the predictability by the atomic ion may be determined by the intrinsic molecular processes involved, such as interatomic Coulombic decay and dissociation. Here the molecular process plays a vital role. It is important to differentiate the role of the observation and molecular dynamics.

Acknowledgments

The experiments were performed at SPring-8 with the approval of JASRI. The author is grateful to all the collaborators listed as coauthors of Refs. [12, 13, 14, 15, 20]. The work was in part supported by Grants-in-Aid for Scientific Researches from the Japan Society for Promotion of Science (JSPS) and by a Grant for Creative Scientific Research from the Ministry of Education, Science, Culture, and Sports of Japan.

References

[1] U. Eichmann *et al.*, Phys. Rev. Lett. *70*, 2359 (1993).
[2] S. P. Walborn *et al.*, Phys. Rev. A *65*, 033818 (2002).
[3] H.D. Cohen and U. Fano, Phys. Rev. **150**, 30 (1966).
[4] N. Stolterfoht, B. Sulik, V. Hoffmann, B. Skogvall, J.Y. Chesnel, J. Rangama, F. Frèmont, D. Hennecart, A. Cassimi, X. Husson, A.L. Landers, J.A. Tanis, M. E. Galassi, and R. D. Rivarola, Phys. Rev. Lett. **87**, 023201 (2001).
[5] D. Misra, U. Kadhane, Y.P. Singh, L.C. Tribedi, P.D. Fainstein, and P. Richard, Phys. Rev. Lett. **92**, 153201 (2004).
[6] D. Akoury *et al.*, Science **318**, 949 (2007).
[7] K. Kreidi *et al.*, Phys. Rev. Letters **100**, 133005 (2008).
[8] R. Dörner, V. Mergel, O. Jagutzki, L. Spielberger, J. Ullrich, R. Moshammer, H. Schmidt-Böcking, Phys. Rep. **330**, 95 (2000).
[9] J. Ullrich, R. Moshammer, A. Dorn, R. Dörner, L.Ph.H. Schmidt, H. Schmidt-Böcking, Rep. Prog. Phys. **66**, 1463 (2003).
[10] D. Rolles *et al.*, Nature **437**, 711 (2005).
[11] B. Zimmermann *et al.*, Nat. Phys. **4**, 649 (2008).
[12] K. Ueda X.-J. Liu, G. Pruemper, T. Lischke, T. Tanaka, M. Hoshino, H. Tanaka, I. Minkov, V. Kimberg, F. Gel'mukhanov, Chem. Phys. **329**, 329 (2006).
[13] X.-J. Liu, N.A. Cherepkov, S.K. Semenov, V. Kimberg, F. Gel'mukhanov, G. Prümper, T. Lischke, T. Tanaka, M. Hoshino, H. Tanaka, and K. Ueda J. Phys. B: At. Mol. Opt. Phys. **39**, 4801 (2006).
[14] X.-J. Liu, G. Prümper, F. Gel'mukhanov, N. A. Cherepkov, H. Tanaka, and K. Ueda, J. Electr. Spectrosc. Relat. Phenom. **156-158**, 73 (2007).
[15] F. Gel'mukhanov, V. Kimberg, X.-J. Liu, G. Prümper, T. Tanaka, M. Hoshino, H. Tanaka, and K. Ueda, J. Electr. Spectrosc. Relat. Phenom. **156-158**, 265 (2007).
[16] A. Thiel, J. Schirmer, and H. Köppel, J. Chem. Phys. **119**, 2088 (2003).
[17] M. Ehara, H. Nakatsuji, M. Matsumoto, T. Hatamoto, X.-J. Liu, T. Lischke, G. Prümper, T. Tanaka, C. Makochekanwa, M. Hoshino, H. Tanaka, J. R. Harries, Y. Tamenori, and K. Ueda, J. Chem. Phys. **124**, 124311 (2006).
[18] U. Hergenhahn, O. Kugeler, A. Rüdel, E.E. Rennie, and A.M. Bradshaw, J. Phys. Chem. A, **105**, 5704 (2001).
[19] S.L. Sorensen, C. Miron, R. Feifel, M.-N. Piancastelli, O. Börneholm, and S. Svensson, Chem. Phys. Lett. **456**, 1 (2008).
[20] R. Püttner, H. Fukuzawa, X.-J. Liu, S. Semenov, N. Cherepkov, T. Tanaka, M. Hoshino, H. Tanaka, and K. Ueda J. Phys. B: At. Mol. Opt. Phys. **41**, 141001 (2008)
[21] S. K. Semenov, N. A. Cherepkov, M. Matsumoto, K. Hatamoto, X.-J. Liu, G. Prümper, T. Tanaka, M. Hoshino, H. Tanaka, F. Gel'mkhanov, and K. Ueda, J. Phys. B: At. Mol. Opt. Phys. **39**, L261 (2006).
[22] K. Ueda, J. Phys. B: At. Mol. Opt. Phys. **36**, R1 (2003); and references cited therein.
[23] H. Ohashi, E. Ishiguro, Y. Tamenori, H. Kishimoto, M. Tanaka, M. Irie T. Tanaka, T. Ishikawa, Nucl. Instrum. Methods A **467-468**, 529 (2001).
[24] H. ohashi, E. Ishiguro, Y. Tamenori, H. Okumura, A. Hiraya, H. Yoshida, Y. Senba, K. Okada, N. Saito, I.H. Suzuki, K. Ueda, T. Ibuki, S. Nagaoka, I. Koyano, T. Ishikawa, Nucl. Instrum. Methods A **467-468**, 533 (2001).
[25] Y. Shimizu, H. Ohashi, Y. Tamenori *et al.*, J. Electr. Spectrosc. Relat. Phenom. **63**, 114 (2001).
[26] S.K. Semenov, N.A. Cherepkov, M. Matsumoto, K. Fujiwara, E. Kukk, H. Yoshida, T. Tanaka, M. Kitajima, H. Tanaka, A. De Fanis, and K. Ueda, J. Phys. B: At. Mol. Opt. Phys. **39**, 375 (2006).
[27] M. Schöffler *et al.*, Science **320**, 892 (2008).
[28] T. Åberg, Phys. Scripta T41, **71** (1992).
[29] T. Åberg and B. Crasemann, in *X-ray Anomalous resonance scattering: theory and experiment*, (eds. K. Fischer, G. Materlik, and C. Sparks, Elsevier, North Holland, Amsterdam, 1994).
[30] G. B. Armen, H. Aksela, T. Åberg and S. Aksela, J. Phys. B: At. Mol. Opt. Phys. **33**, R49 (2000).
[31] M. Kitajima, M. Okamoto, Y. Shimizu, H. Chiba, S. Fritzsche, N. M. Kabachnik, I. P. Sazhina, F. Koike, T Hayaishi, H Tanaka, Y Sato, and K Ueda, J. Phys. B: At. Mol. Opt. Phys. **34**, 3829 (2001).
[32] N.M. Kabachnik, S. Fritzsche, A.N. Grum-Grzhimailo, M. Meyer and K. Ueda Physics Reports **451**, 155-233 (2007).
[33] K. Kreidi *et al.*, J. Phys. B: At. Mol. Opt. Phys. **41**, 101002 (4pp) (2008).
[34] M. Yamazaki, J. Adachi, Y. Kimura, A. Yagishita, M. Stener, P. Decleva, N. Kosugi, H. Iwayama, K. Nagaya, and M. Yao, Phys. Rev. Lett. **101**, 043004 (2008).

Photodetachment of O^- from threshold to 1.2 eV electron kinetic energy using Velocity-Map Imaging

S J Cavanagh, S T Gibson and B R Lewis
Atomic and Molecular Physics Laboratories, Research School of Physics and Engineering, The Australian National University, Canberra, ACT 0200, Australia

E-mail: Steven.Cavanagh@anu.edu.au

Abstract: High-resolution photoelectron imaging from O^- with excess energies between 0.5 meV and 1.2 eV is reported. With electron energy resolutions ranging from 266 μeV to 3 meV, branching ratios and angular-distribution asymmetry parameters for each of the fine-structure transitions were measured. Preliminary data for a subset of these measurements showing possible effects due to electron correlation at low excess energy are presented, in the hope of stimulating further theoretical calculations for this species.

1. Introduction

Photodetachment experiments not only provide valuable data on negative ions relevant to fields such as astrophysics, atmospheric and plasma physics and chemistry, but they also provide the ideal environment to study electron-correlation effects. This is due to the lack of a long-range nuclear Coulombic attraction potential, $\propto r^{-1}$, with the electron instead being bound by a short-range potential of the order r^{-4}. Because of this, subtle electron-correlation and relativistic effects can be studied via photodetachment without being overwhelmed by the strong long-range nuclear Coulombic potential, as is the case in photoionization [1]. Despite this unique environment, experimental photodetachment data, especially angular distributions, are limited in number compared with the situation for photoionization.

To date, most of the measurements undertaken on negative ions have focused on determining cross sections and/or the electron affinity (EA), with only a few measurements additionally determining the photoelectron angular dependence at some finite electron kinetic energy (eKE) [1,2]. Non-threshold photodetachment measurements have been limited to photoelectrons with eKEs from 100's meV to 10's eV. At these eKEs the interaction time of the outgoing electrons with the neutral target is small and any structure (within cross sections or angular distributions) resulting from correlation effects is usually small and overwhelmed by experimental uncertainties, unless near a resonance. Lowering of the eKE increases the interaction time, allowing for an enhancement of the electron correlation effects and the resulting structure.

Velocity-map imaging (VMI) is proving to be a very versatile technique, capable of being used to determine photoelectron spectra and their full angular distributions simultaneously, allowing the extraction of angular-distribution asymmetry parameters, branching ratios and cross sections. VMI also offers the unique ability to map electrons with moderate to very low eKEs while maintaining resolving power, thus increasing the absolute photoelectron energy resolution, with near unity collection efficiency. This latter feature helps to mitigate the dramatic loss of photoelectron signal due to the very small magnitude of cross sections at threshold. However, until recently the relative energy resolutions reported using VMI have been limited to $\Delta E/E \geq 2.0\%$ [3]. With this level of resolution, features such as the fine-structure levels of light atoms, molecular rotational levels and even some vibrational features cannot be resolved.

In 2007, we reported a dramatic improvement in the resolving power of a VMI apparatus [3] achieving a $\Delta E/E = 0.38\%$ ($\Delta v/v = 0.19\%$) for eKE $\approx$ 1 eV. This was not only significant in terms of the resolving power achievable with a VMI spectrometer, but it has allowed the measurement of photoelectron spectra and angular distributions, simultaneously, for light atoms where fine-structure transitions are resolved at eKEs well away from threshold.

In this paper, preliminary results for the photodetachment of O^- are presented. When photodetaching O^- below a photon energy of 3 eV six fine-structure transitions are accessible,

$$O^-(\,^2P_{3/2,1/2}) + h\nu \rightarrow O(\,^3P_{2,1,0}) + e^-, \qquad (1)$$

where the subscripts refer to the total angular momentum J. The data shown includes photoelectron spectra and angular distributions for eKEs between 500 μeV and 1.2 eV, with electron energy resolutions from 200 μeV to several meV. Also presented is the photoelectron branching ratio between two transitions originating from the ground state of the ion, and a detailed plot of the angular-distribution asymmetry parameter for the sub-transition defining the electron affinity. Recently Garand *et al.* [4] claimed a *somewhat higher resolution* than that reported in our publication [3]. This claim is not directly comparable as their reported resolution was achieved at a much longer wavelength than in our publication where the eKEs were 30 – 90 times higher. For a direct comparison of electron energy resolution the values reported in this publication are more relevant.

2. Experimental

The fast anion beam VMI spectrometer has been described elsewhere [3], therefore only a limited description will be given here. The experimental apparatus used in this work (shown in Fig. 1), produces a collimated beam of mass-selected negative ions, which are then interrogated by a laser beam, producing photoelectrons. The detached photoelectrons are imaged using a VMI lens and a multichannel-plate phosphor detector where the photo events are captured with a CMOS camera and each frame is analysed by a PC.

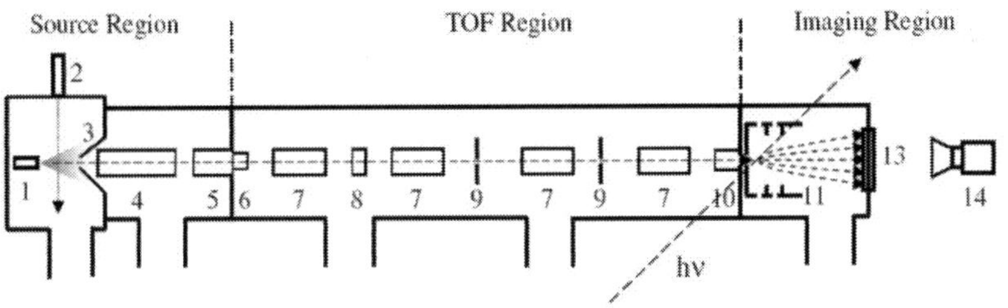

Figure 1. A schematic of the fast anion-beam VMI spectrometer, highlighting its three main regions: Source region (1-5): pulsed ion source and initial acceleration optics; Time-of-Flight (TOF) region (6-10): mass TOF containing ion optics; Imaging region (11-14): VMI spectrometer, MCP/phosphor detector and CMOS camera.

The VMI lens was designed to accommodate a fast ion beam with beam energies from 100's eV to keV and an interaction-region volume of at least 2 mm^3, while achieving high resolution ($\Delta E/E \approx$ 0.3%). Imaging of the photoelectrons is performed using a pair of high dynamic-range image-quality 10 μm-pore microchannel plates (MCPs), a P47 phosphor screen and a 4.2 Mega pixel (2048 × 2048) monochrome Peltier-cooled CMOS camera (PCO AG, PCO2000).

Photoelectrons are produced by crossing the ion beam with a detachment-laser beam generated by a Continuum Sunlite EX optical parametric oscillator (OPO) pumped by a Continuum Powerlite 9010 Nd:YAG laser operated at its third harmonic, 355 nm. For the presented measurements the OPO was operated at 10 Hz with pulse energies in the range of 1 mJ – 3 mJ. Given that the measurements were conducted at a series of wavelengths, the output wavelength of the OPO was measured using a high quality wavemeter (HighFinesse, WS7-UV).

Journal of Physics: Conference Series **212** (2010) 012034 doi:10.1088/1742-6596/212/1/012034

For extraction of the 3D photoelectron distribution from the 2D velocity-map image, an inverse Abel transformation was used, as described by Hansen and Law [5]. Before extraction of the 3D photoelectron distribution, small distortions in the image, resulting mainly from the optical lens system coupling the phosphor to the camera, were removed using a circularisation procedure outlined in [3].

As with all VMI spectrometers, the resolving power of our instrument is maintained for all eKEs, as well as the collection efficiency of the lens itself. With eKEs of 1 eV, instrumental resolving powers of $\Delta E/E = 0.38\%$ are easily obtained, resulting in absolute electron energy resolutions of 2 – 3 meV, full-width of half maximum (FWHM). When operated with eKEs < 100 meV, the absolute electron energy resolution is improved substantially, and for our spectrometer μeV resolutions are obtained. For all measurements, the laser electric-field vector is in the vertical direction, parallel to the MCPs.

3. Results and Discussion

Figure 2 shows a velocity-mapped image (inset) and the corresponding photoelectron spectrum taken at a wavelength corresponding to conditions close to threshold, 832.220(4) nm. The eKEs range from 0.5 – 50 meV with a photoelectron energy resolution in the 100's μeV range. At the lowest eKE achievable where angular distributions can still be extracted, a photoelectron energy resolution of 266 μeV FWHM was achieved.

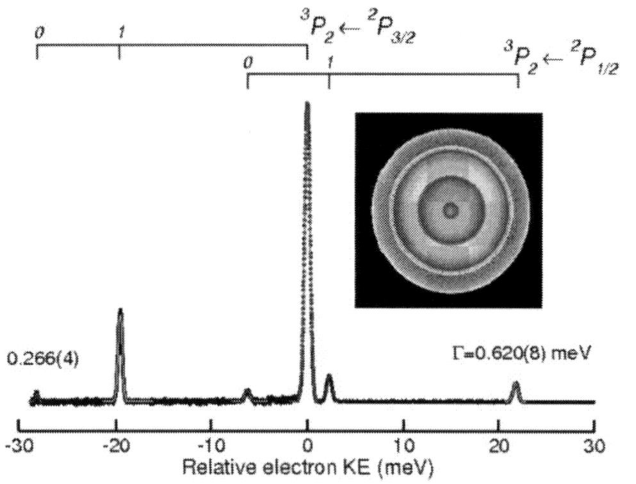

Figure 2. Photoelectron spectrum for photodetachment of O^- ions with 832.220-nm linearly polarized light, together with the corresponding velocity-map photoelectron image (inset). The experimental spectrum (circles) has been fitted with Gaussian functions (lines). The FWHM resolution ranges from ΔE = 0.27 – 0.62 meV as indicated above the first and last feature (left to right). The low intensity for the transitions originating from the $^2P_{1/2}$ anion state is a characteristic of the anion source.

Photodetaching a p electron from O^-, as described by Eq. (1), results in the production of outgoing s and d partial waves. Rau and Fano [6] and Lineberger and Woodward [7] both treated this process theoretically for conditions near threshold, where s waves dominate. They considered the final state as a JJ-coupled (e + atom) complex created after the LS-coupled anion absorbs a photon. Rau [8] showed that under these conditions the fine-structure relative intensities are given by pure geometric factors, which do not involve the full photodetachment dynamics. A generalized theoretical treatment by Pan and Starace [9] for photodetachment showed that by using the approach of term-independent dynamical amplitudes often used in the central-potential model, factorization of the geometric and dynamical factors can be achieved, resulting in intensity ratios given by geometric factors.

The relative intensity ratios between the observed fine-structure transitions agree well with the theoretical values generated using the formalism of [9] and the previous measured values of [10, 11], except for the lowest eKE ($^3P_0 \leftarrow {}^2P_{3/2}$) transition in Fig. 2, which is suppressed. The suppressed intensity for the $^3P_0 \leftarrow {}^2P_{3/2}$ transition is the result of its proximity to threshold and can be accounted for by scaling the transition intensity according to the Wigner law for photodetachment, $\text{eKE}^{l+1/2}$ [12].

Once scaled according to the Wigner law the theoretical values are in excellent agreement with our experimental results.

Figure 3 shows the extracted branching ratio for transitions from the $^2P_{3/2}$ anion ground state to the $^3P_1$ and $^3P_2$ states of the neutral oxygen atom versus the eKE of the transition to $^3P_1$. The eKEs for this ratio range from 500 μeV – 1 eV. The branching ratio exhibits a trend typical of a cross-section for photodetachment, a sharp rise in the cross section from threshold levelling out to a value that is consistent with the intensity ratios given by the geometric factors. Fluctuation in the branching ratio above an eKE = 0.6 eV can be attributed to statistical variations and are not indicative of any physical behaviour. A tentative fit of the Wigner law, expressed as a ratio of each transition, i.e. $\sigma(^3P_1)/\sigma(^3P_2) \propto [eKE(^3P_1)/eKE(^3P_2)]^{1/2}$, has also been plotted for comparison and as a measure for the range of validity. The fit suggests the range of validity of the Wigner law extends from threshold to at least 13 meV. Further analysis of this and the other branching ratios will be published elsewhere.

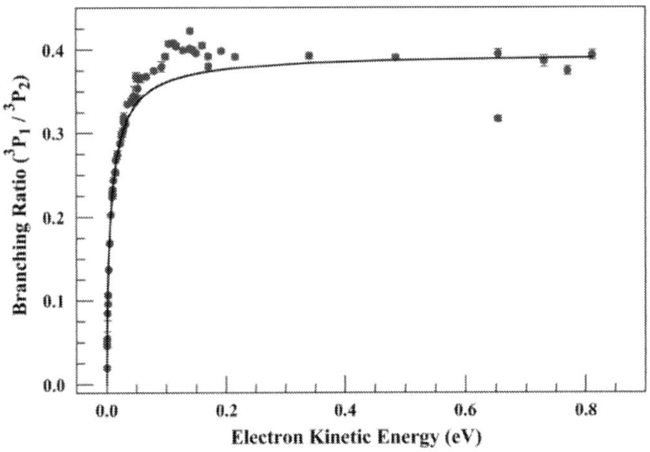

Figure 3. The extracted branching ratio for transitions originating from the $^2P_{3/2}$ state of the anion to the $^3P_1$ and $^3P_2$ states of the neutral oxygen atom. Circles: this work. Solid line: Wigner Law for photodetachment of an $l = 1$ orbital.

The angular anisotropy described by the β parameter contains all the dynamics of the photodetachment process. This parameter was extracted from the photoelectron images by plotting electron intensity against $P_2(\cos\theta)$ and fitting to the well-known equation [13]:

$$\frac{d\sigma(E)}{d\Omega} = \frac{\sigma_{\text{total}}(E)}{4\pi}\left[1 + \beta(E)\,P_2(\cos\theta)\right], \qquad (2)$$

where E is the eKE, θ is the angle between the photon electric vector and the electron momentum vector, σ_{total} is the total cross section for photodetachment, β is the angular-distribution asymmetry parameter and $P_2(\cos\theta)$ is a Legendre polynomial of second order. Plotting intensity against $P_2(\cos\theta)$ produces a straight line where the slope of the line equals β [3]. Photodetachment of a p orbital from atomic oxygen gives rise to s and d outgoing partial waves with the value of β dependent on the intensity and relative phase shift between each partial wave.

Figure 4 shows the experimental β values versus eKE for the $^3P_2 \leftarrow {}^2P_{3/2}$ transition (circles) compared with the central-potential Cooper-Zare model [13] (solid curve) and a parameterised version of this model by Hanstorp *et al.* [14] (dashed and dotted curves). Also shown for comparison is the experimental data of Hanstorp *et al.*, [14] (squares), Hall and Siegel [15] (diamond) and Breyer *et al.* [10] (triangles). It should be noted that the β estimates for [14,15] are for unresolved fine-structure transitions and therefore are not strictly comparable to the current data set. They are shown as a guide to the relative accuracy of our β values for eKEs higher than 0.24 eV. The experimental range of eKE in Fig. 4 is 480 μeV – 1.26 eV.

The parameterisation by Hanstorp *et al.* [14] provides an analytical expression for β versus eKE with two adjustable parameters. With a simple assumption regarding the energy dependence of dipole radial matrix elements, $R_{l\pm1}$, their parameterization yielded:

$$\beta(E) = \frac{2A_2^2E^2 - 4A_2Ec}{1 + 2A_2^2E^2}, \tag{3}$$

where $c = \cos(\delta_2 - \delta_0)$ is the relative phase shift between the *s* and *d* outgoing waves, $A_2E = R_2/R_0$ is the ratio of radial matrix elements and E is the photoelectron eKE.

The two parameterised curves in Fig. 4 have the same value for the ratio between the radial matrix elements, set by fitting to the Cooper-Zare model [13], but different values for the relative phase shift between the two outgoing partial waves [1.0 (dotted curve), 0.925 (dashed curve)] [3]. The experimental values for β, with eKEs above 0.24 eV, lie closest to the dashed curve, the curve with the smaller phase-shift value, and then converge to a value common to all curves below this point until an eKE of 40 meV. The magnitudes of our data points are also consistent with the values of [14,15]. This suggests that, for eKEs higher than 0.24 eV, there may be slightly less phase shift between the two partial waves or a greater *s*-wave contribution than predicted by theory [13].

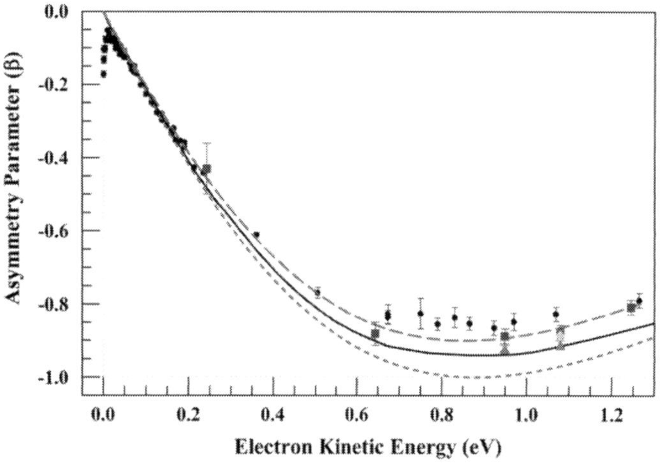

Figure 4. A comparison of the experimental and theoretical β values for the $^3P_2 \leftarrow {}^2P_{3/2}$ transition. Experimental data: (Circles) this work, (triangles) Breyer *et al.* [10], (squares) Hanstorp *et al.* [14] and (diamond) Hall and Siegel [15]. Theory curves: (Solid line) Cooper and Zare [13], (dotted and dashed lines) model of Hanstorp *et al.* [14] with phase shifts of 1.0 and 0.925, respectively.

The convergence of data and theory for eKE between 40 meV and 0.24 eV clearly shows the decreasing *d*-wave contribution and is consistent with the Cooper-Zare formalism [13] which predicts that, as the eKE decreases, the outgoing electrons become more *s*-wave in nature and should tend to a completely isotropic distribution at zero eKE. However, below an eKE of 40 meV a slow and increasing turnaround towards lower values of β is evident. Recent theoretical calculations by Kutzner *et al.* [16] and Roberston *et al.* [17] for photodetachment of atomic halogen anions, where correlation effects such as polarization and relaxation are taken into account, show a similar turnaround in the value of β for outgoing electrons as eKE $\rightarrow$ 0. Their conclusion is that, as the outgoing electron's eKE becomes lower, electron correlation effects become more pronounced. Our results may show the effect of electron correlations at low eKE and, if so, would serve as a good test for the inclusion of electron correlation-effects into relativistic theoretical models for open-shell atoms.

4. Conclusions and future work

Preliminary results for the photodetachment of the atomic oxygen anion have been presented. They include the $^3P_1$ and $^3P_2$ branching ratio [i.e., $\sigma(^3P_1)/\sigma(^3P_2)$] for both transitions originating from the

$^2P_{3/2}$ state of the anion and a measurement of the angular-distribution asymmetry parameter for eKEs across a wide range. The results were collected using a very high-resolution VMI spectrometer, which is coupled to an anion source and mass TOF spectrometer. The measurements of the β parameter represent a new standard, not only due to the extensive range of the measurements, but more importantly, the lowest eKE for which a value of β has been obtained. These measurements should prove to be a substantial test for theory and will hopefully act as an incentive for new and detailed calculations for the photodetachment of atomic oxygen, particularly relativistic calculations that include fine-structure detail, polarization and relaxation effects. The finalized data for O^-, including further results for the remaining fine-structure transitions, will be published elsewhere.

In the future, measurements on atomic sulfur will be undertaken across the same eKE range as undertaken on atomic oxygen. These measurements should not only add to the collective spectroscopic knowledge of a significant species but also provide a stronger test for relativistic theory, especially at threshold, given sulfur's stronger dipole-polarizability.

Acknowledgements

The authors would like to thank the Australian Research Council Discovery Grants No. DP0666267 and DP0880850, for supporting this research and would also like to thank Colin Dedman, Graeme Cornish, Stephen Battisson, Ross Tranter and Ron Cruikshank for providing technical support.

References

[1] Covington A M, Calabrese D, Williams W W and Thompson J S (1997) *Phys. Rev. A* **56** 4746
[2] Andersen T (2004) *Phys. Rep.* **394** 157
[3] Cavanagh S J, Gibson S T, Gale M N, Dedman C J, Roberts E H and Lewis B R (2007) *Phys. Rev. A* **76** 052708
[4] Garand E, Buchachenko A A, Yacovitch T I, Szczęśniak M M, Chalasinski G and Neumark D M (2009) *J. Phys. Chem. A* **113** 4631
[5] Hansen E W and Law P -L (1985) *J. Opt. Soc. Am. A* **2** 510
[6] Rau A R P and Fano U (1971) *Phys. Rev. A* **4** 1751
[7] Lineberger W C and Woodward B W (1970) *Phys. Rev. Lett.* **25** 424
[8] Rau A R P, *Electron and Photon Interaction with Atoms*, edited by Kleinpoppen H and McDowell M R C (Plenum, New York, 1976), Chap. 12
[9] Pan C and Starace A F (1993) *Phys. Rev. A* **47** 295
[10] Breyer F, Frey P and Hotop H (1978) *Z. Phys. A* **286** 133
[11] Suzuki T and Kasuya T (1987) *Phys. Rev. A* **36** 2129
[12] Wigner E P (1943) *Phys. Rev.* **73** 1002
[13] Cooper J and Zare R N (1968) *J. Chem. Phys.* **48** 942
[14] Hanstorp D, Bengtsson C and Larson D J (1989) *Phys. Rev. A* **40** 670
[15] Hall J H and Siegel M W (1968) *J. Chem. Phys.* **48** 943
[16] Kutzner M, Robertson J A and Pelley P (2000) *Phys. Rev. A* **62** 062717
[17] Robertson J A, Kutzner M and Pelley P (2001) *Phys. Rev. A* **63** 042715

Non-linear photointeractions under FEL intense short-wavelength radiation: case study

A Miheli\v{c}[1], P Lambropoulos[2,3] and M G Makris[2]

[1] Jožef Stefan Institute, Jamova cesta 39, SI-1000 Ljubljana, Slovenia
[2] Institute of Electronic Structure and Laser, Foundation for Research and Technology – Hellas, P. O. Box 1527, GR-71110 Heraklion, Greece
[3] Department of Physics, University of Crete, P. O. Box 2208, GR-71003 Heraklion, Greece

E-mail: andrej.mihelic@ijs.si

Abstract. We present a theoretical study of a two-photon single-colour and two-colour excitation of ground-state krypton atoms to the final $3d^{-1}_{5/2}4d$ autoionising states.

1. Introduction

Experimental studies of multiphoton excitation or ionisation of matter with short-wavelength radiation have become feasible only recently: Free-electron laser (FEL) sources have opened ways to new experiments in which non-linear processes can be studied with radiation in the extreme ultraviolet and x-ray regime. In order to understand such experimental results on more complex systems, quantitative experimental and theoretical studies on simpler systems, such as noble gases, have been undertaken.

Depending on the intensity and pulse duration of the radiation, and the system under investigation, various theoretical approaches may need to be employed [1, 2, 3, 4]. If the intensity is sufficiently low, so that the ponderomotive energy is much smaller than the photon energy, and the pulse duration, in terms of cycles, is sufficiently long, a description in terms of appropriate cross sections and rate equations may be sufficient. If several ionic species are involved, the relevant ionisation yields can be obtained through the solution of coupled differential (kinetic or rate) equations describing their evolution during the laser pulse [3, 4]. Generalised multiphoton cross sections [1, 2], which enter these differential equations as parameters, can be calculated in the appropriate lowest non-vanishing order perturbation theory. The necessary accuracy in the calculation of such cross sections is dictated by the type and accuracy of the experimental data under consideration. In certain cases that may involve more detailed data, the use of density matrix techniques, within the space of the necessary atomic states, may be in order.

In this paper, we present a brief outline of the theoretical treatment of two-photon, single-colour and two-colour resonant excitation of krypton atoms from the ground state to the $3d^{-1}_{5/2}4d$ autoionising states with the total angular momentum $J = 0$ and $J = 2$. The motivation for this work came from ongoing experiments at the FEL facility FLASH in Hamburg [5]. Our aim here is to set up a simple theoretical framework for the experiment, providing an estimate of the expected ion yields and their dependence on the laser parameters, which can serve the purpose of optimisation of the experimental procedure.

2. Two-photon excitation of the krypton atom

We describe the atom with a 5-level atom model within the density matrix formalism. The values of the laser parameters employed in the calculation are chosen so as to correspond to the experimental conditions.

We study two different ways of exciting the atom. In the first case, to be referred to as the *two-colour* case, we consider the krypton atoms exposed simultaneously to the FEL and Ti:sapphire (IR) laser pulses. The energy of the IR photons is fixed to 1.55 eV, and the energy of the FEL photons is chosen so that the sum of the photon energies $\hbar\omega_1 + \hbar\omega_2$ lies close to the energy difference between the ground state and the higher of the two $3d^{-1}_{5/2}4d$ $J = 2$ final states. In the second case, to be referred to as the *single-colour* case, we consider the excitation with two FEL photons, each of them having the energy of 46 eV. In both of the above cases, the two-photon excitation proceeds through the $3d^{-1}_{5/2}5p$ $J = 1$ intermediate state.

The $3d^{-1}4d$ and $3d^{-1}5p$ states form two groups of states separated in energy by approximately 1.2 eV: the lower-lying group is characterised by the coupling of the $3d$ electrons to the total angular momentum $J_h = 5/2$ and the higher-lying group by $J_h = 3/2$. On the basis of known experiments [6], the widths of $3d^{-1}5p$ states are known to be about 100 meV, stemming predominantly from the Auger decay of the $3d$ hole, and this is also the value we assume for the $3d^{-1}_{5/2}4d$ and $3d^{-1}_{5/2}5p$ states. For the two-colour case, the bandwidth of the FEL operating at ~ 90 eV is about 0.9 eV, whereas the bandwidth of the IR laser is much smaller [5]. For 46-eV incident photons, the FEL bandwidth is further reduced to approximately one half of the above value. Although, in all generality, the present model should also include the $J_h = 3/2$ states, they have not been taken into account for the purposes of the present preliminary calculations. Their inclusion should not substantially change the results or affect the main ideas. Thus, we focus on the excitation to the final states with $J_h = 5/2$ only.

2.1. Theoretical model

The system under investigation is depicted in Fig. 1. We describe it with the Hamilton operator

$$H = \sum_i E_i|i\rangle\langle i| - \frac{\hbar}{2}\left(\Omega_{21}(t)\,e^{-i\omega_1 t}|2\rangle\langle 1| + \sum_{k\geq 3}\Omega_{k2}(t)\,e^{-i\omega_2 t}|k\rangle\langle 2| + \mathrm{H.c.}\right), \tag{1}$$

where the first term represents the free atom, and the second the atom-field interaction. The energies of the atomic states $|i\rangle$ are denoted by E_i. The frequencies ω_1 and ω_2 are associated with the first (FEL) and the second (IR or FEL) photon, respectively. The Rabi frequencies $\Omega_{12} = \Omega_{21}$, $\Omega_{2k} = \Omega_{k2}$, $k = 3, \ldots, 5$, are assumed real. In the semi-classical treatment, in which the field is treated classically, the Rabi frequency connecting the states $|i\rangle$ and $|j\rangle$ is defined as:

$$\Omega_{ij}(t) = \hbar^{-1}\,2\mathcal{F}(t)\,D_{ij} = \hbar^{-1}\,2\mathcal{F}_0\,\exp\left(-\frac{4t^2\ln 2}{T^2}\right)D_{ij}, \tag{2}$$

where $2\mathcal{F}_0 = 2\mathcal{E}_1$ ($2\mathcal{F}_0 = 2\mathcal{E}_2$) is the electric field amplitude associated with the first (second) radiation field, $T = \tau_1$ ($T = \tau_2$) the duration of the first (second) pulse, and D_{ij} the electric dipole matrix element. Note that the intensity of the radiation is connected to the field strength through the equality $I(t) = 2\epsilon_0 c\,\mathcal{F}^2(t)$ (in SI units, c is the speed of light *in vacuo*, and ϵ_0 is the vacuum permittivity).

Since under the conditions relevant to the experiment, the rotating-wave approximation (RWA) is valid, the time-evolution of the system is described by the following set of equations:

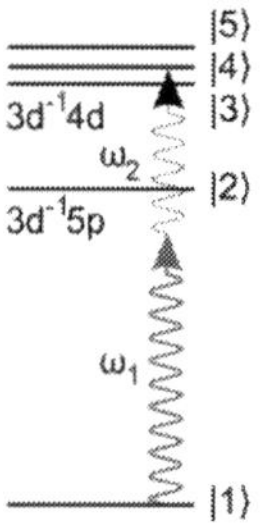

Figure 1. Two-photon excitation of the krypton atom.

$$\begin{aligned}
\dot{\sigma}_{11} &= -\frac{i}{2}\big(\Omega_{12}\sigma_{12} - \Omega_{12}\sigma_{12}^*\big), &(3)\\
\dot{\sigma}_{22} &= -\Gamma_2/\hbar\,\sigma_{22} - \frac{i}{2}\big(- \Omega_{12}\sigma_{12} + \Omega_{12}\sigma_{12}^* + \Omega_{23}\sigma_{23} - \Omega_{23}\sigma_{23}^* &(4)\\
&+ \Omega_{24}\sigma_{24} - \Omega_{24}\sigma_{24}^* + \Omega_{25}\sigma_{25} - \Omega_{25}\sigma_{25}^*\big),\\
\dot{\sigma}_{kk} &= -\Gamma_k/\hbar\,\sigma_{kk} - \frac{i}{2}\big(- \Omega_{2k}\sigma_{2k} + \Omega_{2k}\sigma_{2k}^*\big), \qquad k = 3, \ldots, 5, &(5)
\end{aligned}$$

$$\begin{aligned}
\dot{\sigma}_{12} &= \hbar^{-1}\big[- \Gamma_2/2 - \gamma_1 - i(E_1 + S_1 - E_2 - S_2 + \hbar\omega_1)\big]\sigma_{12} &(6)\\
&- \frac{i}{2}\big(\Omega_{12}\sigma_{11} - \Omega_{12}\sigma_{22} + \Omega_{23}\sigma_{13} + \Omega_{24}\sigma_{14} + \Omega_{25}\sigma_{15}\big),\\
\dot{\sigma}_{1k} &= \hbar^{-1}\big[- \Gamma_k/2 - i(E_1 + S_1 - E_k - S_k + \hbar\omega_1 + \hbar\omega_2)\big]\sigma_{1k} &(7)\\
&- \frac{i}{2}\big(- \Omega_{12}\sigma_{2k} + \Omega_{2k}\sigma_{12}\big), \qquad k = 3, \ldots, 5,
\end{aligned}$$

$$\begin{aligned}
\dot{\sigma}_{23} &= \hbar^{-1}\big[- \Gamma_2/2 - \Gamma_3/2 - \gamma_2 - i(E_2 + S_2 - E_3 - S_3 + \hbar\omega_2)\big]\sigma_{23}\\
&- \frac{i}{2}\big(- \Omega_{12}\sigma_{13} + \Omega_{23}\sigma_{22} - \Omega_{23}\sigma_{33} - \Omega_{24}\sigma_{34}^* - \Omega_{25}\sigma_{35}^*\big), &(8)\\
\dot{\sigma}_{24} &= \hbar^{-1}\big[- \Gamma_2/2 - \Gamma_4/2 - \gamma_2 - i(E_2 + S_2 - E_4 - S_4 + \hbar\omega_2)\big]\sigma_{24}\\
&- \frac{i}{2}\big(- \Omega_{12}\sigma_{14} - \Omega_{23}\sigma_{34} + \Omega_{24}\sigma_{22} - \Omega_{24}\sigma_{44} - \Omega_{25}\sigma_{45}^*\big), &(9)\\
\dot{\sigma}_{25} &= \hbar^{-1}\big[- \Gamma_2/2 - \Gamma_5/2 - \gamma_2 - i(E_2 + S_2 - E_5 - S_5 + \hbar\omega_2)\big]\sigma_{25}\\
&- \frac{i}{2}\big(- \Omega_{12}\sigma_{15} - \Omega_{23}\sigma_{35} - \Omega_{24}\sigma_{45} + \Omega_{25}\sigma_{22} - \Omega_{25}\sigma_{55}\big), &(10)
\end{aligned}$$

$$\begin{aligned}
\dot{\sigma}_{3k} &= \hbar^{-1}\big[- \Gamma_3/2 - \Gamma_k/2 - i(E_3 + S_3 - E_k - S_k)\big]\sigma_{3k} &(11)\\
&- \frac{i}{2}\big(- \Omega_{23}\sigma_{2k} + \Omega_{2k}\sigma_{23}^*\big), \qquad k = 4, 5,\\
\dot{\sigma}_{45} &= \hbar^{-1}\big[- \Gamma_4/2 - \Gamma_5/2 - i(E_4 + S_4 - E_5 - S_5)\big]\sigma_{45} &(12)\\
&- \frac{i}{2}\big(- \Omega_{24}\sigma_{25} + \Omega_{25}\sigma_{24}^*\big).
\end{aligned}$$

The elements of the density matrix in the frame rotating with the frequency of the external fields are denoted by σ_{ij}, the Auger energy widths of the intermediate ($i = 2$) and final states ($i = 3, \ldots, 5$) by Γ_i, and the bandwidths (half-width at half-maximum) of the first and the second radiation field by γ_1 and γ_2. Strictly speaking, the widths γ_1 and γ_2 describe *phase* fluctuations of the driving fields. However, if no bound-bound transition is saturated, the present treatment is also valid in the case of *amplitude* fluctuations of the fields. The field-induced coupling between the levels of the model results in energy shifts of these levels. In addition, we also introduce shifts S_i due to *all other levels not included in the 5-level model.* It turns out that the shifts of the levels are dominated by the ponderomotive shift. It is well known that the ponderomotive shift U_p is inversely proportional to the square of the photon frequency. In the two-colour case, the ponderomotive shift due to the IR laser is thus much larger than the shift due to the FEL, and we may write $S_i(t) \approx e^2 I_2(t)/(2m_e\epsilon_0 c\,\omega_2^2)$, $i = 2, \ldots, 5$. For the single-colour case, we write $S_i(t) = e^2 I_1(t)/(2m_e\epsilon_0 c\,\omega_1^2)$. The shift of the initial state has been set to zero.

The parameter values used in the calculation are given in Table 1.

Table 1. Parameter values used in the calculation. The values of the dipole matrix elements (in Babushkin gauge) have been obtained using a Dirac-Fock calculation [7] for polarisation of the incident light along the quantisation (z) axis. The total angular momenta of the states are: $J_1 = J_5 = 0$, $J_2 = 1$, and $J_3 = J_4 = 2$. The Auger widths are $\Gamma_1 = 0$ and $\Gamma_i = 100$ meV ($i \geq 2$).

i	1	2	3	4	5
E_i [eV]	$\equiv 0$	91.245	92.033	92.049	92.111
D_{2i} [a.u.]			0.857	2.234	-2.855
D_{12} [a.u.]	-0.027				

2.2. Two-colour excitation

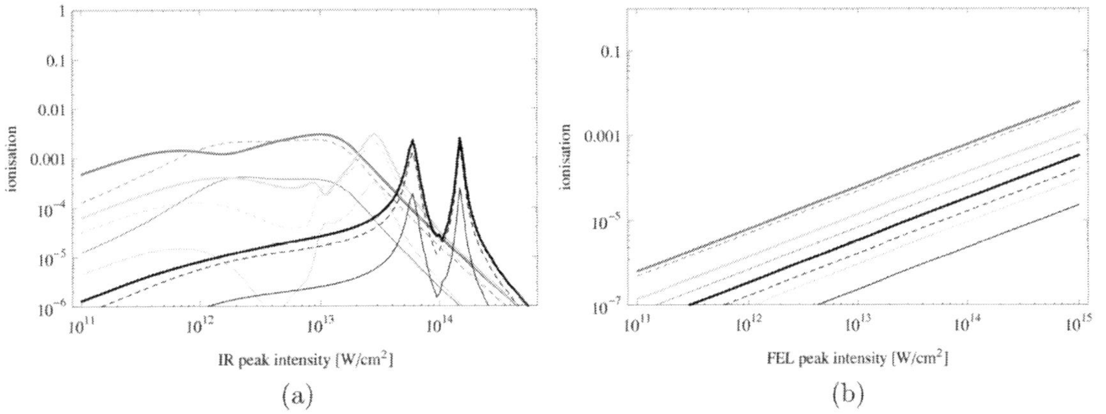

Figure 2. Two-photon two-colour excitation. Left: the peak intensity of the IR laser is varied. Right: the peak intensity of the FEL is varied. Thin solid, dashed, and thick solid patterns correspond to the final states $|3\rangle$, $|4\rangle$, and $|5\rangle$, respectively. See text for details.

We have calculated the ion yields resulting from various $3d_{5/2}^{-1}4d$ ($J = 0, 2$) final states. The photon energy of the IR laser is fixed to $\hbar\omega_2 = 1.55$ eV ($\lambda_2 = 800$ nm). The energy of the FEL photons $\hbar\omega_1$ is chosen in such a way that the sum of the photon energies $\hbar\omega_1 + \hbar\omega_2$ is equal to the energy of level 4 shifted by S_4, i.e., $\hbar\omega_1 = (E_4 + S_4) - (E_1 + S_1) - \hbar\omega_2$, *at a selected value of the peak intensity of the IR laser*. The pulse durations are $\tau_1 = 30$ fs for the FEL and $\tau_2 = 120$ fs for the IR laser. The bandwidth (half-width at half-maximum) of the FEL is set to $\gamma_1 = 0.45$ eV, whereas the bandwidth of the IR laser has been neglected because it is much smaller than γ_1 and would not affect the results.

In Figure 2(a) we show the variation of the yields with the peak intensity of the IR laser ($I_{2,\max} = 2\epsilon_0 c\,\mathcal{E}_2^2$). The peak intensity of the FEL is $I_{1,\max} = 2 \times 10^{13}$ W/cm^2. Red (dark grey), green (light grey), and black curves correspond to the energy of the first photon $\hbar\omega_1$ chosen to reach the (shifted) level 4 at the IR peak intensities 1×10^{12} W/cm^2 ($\hbar\omega_1 = 90.559$ eV), 1×10^{13} W/cm^2 ($\hbar\omega_1 = 91.097$ eV), and 1×10^{14} W/cm^2 ($\hbar\omega_1 = 96.473$ eV). Especially interesting is the behaviour of the ion yields for intensities above 10^{12} W/cm^2: it should be pointed out that the shape of the yields is the consequence of a rather intricate interplay between the intensity dependence of the driving (Ω_{ij}) and the induced shifts. In Figure 2(b) we show the dependence of the ion yields on the FEL peak intensity ($I_{1,\max} = 2\epsilon_0 c\,\mathcal{E}_1^2$). Red (dark grey), green (light grey), and black curves correspond to the IR peak intensity set to 2×10^{13} W/cm^2 ($\hbar\omega_1 = 91.694$ eV), 1×10^{14} W/cm^2 ($\hbar\omega_1 = 96.473$ eV), and 5×10^{14} W/cm^2 ($\hbar\omega_1 = 120.37$ eV).

2.3. Single-colour excitation

As in the two-colour case, we have calculated the single-colour ion yields (Figure 3) resulting from the Auger decay of the final states. The photon energy of the FEL has been fixed to $\hbar\omega_1 = \hbar\omega_2 = 46.0$ eV, and the pulse duration set to $\tau_1 = \tau_2 = 30$ fs. The bandwidth of the FEL is set to $\gamma_1 = \gamma_2 = 0.25$ eV.

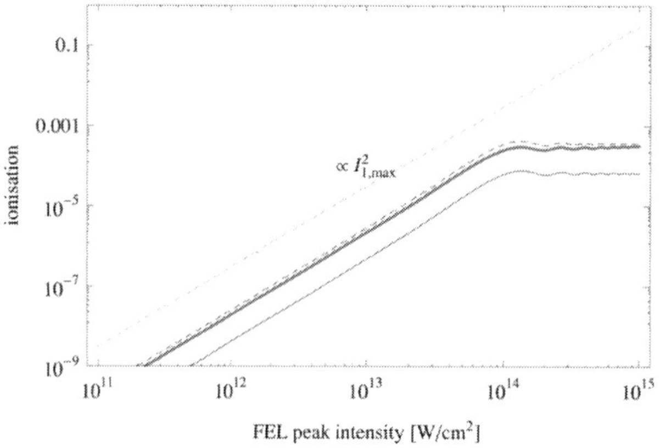

Figure 3. Two-photon single-colour excitation. Thin solid, dashed, and thick solid lines correspond ion yields from the final states $|3\rangle$, $|4\rangle$, and $|5\rangle$, respectively. The dashed line (top) shows a quadratic dependence of the yield on the peak intensity (i.e., $Y \propto I_{1,\max}^2$) to guide the eye.

For low peak intensities of the FEL (below 10^{14} W/cm^2), the yields exhibit a quadratic behaviour, as can be seen by comparing the slopes of the calculated yields with the (quadratic) power dependence shown in Figure 3, top. At higher intensities, for $I_{1,\max} \gtrsim 10^{14}$ W/cm^2, a deviation of the yields from the quadratic behaviour can be observed: although the field-induced shifts may be small for lower intensities, they are no longer negligible above 10^{14} W/cm^2. The final $3d^{-1}4d$ states move away from the energy $E_1 + \hbar\omega_1 + \hbar\omega_2$, and the yields do not grow proportionally to $I_{1,\text{max}}^2$ any longer.

3. Conclusion

We have studied two-photon excitation of the ground-state krypton atoms exposed to: (i) two-colour FEL + IR laser pulses, and (ii) single-colour FEL pulses. The present calculations serve as a theoretical framework for the interpretation of the recent experimental data. We have used the density-matrix formalism to calculate population of the $3d_{5/2}^{-1}4d$ $J = 0, 2$ final states and the ion yields resulting from the Auger decay. We have used realistic parameter values to approach the experimental situation. One of the surprises in the course of this work has been the crucial role of the ponderomotive shifts in the single-colour case, in which the photon energy of 46 eV is rather large. Although such shifts are still smaller than the photon energy, they are comparable to the laser bandwidth and they need to be taken into consideration in order to optimise the two-photon resonance condition at the peak of the pulse.

The support of the European COST Action CM0702 is gratefully acknowledged. One of us (A. M.) also acknowledges the support of the Marie Curie RTN EMALI.

References

[1] Lambropoulos P and Tang X 1987 *J. Opt. Soc. Am.* B **4** 821
[2] Lambropoulos P, Maragakis P and Zhang J 1998 *Phys. Rep.* **305** 203
[3] Makris M G, Lambropoulos P 2008 *Phys. Rev.* A **77** 023401
[4] Makris M G, Lambropoulos P and Mihelič A 2009 *Phys. Rev. Lett.* **102** 033002
[5] Meyer M 2009 private communication
[6] King G C, Tronc M, Read F H and Bradford R C 1977 *J. Phys.* B **10** 2479
[7] Dyall K G, Grant I P, Johnson C T, Parpia F A and Plummer E P 1989 *Comput. Phys. Commun.* **55** 425

AUTHOR INDEX

AUTHOR INDEX

Institute of Physics
Dirac House, Temple Back
Bristol BS1 6BE UK

CURRAN ASSOCIATES INC
proceedings.com